VOLUME ONE HUNDRED AND TWENTY TWO

ADVANCES IN CANCER RESEARCH

Redox and Cancer Part A

VOLUME ONE HUNDRED AND TWENTY TWO

ADVANCES IN CANCER RESEARCH

Redox and Cancer Part A

Edited by

DANYELLE M. TOWNSEND

Assistant Professor,
Department of Pharmaceutical and Biomedical Sciences,
Medical University of South Carolina,
Charleston, South Carolina, USA

KENNETH D. TEW

Professor and Chairman,
Department of Cell and Molecular Pharmacology,
John C. West Chair of Cancer Research,
Medical University of South Carolina,
Charleston, South Carolina, USA

ELSEVIER

AMSTERDAM • BOSTON • HEIDELBERG • LONDON
NEW YORK • OXFORD • PARIS • SAN DIEGO
SAN FRANCISCO • SINGAPORE • SYDNEY • TOKYO
Academic Press is an imprint of Elsevier

Academic Press is an imprint of Elsevier
525 B Street, Suite 1800, San Diego, CA 92101-4495, USA
225 Wyman Street, Waltham, MA 02451, USA
32 Jamestown Road, London, NW1 7BY, UK
The Boulevard, Langford Lane, Kidlington, Oxford, OX5 1GB, UK
Radarweg 29, PO Box 211, 1000 AE Amsterdam, The Netherlands

First edition 2014

ISBN: 978-0-12-420117-0
ISSN: 0065-230X

For information on all Academic Press publications
visit our website at store.elsevier.com

14 15 16 12 11 10 9 8 7 6 5 4 3 2 1

CONTENTS

CONTRIBUTORS

Ningfei An
Division of Hematology and Oncology, Department of Medicine, Medical University of South Carolina, Charleston, South Carolina, USA

Joseph J. Barycki
Department of Biochemistry and the Redox Biology Center, University of Nebraska, Lincoln, Nebraska, USA

Arthur J.L. Cooper
Department of Biochemistry and Molecular Biology, New York Medical College, Valhalla, New York, USA

Christina Grek
Department of Cell and Molecular Pharmacology and Experimental Therapeutics, Medical University of South Carolina, Charleston, South Carolina, USA

Marie H. Hanigan
Department of Cell Biology, University of Oklahoma Health Sciences Center, Oklahoma City, Oklahoma, USA

Merja Heinäniemi
Department of Biotechnology and Molecular Medicine, A.I. Virtanen Institute for Molecular Sciences, and Institute of Biomedicine, School of Medicine, University of Eastern Finland, P.O. Box 1627, Kuopio, Finland

Annastasia S. Hyde
Department of Biochemistry and the Redox Biology Center, University of Nebraska, Lincoln, Nebraska, USA

Yubin Kang
Division of Hematology and Oncology, Department of Medicine, Medical University of South Carolina, Charleston, South Carolina, USA

Emilia Kansanen
Department of Biotechnology and Molecular Medicine, A.I. Virtanen Institute for Molecular Sciences, University of Eastern Finland, P.O. Box 1627, Kuopio, Finland

Hanna M. Leinonen
Department of Biotechnology and Molecular Medicine, A.I. Virtanen Institute for Molecular Sciences, University of Eastern Finland, P.O. Box 1627, Kuopio, Finland

Anna-Liisa Levonen
Department of Biotechnology and Molecular Medicine, A.I. Virtanen Institute for Molecular Sciences, University of Eastern Finland, P.O. Box 1627, Kuopio, Finland

Yilin Liu
Department of Biochemistry and the Redox Biology Center, University of Nebraska, Lincoln, Nebraska, USA

Volodymyr I. Lushchak
Department of Biochemistry and Biotechnology, Vassyl Stefanyk Precarpathian National University, Ivano-Frankivsk, Ukraine

Yefim Manevich
Department of Cell and Molecular Pharmacology and Experimental Therapeutics, Medical University of South Carolina, Charleston, South Carolina, USA

Bengt Mannervik
Department of Chemistry-BMC, Uppsala University, Uppsala, and Department of Neurochemistry, Stockholm University, Stockholm, Sweden

Olof Modén
Department of Chemistry-BMC, Uppsala University, Uppsala, Sweden

Petri Pölönen
Department of Biotechnology and Molecular Medicine, A.I. Virtanen Institute for Molecular Sciences, and Institute of Biomedicine, School of Medicine, University of Eastern Finland, P.O. Box 1627, Kuopio, Finland

Lijian Shao
Division of Radiation Health, Department of Pharmaceutical Sciences, Winthrop P. Rockefeller Cancer Institute, University of Arkansas for Medical Sciences, Little Rock, Arkansas, USA

Melanie A. Simpson
Department of Biochemistry and the Redox Biology Center, University of Nebraska, Lincoln, Nebraska, USA

Douglas R. Spitz
Free Radical and Radiation Biology Program, Department of Radiation Oncology, Holden Comprehensive Cancer Center, Carver College of Medicine, The University of Iowa, Iowa City, Iowa, USA

Kenneth D. Tew
Professor and Chairman, Department of Cell and Molecular Pharmacology, John C. West Chair of Cancer Research, Medical University of South Carolina, Charleston, South Carolina, USA

Danyelle M. Townsend
Department of Pharmaceutical and Biomedical Sciences, Medical University of South Carolina, Charleston, South Carolina, USA

Zhi-Wei Ye
Department of Cell and Molecular Pharmacology and Experimental Therapeutics, Medical University of South Carolina, Charleston, South Carolina, USA

Jie Zhang
Department of Cell and Molecular Pharmacology and Experimental Therapeutics, Medical University of South Carolina, Charleston, South Carolina, USA

Daohong Zhou
Division of Radiation Health, Department of Pharmaceutical Sciences, Winthrop P. Rockefeller Cancer Institute, University of Arkansas for Medical Sciences, Little Rock, Arkansas, USA

PREFACE

Because solid tumors frequently have reduced oxygen and altered energy metabolism, they also demonstrate altered redox homeostasis, both creating and tolerating higher levels of reactive oxygen and nitrogen species. Within this first Redox and Cancer volume, authors variously discuss those enzyme systems within a cell that help regulate ROS and contribute to the unbalanced growth that characterizes tumor formation. The nucleophilicity of sulfur helps to counteract the ROS threat, and tumors can have quite aberrant small molecule thiol homeostasis, the pathways for which are complex and interreactive, frequently accompanied by degrees of redundancy. This complexity provides many avenues to interrogate specific enzymes and pathways. For example, glutathione as the most prevalent nonprotein source of sulfhydryls can be synthesized *de novo* or through salvage of constituent amino acids. The balance of these processes depends upon numerous variables, but can be a distinguishing feature of many types of cancer cells. Moreover, progress in understanding recycling of GSH amino acids is facilitated by recent progress in the crystal structure of γ-glutamyl transpeptidase (GGT), and two of the present contributions focus on the biology of this enzyme.

The addition of glutathione to acceptor cysteines in target proteins is a posttranslational modification referred to as S-glutathionylation. This alters the size and charge of a number of proteins that can alter their structure and/or function. Two chapters discuss the importance of reversible and irreversible S-glutathionylation, processes that yield a cycle that controls many critical signaling events, perhaps differentially in normal and tumor cells. Glutathione *S*-transferases (GST) have been the focus of oncology-based studies for a number of decades. Their importance in detoxification of small molecule drugs is discussed, as is the role of a specific isozyme (GSTP) in regulating, among other things, cellular kinase signaling and the forward reaction of the S-glutathionylation cycle.

Thioredoxin also contributes reducing equivalents that stabilize redox homeostasis. It can reactivate a number of enzyme systems and more recently has found a role in the physiological functioning of bone marrow, a tissue of high relevance in determining response to anticancer drug therapies. This area of research is also covered in this volume.

Finally, Nrf2 is a master transcriptional regulator and reacts to changes in redox environment. As will be seen, its regulation is functionally critical to controlling expression of many of the gene products discussed in the other chapters of this volume.

Subsequent volumes of *Advances in Cancer Research* will expand the redox horizon in ways that will build upon the current contents.

CHAPTER ONE

Reactive Oxygen Species in Normal and Tumor Stem Cells

Daohong Zhou[*,1], **Lijian Shao**[*], **Douglas R. Spitz**[†,1]
[*]Division of Radiation Health, Department of Pharmaceutical Sciences, Winthrop P. Rockefeller Cancer Institute, University of Arkansas for Medical Sciences, Little Rock, Arkansas, USA
[†]Free Radical and Radiation Biology Program, Department of Radiation Oncology, Holden Comprehensive Cancer Center, Carver College of Medicine, The University of Iowa, Iowa City, Iowa, USA
[1]Corresponding authors: e-mail address: dzhou@uams.edu; douglas-spitz@uiowa.edu

Contents

Abstract

Reactive oxygen species (ROS) play an important role in determining the fate of normal stem cells. Low levels of ROS are required for stem cells to maintain quiescence and self-renewal. Increases in ROS production cause stem cell proliferation/differentiation, senescence, and apoptosis in a dose-dependent manner, leading to their exhaustion. Therefore, the production of ROS in stem cells is tightly regulated to ensure that they have the ability to maintain tissue homeostasis and repair damaged tissues for the life span of an organism. In this chapter, we discuss how the production of ROS in normal stem cells is regulated by various intrinsic and extrinsic factors and how the fate of these cells is altered by the dysregulation of ROS production under various pathological conditions. In addition, the implications of the aberrant production of ROS by tumor stem cells for tumor progression and treatment are also discussed.

Advances in Cancer Research, Volume 122
ISSN 0065-230X
http://dx.doi.org/10.1016/B978-0-12-420117-0.00001-3

1. INTRODUCTION

About 2.5 million years ago, cyanobacteria evolved to gain the ability to produce oxygen (O_2) as a by-product of photosynthesis. O_2 is a paramagnetic gas that readily reacts with other elements like hydrogen, carbon, copper, and iron. As O_2 accumulated, it is thought to have converted the early reducing atmosphere into an atmosphere more conducive to oxidation reactions. Also, as atmospheric O_2 levels rose, many new organisms evolved and flourished after developing antioxidant defense systems to protect against the toxicity of by-products related to O_2 metabolism. Moreover, early aerobic organisms continued evolving to become multicellular organisms by taking selective advantage of efficient O_2 utilization in various vital metabolic processes, such as employing O_2 as the terminal electron acceptor for mitochondrial electron transport chain (ETC) activity during oxidative phosphorylation (OXPHOS), allowing for the efficient production of energy (Halliwell & Gutteridge, 2007).

However, utilizing O_2 in many essential metabolic processes by living systems came at an evolutionary price, because O_2 metabolism can lead to the production of reactive oxygen species (ROS) (Boveris, 1977; Buettner, 1993; Chance, Sies, & Boveris, 1979; Forman & Kennedy, 1974, 1975; Fridovich, 1978). Fortunately, living systems are normally maintained in a nonequilibrium steady-state that is highly reducing and is exemplified by the reduced glutathione (GSH)/glutathione disulfide (GSSG) redox couple that oscillates between about −200 and −240 mV (Schafer & Buettner, 2001). This highly reducing intracellular environment keeps steady-state ROS at relatively low levels that oscillate with changes in metabolic activity, which can communicate normal shifts in oxidative metabolism to signaling and gene expression pathways that control many diverse cellular functions including cell proliferation, circadian rhythms, differentiation, immunological functions, tissue remodeling, and vascular reactivity (Beckman & Koppenol, 1996; Kessenbrock, Plaks, & Werb, 2010; Menon & Goswami, 2007; Oberley, Oberley, & Buettner, 1980, 1981; Reuter, Gupta, Chaturvedi, & Aggarwal, 2010; Rutter, Reick, Wu, & McKnight, 2001). If the metabolic production of ROS exceeds the capacity of the endogenous antioxidant defense systems, oxidative stress can occur (Sies, 1991; Spitz, Azzam, Li, & Gius, 2004). Depending on the severity of oxidative stress, an organism may adapt by increasing its antioxidant capacity, increasing the capacity to repair oxidative damage, or shifting metabolic processes away from oxidative metabolism towards glycolytic metabolism. If the cellular adaptive processes that are induced in response to chronic metabolic oxidative stress cannot mitigate the accumulation of oxidative damage to critical biomolecules, potentially pathological conditions can develop due to increasing oxidative damage to DNA, proteins, and lipids. It is this gradual accumulation of oxidative damage to critical biomolecules that is believed to contribute to most if not all degenerative diseases associated with aging and cancer (Droge, 2002; Finkel, 2005).

Although all cells in an organism can be affected by the accumulation of oxidative damage, the effects of ROS on stem cells (or pluripotent cells) in most self-renewing tissues are of particular interest to the processes of aging and cancer development because of their undifferentiated state and longevity of replicative potential (Kobayashi & Suda, 2012; Oberley et al., 1980, 1981; Shyh-Chang, Daley, & Cantley, 2013). Stem cells can exist in a completely undifferentiated state, such as pluripotent embryonic stem cells (ESCs), or can be more committed to a particular lineage in a tissue as tissue stem cells or adult stem cells (ASCs). All normal stem cells appear to be highly sensitive to oxidative stress because of their relatively undifferentiated state with a long division potential for accumulating genetic damage. Accumulation of oxidative damage in normal stem cells can lead to cell transformation and tumorigenesis or cause tissue injury, loss of function, enhanced senescence, and loss of division potential associated with degenerative diseases associated with aging (Shyh-Chang, Daley, et al., 2013). Therefore, in this chapter, we will focus our discussions on the role of metabolic ROS in stem cell physiology and pathology and discuss strategies to exploit the differences in normal and tumor stem cell (TSC) sensitivities to oxidative stress for selectively protecting normal ASCs while sensitizing TSCs including leukemia stem cells (LSCs) and cancer stem cells (CSCs) to oxidative damage induced during leukemia and cancer therapy.

2. ROS

2.1. Common biological ROS

ROS is a collective term for oxygen-containing species that are more reactive than molecular O_2. The most likely ROS to be produced initially during the metabolism of O_2 by living systems were proposed to derive from the superoxide anion ($O_2^{\bullet-}$) because it represents the one-electron reduction product of O_2 (Boveris, 1977; Buettner, 1993; Chance et al., 1979; Fridovich, 1978). $O_2^{\bullet-}$ is a relatively weak oxidant but is an excellent reductant for transition metal ions such as Fe^{+3}, Mn^{+3}, and Cu^{+2} leading to the formation of Fe^{+2}, Mn^{+2}, and Cu^{+1}. Interestingly, approximately 0.3% of $O_2^{\bullet-}$ in cellular aqueous environments is believed to exist in the protonated form as the hydroperoxyl radical ($HO_2^{\bullet}$). In contrast to $O_2^{\bullet-}$, $HO_2^{\bullet}$ can diffuse more readily through hydrophobic environments because it is uncharged. In addition, $HO_2^{\bullet}$ is a much more aggressive oxidant than $O_2^{\bullet-}$, capable of damaging a wide variety of potentially critical biomolecules such as polyunsaturated fatty acids to initiate lipid peroxidation chain reactions forming ROOH and aldehydic by-products of lipid oxidation (Buettner, 1993).

The other common ROS derived from reactions of $O_2^{\bullet-}$ and $HO_2^{\bullet}$ in biological matrices include hydrogen peroxide (H_2O_2), organic hydroperoxides (ROOH), organic hydroperoxyl radicals ($RO_2^{\bullet}$), alkoxyl radicals ($RO^{\bullet}$), and the hydroxyl radical ($HO^{\bullet}$) (Buettner, 1993). H_2O_2 forms from the spontaneous dismutation of $O_2^{\bullet-}$ with a rate constant of 10^5 mol^{-1} s^{-1} or through enzymatic dismutation of $O_2^{\bullet-}$ at a rate of 10^9 mol^{-1} s^{-1} in the presence of the superoxide dismutase (SOD) enzymes (Fridovich, 1978). ROOH are formed from the oxidation of organic molecules such as during lipid peroxidation reactions (Buettner, 1993; Buettner, Ng, Wang, Rodgers, & Schafer, 2006; Fridovich, 1978). More highly reactive O_2 radicals are also formed from Fenton-type reactions involving transition metals such as Fe^{+2} and Cu^{+1} with H_2O_2 and ROOH, respectively. These reactions result in the oxidation of the transition metals (to Fe^{+3} and Cu^{+2}) and the formation of $HO^{\bullet}$ and organic hydroperoxyl and alkoxyl radicals ($RO_2^{\bullet}$). $HO^{\bullet}$, $RO_2^{\bullet}$, and $RO^{\bullet}$ are potentially very damaging species because they are formed in close proximity to sites of metal ion binding to biomolecules and they readily react with many critical biological macromolecules including lipids, proteins, and nucleic acids. For this reason, $HO^{\bullet}$, $RO^{\bullet}$, and $RO_2^{\bullet}$ are thought to significantly contribute to the accumulation of oxidative damage to critical biomolecules leading to cytotoxic and mutagenic responses involved in carcinogenesis and aging (Droge, 2002; Fridovich, 1978; Gius & Spitz, 2006; Jorgenson, Zhong, & Oberley, 2013; Oberley et al., 1980, 1981; Spitz et al., 2004).

2.2. Sources of ROS

2.2.1 Mitochondria

The mitochondria have been considered as a major source of cellular-derived ROS based on the fact that most cellular O_2 consumption occurs as a result of mitochondrial ETC activity in mammalian cells (Balaban, Nemoto, & Finkel, 2005; Boveris, 1977; Chance et al., 1979; Finkel & Holbrook, 2000;

Forman & Kennedy, 1974, 1975; Fridovich, 1978). Mitochondrial ETCs consist of four multiprotein complexes (complexes I–IV) assembled on the inner mitochondrial membrane that contains a series of redox catalysts (i.e., pyridine nucleotides, flavoproteins, iron sulfur proteins, ubiquinone, and cytochromes) that utilize electrons obtained from glycolysis and the tricarboxylic acid cycle to generate energy via the process of OXPHOS (Lehninger, 1976; Voet, Voet, & Pratt, 1999) (Fig. 1.1). The ETC proteins are arranged according to their redox potentials, and the flow of electrons down the ETC chains is accompanied by the pumping of protons out of the inner mitochondrial membrane creating a proton gradient (−140 to −180 mV) across the inner membrane that is coupled to the production of ATP from ADP + Pi as the protons reequilibrate through the ATP synthase (also known as ETC complex V) (Fig. 1.1). O_2 is the terminal electron acceptor at complex IV (cytochrome oxidase) where it undergoes a 4-electron reduction to form two water molecules in a tightly regulated process that is not believed to result in the production of $O_2^{\bullet -}$ or H_2O_2. Despite the efficiency of the 4-electron reduction of O_2 to form H_2O at complex IV (≈99% of all mitochondrial O_2 consumption), there is believed to be a finite probability that 0.1–1.0 % of all O_2 consumed by mitochondria can undergo 1-electron reductions at sites in complexes I, II, and III of the ETCs to form $O_2^{\bullet -}$ (Ahmad et al., 2005; Boveris, 1977; Boveris & Cadenas, 1982) (Fig. 1.1). Once formed, $O_2^{\bullet -}$ is converted to H_2O_2 via the spontaneous or enzymatically driven dismutation

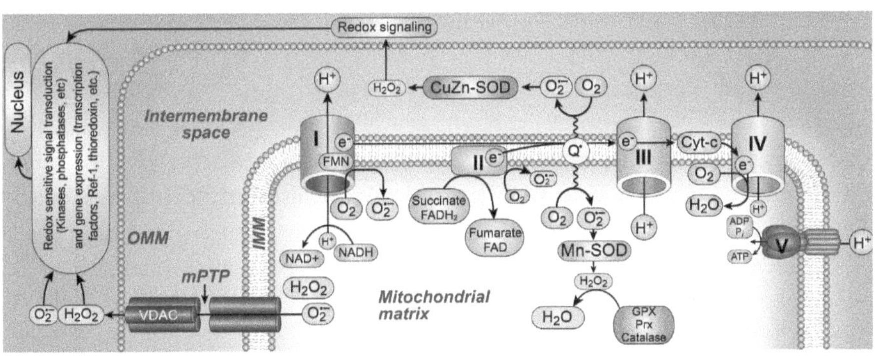

Figure 1.1 Mitochondrial electron transport chains can form superoxide and hydrogen peroxide that can act as signaling molecules transducing redox signals from metabolic processes to the nucleus. OMM, outer mitochondrial membrane; IMM, inner mitochondrial membrane; VDAC, voltage-dependent anionic channel; Ref-1, redox factor 1; FMN, flavin mononucleotide; mPTP, mitochondrial permeability transition pore; MnSOD, manganese superoxide dismutase; CuZnSOD, copper/zinc superoxide dismutase; Cyt-C, cytochrome C. (See the color plate.)

of $O_2^{\bullet-}$. Results using isolated mitochondria from normal tissues suggest that the majority of $O_2^{\bullet-}$ and H_2O_2 originate from complexes I and III (Boveris, 1977). Again, based on the data using isolated mitochondria from normal tissues, complex II is typically found to produce small amounts of $O_2^{\bullet-}$ (1% or less of the total) (Boveris, 1977) and was thought to be a minor contributor to overall $O_2^{\bullet-}$ and H_2O_2 production in normal cells (Fig. 1.1).

Recently, evidence is accumulating to suggest that cancer cell mitochondria may produce greater steady-state levels of $O_2^{\bullet-}$ and H_2O_2, relative to normal cells (Aykin-Burns, Ahmad, Zhu, Oberley, & Spitz, 2009). Several recent studies have also shown that mutations in complex II subunits C and D increase $O_2^{\bullet-}$ and H_2O_2 production as well as causing genomic instability and cancer induction (Aykin-Burns et al., 2011; Ishii et al., 1998; Owens et al., 2012; Slane et al., 2006). Furthermore, familial forms of two human cancers (paraganglioma and pheochromocytoma) have also been found to be associated with mutations in genes coding for subunits B, C, and D in complex II (Fliedner et al., 2012; Gimm, Armanios, Dziema, Neumann, & Eng, 2000). In addition, the mitochondria of malignant human tumor cells and rodent tumors have been shown to exhibit histological abnormalities characterized by unusual arrangements of mitochondrial cristae, mitochondrial hypertrophy, and mitochondrial fragmentation when compared to normal human cells (Bize, Oberley, & Morris, 1980; Springer, 1980). Furthermore, many tumors, including human epithelial cancers (i.e., colon and breast), have been shown to have high rates of mitochondrial DNA (mtDNA) mutations (relative to normal human tissues), and this has been suggested to contribute to increased $O_2^{\bullet-}$ and H_2O_2 production (Penta, Johnson, Wachsman, & Copeland, 2001; Yankovskaya et al., 2003). Furthermore, there are now several reports measuring $O_2^{\bullet-}$ and H_2O_2 in normal versus cancerous human cells, suggesting that cancer cell mitochondria produce higher steady-state levels of $O_2^{\bullet-}$ and H_2O_2, relative to normal cell mitochondria (Ahmad et al., 2005; Aykin-Burns et al., 2009).

Given this mounting evidence of increased steady-state levels of $O_2^{\bullet-}$ and H_2O_2 in cancer versus normal cells (Aykin-Burns et al., 2009; Dhar & St Clair, 2012; Droge, 2002; Yankovskaya et al., 2003), it is reasonable to hypothesize that mitochondria from cancer cells demonstrate alterations in the proper assembly of ETC complexes such that there is an increase in the residence time and/or accessibility of electrons on sites capable of mediating increased 1-electron reductions of O_2 (relative to normal cells) resulting in increased steady-state levels of $O_2^{\bullet-}$ and H_2O_2. This increased flux of ROS from cancer cell mitochondria could contribute to uncontrolled growth, the inability of cancer cells to differentiate, genomic instability, and disease progression. This hypothesis was originally proposed by Oberley et al. and forms the basis for the continuously evolving free radical theory of cancer (Oberley & Buettner, 1979; Oberley et al., 1980, 1981; Spitz, Sim, Ridnour, Galoforo, & Lee, 2000).

2.2.2 Nicotinamide adenine dinucleotide phosphate (NADPH) oxidases (NOXs)

An increasing body of evidence indicates that another major cellular source of ROS can be generated from a family of tightly regulated NOXs that are homologues of the phagocyte NOX (e.g., Phox or NOX2) originally discovered in neutrophils (Bedard & Krause, 2007; Lambeth, 2007). The NOX family of enzymes (NOXs 1–5, DUOX1, and DUOX2) represent a diverse and widely distributed group of redox signaling proteins that play physiological and pathophysiological roles in biology and medicine by transferring electrons from NAD(P)H to molecular O_2 to form $O_2^{\bullet -}$ and H_2O_2. In this regard, NOX enzymes can potentially transduce redox signals regarding the availability of NAD(P)H, which is necessary for many biosynthetic and detoxification processes necessary for maintaining redox homeostasis, to redox-sensitive signaling pathways that govern growth and survival. In addition to their originally proposed role in phagocytic cells and inflammatory cells, NOXs have also been shown to mediate redox signaling for a wide variety of normal cellular functions such as cell division, differentiation, wound healing, fibrosis, induction of progrowth/prosurvival pathways, and angiogenesis (Bonner & Arbiser, 2012; Coso et al., 2012; Kamata, 2009). Recently, it has also been suggested that the activation of NOXs in cancer cells could play an important role in oncogene-mediated cancer cell survival that could be targeted to inhibit prosurvival/progrowth pathways in order to inhibit cancer cell growth and progression (Weyemi, Redon, Parekh, Dupuy, & Bonner, 2013). In this regard, NOX enzymes appear to represent a major source of ROS-producing enzymes that can govern biological responses relevant to both cancer biology and therapy including cancer cell survival and normal tissue responses.

2.2.3 Other sources

In addition to mitochondria and NOX enzymes, there are a number of different pathways by which ROS can be formed during oxidative metabolism including peroxisomal metabolism, cytochrome P450 metabolism, inflammatory reactions, and the metabolism of lysyl oxidases (LOXs) that may be relevant to cancer biology (Barker, Cox, & Erler, 2012; Gonda, Tu, & Wang, 2009; Nordgren & Fransen, 2014; Pani, Galeotti, & Chiarugi, 2010; Rusyn et al., 2004; Seitz & Stickel, 2006). Peroxisomes are the site of many fatty acid and other enzymatic oxidation reactions that generate ROS containing antioxidant enzymes that metabolize ROS including catalasing that is very efficient at dealing with high levels of H_2O_2 (Antonenkov, Grunau, Ohlmeier, & Hiltunen, 2010). In fact, it appears that peroxisomes are highly compartmentalized for the specific purpose of keeping the enzymes that generate large amounts of H_2O_2 sequestered from the rest of the cellular compartments. Peroxisomal enzymes that produce ROS include many FAD-dependent (or FMN-dependent) oxidases that generate $O_2^{-\bullet}$ and/or H_2O_2 as reaction by-products including xanthine oxidase, acyl-CoA oxidases, urate oxidase, 2-hydroxyacid oxidases, polyamine oxidases, pipecolic acid and sarcosine oxidases, D-amino acid and D-aspartate oxidases, palmitoyl-CoA oxidase, and pristanoyl-CoA oxidase. These enzymes oxidize fatty acids, amino acids, purines, and nitrites and are important sources of breakdown products that represent building blocks for salvage pathways in metabolism. These enzymes necessarily generate high levels of H_2O_2 as a by-product of their enzymatic activity, which is believed to be the reason why catalase is peroxisomally located. Compounds that cause an increase in peroxisomal activity (termed peroxisomal proliferators) are thought to contribute to hepatocellular carcinoma development by increasing H_2O_2-mediated oxidative stress and DNA damage, leading to disturbances in redox signaling causing inflammation and aberrant cell proliferation and neoplastic transformation (Bosgra, Mennes, & Seinen, 2005).

Cytochrome P450 enzymes and their partner P450 reductases are xenobiotic-metabolizing enzymes that have also been associated with increased production of $O_2^{-\bullet}$ and/or H_2O_2 and carcinogenesis (Imaoka et al., 2004). Phenobarbital, 3-methylcholanthrene, and ethanol are strong inducers of cytochrome P450 enzymes that have been associated with carcinogenesis and tumor promotion. The induction of cytochrome P450 and P450 reductase by phenobarbital has also been shown to induce ROS production from liver microsomes that is capable of causing oxidative damage to DNA and lipids. This cytochrome P450 reductase-mediated ROS and subsequent oxidative stress have been suggested to account for the tumor-promoting activity of phenobarbital. It has also been proposed that genetic polymorphisms of cytochrome P450 enzymes associated with leukemia may also result in elevated ROS, which could contribute to genomic instability and uncontrolled growth signaling, but the causal involvement of ROS in these phenomena needs further confirmation (Irwin, Rivera-Del Valle, & Chandra, 2013).

The hypothesis that oxidative stress induced by chronic inflammation could mediate neoplastic transformation of stem cells arose from the realization that chronic inflammation was accompanied by the copious production of ROS, nitric oxide ($^{\bullet}NO$), and hypohalous acids (such as HOCl) by immune-competent cells that had become activated by a variety of stressors including autoimmune disease, xenobiotic exposure, viral infection, radiation exposure, obesity, asbestos exposure, and smoking (Archer, 1979; De Marco, 2013; Nathan & Cunningham-Bussel, 2013; Ohshima, Tatemichi, & Sawa, 2003; Saeidnia & Abdollahi, 2013). In this regard, inflammatory reactions may contribute to the induction of cancer caused by many if not all environmental stresses. Furthermore, ROS associated with inflammation are also causally associated with all the steps of carcinogenesis including initiation, promotion, and progression (De Flora & Ferguson, 2005; De Marco, 2013; Gukovsky, Li, Todoric, Gukovskaya, & Karin, 2013; Nathan & Cunningham-Bussel, 2013; Ohshima et al., 2003). In this regard, chronic inflammation appears to be a complete carcinogen that is capable of contributing to stress-induced and aging-associated carcinogenesis (Cannizzo, Clement, Sahu, Follo, & Santambrogio, 2011).

The extracellular LOX enzymes are regulated by hypoxia-inducible factor-1 (HIF-1)/HIF-2 transcription factors and utilize the lysyl tyrosylquinone cofactor to convert lysine residues to the aldehydic product, allysine, that is required for cross-link formation in the stabilization of collagen and elastin. The LOX catalytic cycle also produces H_2O_2 as a by-product. LOX activity is believed to contribute to metastasis by facilitating the cross-linking of collagen fibers, which is required for the recruitment of bone marrow (BM)-derived cells believed to be required for the formation of metastatic niches for CSCs to colonize distant organs (Semenza, 2013). In addition, the H_2O_2 produced as a by-product of LOX activity has been hypothesized to lead to enhanced migration and cell-matrix adhesion formation of invasive breast cancer cells via activation of the FAK/Src signaling pathway. These data suggest that inhibition of pathways leading to the activation of LOXs may represent viable targets for limiting the progression of aggressive breast cancers (Payne et al., 2005).

2.3. Antioxidants and antioxidant enzymes

2.3.1 Cellular antioxidants

An intricate system of nutritional antioxidants (i.e., vitamin C, vitamin E, lycopenes, and selenium), low-molecular-weight reducing cofactors and peptides (i.e., GSH, thioredoxin (Trx), and NADPH), and antioxidant enzymes [i.e., SOD enzymes, catalase, glutathione peroxidases (GPx), and peroxiredoxins (Prx)] regulates the oscillations in intra- and extracellular redox state that occur normally during oxidative metabolism and provides protection from ROS-induced damage during conditions of oxidative stress (Fig. 1.2) (Fridovich, 1978; Halliwell & Gutteridge, 2007; Rhee, Chae, & Kim, 2005; Sies, 1991; Spitz et al., 2004). During normal steady-state oxidative metabolism, periodic fluctuations in the intra- and extracellular redox environment are believed to coordinately regulate signal transduction and gene expression through redox-sensitive kinases, phosphatases, and transcription factors to maintain the cell in a nonequilibrium steady-state supporting most normal cellular functions including proliferation,

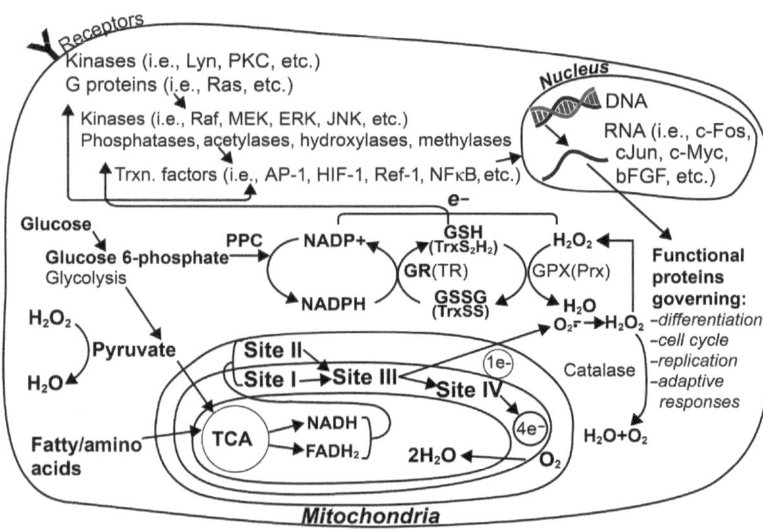

Figure 1.2 Theoretical model outlining redox couples and antioxidant pathways that coordinately regulate the flow of electrons from metabolism to redox-sensitive signaling and gene expression pathways contributing to the maintenance of the non-equilibrium steady state necessary for normal cellular functions during growth and development. TCA, tricarboxylic acid cycle; PPC, pentose phosphate cycle; GR, glutathione reductase; TR, thioredoxin reductase; GSH/GSSG, glutathione/glutathione disulfide; $TrxS_2H_2$/TrxSS, thioredoxin reduced and oxidized; GPX, glutathione peroxidase; Prx, peroxiredoxin; sites I–IV, electron transport chain complexes I–IV; Ref-1, redox factor 1.

differentiation, epigenetic regulation, and quiescence (Fig. 1.2) (Cyr & Domann, 2011; Cyr, Hitchler, & Domann, 2013; Droge, 2002; Gius & Spitz, 2006; Hitchler & Domann, 2012; Jorgenson et al., 2013; Li, Wicha, Schwartz, & Sun, 2011; Monfort & Wutz, 2013; Oberley et al., 1980, 1981; Ogasawara & Zhang, 2009; Sarsour, Kumar, Chaudhuri, Kalen, & Goswami, 2009; Spitz et al., 2000, 2004). Since cancer cells, relative to normal cells, are believed to demonstrate increased steady-state fluxes of ROS due to disruptions in oxidative metabolism, it is logical to propose that they would adapt by upregulating fluxes through reductive antioxidant pathways to avoid the potentially growth-inhibitory and cytotoxic accumulation of oxidative damage (Blackburn et al., 1999; Spitz et al., 2000). This proposition has led to the hypothesis that glucose metabolism is increased in cancer versus normal cells to (1) increase levels of pyruvate from glycolysis because pyruvate reacts with H_2O_2 through a deacetylation reaction to form H_2O and (2) regenerate NADPH from NADP+ in the pentose cycle to provide reducing equivalents for GPx and Prx enzymes to detoxify hydroperoxides (Fig. 1.2) (Blackburn et al., 1999; Spitz et al., 2000). This logic has also led to the hypothesis that increased fluxes of electrons through both prooxidant and antioxidant pathways in cancer cells, relative to normal cells, might represent a significant target for selectively manipulating cancer cell defects in oxidative metabolism to limit the growth and progression to the malignant phenotype while sparing sensitive normal tissues.

2.3.2 SOD and catalase

The first indication of an altered antioxidant profile in cancer versus normal cells was observed when the activity of the mitochondrial matrix form of the SOD enzyme, MnSOD, and sometimes catalase was found to be decreased in many transformed versus normal cells (Dionisi, Galeotti, Terranova, & Azzi, 1975; Oberley, Bize, Sahu, Leuthauser, & Gruber, 1978; Oberley & Buettner, 1979; Peskin, Koen, Zbarsky, & Konstantinov, 1977; Sahu, Oberley, Stevens, & Riley, 1977). The significance of fluctuations in the expression of MnSOD in cancer cells is still controversial (Dhar & St Clair, 2012), but a preponderance of recent evidence has supported the conclusion that during the initial stages of tumor formation, MnSOD is generally downregulated, but as cancer cells progress, the enzyme is upregulated contributing to the enhanced ability to achieve the metastatic potential seen in later-stage disease (Dhar & St Clair, 2012). This transition between lowered and increased expression of MnSOD during cancer progression has been shown to correlate with an apparently metabolic reprogramming of cancer cells that may relate to differences in steady-state H_2O_2 levels caused by the activity of MnSOD (Buettner, 2011; Dhar & St Clair, 2012; Li, Yan, Yang, Oberley, & Oberley, 2000).

Increased fluxes of H_2O_2 in cancer cells could be driven by the MnSOD-catalyzed conversion of $2O_2^{\bullet-} + 2H^+ \rightarrow H_2O_2 + O_2$, since the rate constant for this reaction is nearly diffusion-limited at $\approx 10^9$ $mol^{-1}s^{-1}$. If the forward rate constant (k_f) for a given protein (such as ubisemiquinone, $Q^\bullet$, in complex III; Fig. 1.1) to reduce O_2 to form $O_2^{\bullet-}$ ($Q^\bullet + O_2 \rightarrow Q + O_2^{\bullet-}$) is less than the reverse rate constant (k_r) of back reaction ($O_2^{\bullet-} + Q \rightarrow Q^\bullet + O_2$), then having high levels of MnSOD could actually lead to increased steady-state H_2O_2 formation by pulling the reaction of $Q^\bullet + O_2$ in the forward direction towards $O_2^{\bullet-}$ being rapidly converted to H_2O_2 (Buettner, 2011; Buettner et al., 2006; Kaewpila, Venkataraman, Buettner, & Oberley, 2008).

It has been postulated that the growth-inhibitory effects of MnSOD overexpression in cancer cells can be explained in part by increased fluxes of H_2O_2 (Dhar & St Clair, 2012; Oberley, 2005; Wang et al., 2005). Also, from this line of thinking, it is logical to hypothesize that once cancer cells have adapted to higher fluxes of H_2O_2 by increasing their capacity to detoxify hydroperoxides through increasing glycolytic metabolism and the activities of catalase, GPx, and Prx enzymes (Fig. 1.2), then the H_2O_2-mediated growth-inhibitory effects of MnSOD may no longer be a selective pressure against progression to metastasis (Dhar & St Clair, 2012; Oberley, 2005). Furthermore, increasing MnSOD activity in cancer cells when they have adequate peroxide removal systems in place may actually help to accelerate cancer cell progression to metastasis phenotype by limiting the reactions of $O_2^{\bullet-}$ with redox-active metals such as Fe^{+3} and Cu^{+2}, which would in turn limit the ability of these metal ions and $O_2^{\bullet-}$ to participate in free radical production from hydroperoxides through Haber–Weiss and Fenton chemistry (Koppenol, 2001).

Throughout the 30- to 40-year history of the paradoxical reports of alterations in SOD and catalase activities in cancer cells, the clear pattern that has emerged is that the regulation of SODs, catalase, GPx, and Prx is nearly universally altered in cancer versus normal cells (Buettner, 2011; Dhar & St Clair, 2012; Oberley, 2005). These alterations in $O_2^{\bullet-}$– and H_2O_2-metabolizing enzymes in cancer versus normal cells are thought to be indicative of disruptions in cancer cell oxidative metabolism leading to the inability of cancer cells to coordinately regulate one-electron signaling (directed by reactions of $O_2^{\bullet-}$ with metal ions in ETC chains and metal ions required for governing HIFs and Jumonji protein activity) with two-electron signaling pathways (directed by reactions of H_2O_2 with sulfhydryl and methionine residues on kinases, phosphatases, and acetylases) (Blackburn et al., 1999; Buettner, 2011; Cyr & Domann, 2011; Cyr et al., 2013; Hitchler & Domann, 2009, 2012; Spitz et al., 2000). In this regard, if CSCs are experiencing higher fluxes of ROS, relative to normal stem cells, in a background of an immortalized phenotype, the inability to regulate one-electron versus two-electron signaling could represent a fundamental metabolic abnormality rendering CSCs unable to properly control cell division and normal differentiation leading to the inevitable progression to malignancy.

2.3.3 GSH

GSH is the most abundant (1–10 mM) intracellular low-molecular-weight soluble thiol and is thought to act as a major intracellular redox buffer. GSH is a tripeptide containing a redox-active cysteine residue capable of providing reducing equivalents to the GPx enzymes to detoxify a wide variety of hydroperoxides including H_2O_2 and organic hydroperoxides (Fig. 1.2). Once oxidized to GSSG, GSH can be regenerated using electrons from NADPH by the glutathione reductase enzyme (Fig. 1.2). NADPH is then regenerated by metabolizing glucose in the pentose phosphate cycle (PPC) (Fig. 1.2). GSH can also be used by the glutathione transferase enzymes in conjugation reactions to detoxify a wide variety of electrophiles formed during oxidative stress (Spitz, Malcolm, & Roberts, 1990; Spitz, Sullivan, Malcolm, & Roberts, 1991). GSH is synthesized by transporting cystine into the cell via the xCT transporter; the cystine is then reduced to cysteine, which is conjugated to glutamate by the action of glutamate–cysteine ligase to form γ-glutamyl cysteine. γ-Glutamyl cysteine is then conjugated to glycine to form GSH by the action of glutathione synthase.

Given that cancer cells demonstrate increased steady-state fluxes of hydroperoxides relative to normal cells, it is not surprising that cancer cell GSH metabolism is usually dramatically upregulated (Spitz et al., 2000). Interestingly, CSCs have also been suggested to upregulate glutathione-dependent metabolism. (Herault et al., 2012; Ishimoto et al., 2011; Nagano, Okazaki, & Saya, 2013; Pei et al., 2013; Sato et al., 2013; Tamada et al., 2012). Since CSCs appear to rely heavily on the reducing equivalents in GSH to maintain redox homeostasis in the face of metabolic oxidative stress, manipulation of GSH by limiting its synthesis and/or inhibiting its redox recycling has been shown to be an effective means of sensitizing CSC clonogens to cell killing using a number of different pharmacological and genetic manipulations (Fath, Ahmad, Smith, Spence, & Spitz, 2011; Hadzic et al., 2010; Herault et al., 2012; Ishimoto et al., 2011; Nagano et al., 2013; Pei et al., 2013; Sato et al., 2013; Scarbrough et al., 2012; Simons, Ahmad, Mattson, Dornfeld, & Spitz, 2007; Simons et al., 2009; Sobhakumari et al., 2012; Tamada et al., 2012). Many of these pharmacological approaches are well tolerated in animals and show great promise for the development of combined-modality cancer therapies designed to selectively enhance metabolic oxidative stress in cancer versus normal stem cells.

2.3.4 Trx

Trx1 and Trx2 are 12 kDa proteins with a vicinal thiol moiety at their active site that acts as an electron donor in many redox signaling/gene expression pathways (Wei et al., 2000), ROS detoxification pathways (Rhee et al., 2005), and protein activation pathways where maintaining protein thiols in the reduced state is necessary for proper function (Arner & Holmgren, 2006; Du, Zhang, Zhang, Lu, & Holmgren, 2013; Lee, Kim, & Lee, 2013; Lu & Holmgren, 2014; Nakamura, Nakamura, & Yodoi, 1997; Ueda et al., 2002). Trx1 is primarily cytosolic and can be transported to the nucleus to activate redox-sensitive transcription, and Trx2 is located in the mitochondria and is believed to keep functional proteins in the reduced state. The vicinal thiol moiety of Trx contains two redox-active cysteine residues capable of providing reducing equivalents to the Prx enzymes (aka Trx peroxidases) to detoxify a wide variety of hydroperoxides including H_2O_2 and organic hydroperoxides (Fig. 1.2). In this regard, the Trx system in many cellular compartments is redundant with the GSH/GPx system, highlighting the importance of controlling hydroperoxide levels in cancer and normal cells. Once oxidized from $Trx(SH)_2$ to $Trx(S)_2$, $Trx(S)_2$ can be regenerated using electrons from NADPH via the action of Trx reductases (TrxRs) (Fig. 1.2). NADPH is again regenerated by metabolizing glucose in PPC (Fig. 1.2) (Arner & Holmgren, 2006; Lee et al., 2013; Nakamura et al., 1997). Given that cancer cells demonstrate increased steady-state fluxes of hydroperoxides, relative to normal cells, it is logical to hypothesize that cancer cell Trx metabolism is generally significantly upregulated (Arner & Holmgren, 2006; Du et al., 2013; Lee et al., 2013; Lu & Holmgren, 2014; Powis, Mustacich, & Coon, 2000). Since CSCs are believed to require the reducing equivalents in Trx and GSH to maintain redox homeostasis in the face of metabolic oxidative stress, manipulation of Trx activity by inhibiting its redox recycling has been shown to be an effective means of sensitizing CSC clonogens to cell killing using a number of different pharmacological manipulations including TrxR inhibitors such as auranofin combined with inhibitors of glutathione metabolism such as buthionine sulfoximine (BSO) (Du, Zhang, Lu, & Holmgren, 2012; Fath et al., 2011; Scarbrough et al., 2012; Simons et al., 2009; Sobhakumari et al., 2012). Again as with manipulation of GSH, many of the pharmacological approaches used to manipulate the Trx system, such as auranofin, are well-tolerated in humans and show promise for the development of combined-modality cancer therapies designed to selectively enhance metabolic oxidative stress-induced cell killing in cancer versus normal stem cells (Nguyen, Awwad, Smart, Spitz, & Gius, 2006; Pennington et al., 2007; Scarbrough et al., 2012).

2.3.5 Coordinate regulation of intracellular redox environment

A functionally interdependent relationship exists between metabolic and genetic processes necessary for maintaining living systems in the non-equilibrium steady state within which they propagate and reproduce their species (Fig. 1.3). Mammals are O_2-breathing organisms that represent complex higher-order biological structures that derive their life force from the ability to

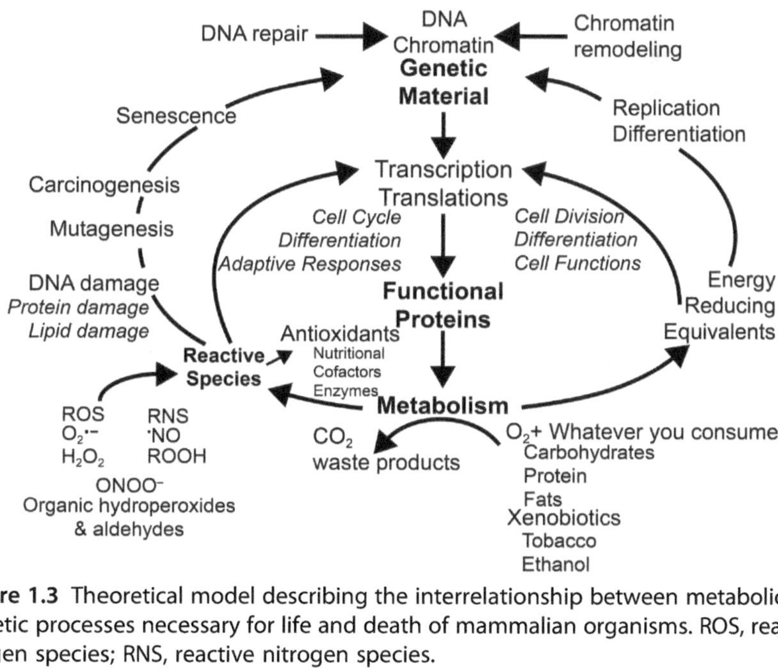

Figure 1.3 Theoretical model describing the interrelationship between metabolic and genetic processes necessary for life and death of mammalian organisms. ROS, reactive oxygen species; RNS, reactive nitrogen species.

extract, store, and move high-energy electrons. The genetic material coded for in the DNA of mammalian cells maintains the blueprint for transcription and translation of the functional proteins necessary for replicating and maintaining the living biological structure of each organism (Fig. 1.3). Oxidative metabolism by the biological structures removes high-energy electrons from the food sources and moves them onto electron carriers such as NADH, NADPH, and $FADH_2$. These high-energy electron carriers are used to generate energy via OXPHOS where O_2 acts as the terminal acceptor of the electrons once the energy from the electrons, in the form of proton motive force across the inner mitochondrial membrane, has been harnessed to synthesize ATP (Figs. 1.1 and 1.3). The high-energy electron carriers are also utilized as reducing equivalents for biosynthetic enzymes to make and break chemical bonds in organic molecules to maintain the differentiated structure of the living system, accomplish normal cellular functions, and drive the replication and maintenance of the living system (Fig. 1.3).

The biochemical basis of living systems that are based on O_2 metabolism is very efficient, and greater than 99% of the electrons fluxing through the biological structure leave the system as H_2O via the four-electron reduction of O_2. However, small amounts of one-electron reduction products of O_2 in the form of $O_2^{\bullet -}$ can be produced during metabolism that together with redox-active metal ions and $^{\bullet}NO$ can form a variety of highly reactive ROS and RNS (Fig. 1.3). The levels of these potentially damaging reactive species are regulated by cellular and nutritional antioxidants and low-molecular-weight redox cofactors necessary for antioxidant enzymes to function. The normal fluctuations in the oxidized to reduced ratios of electron carriers and thiol redox couples and the fluctuations in the reactive species act as redox signaling molecules that allow signal transduction proteins and gene expression pathways to "taste" the fluxes of electrons moving through all the metabolic pathways in order to coordinately regulate metabolism and gene expression for the execution of normal cell division, differentiation, and adaptive responses necessary to maintain homeostasis of the living system (Figs. 1.1–1.3). There is believed to be a small but finite probability that the reactive species produced as by-products of oxidative metabolism can evade the antioxidant pathways to cause covalent alterations in the genetic material in the form of base damage, adducts, and deletions to the DNA. This damage is usually repaired with high fidelity, but slowly, as a function of time, some of this damage escapes the high-fidelity repair pathways and can lead to heritable changes in the genetic material of the cell. When this damage accumulates to a level where the assembly and efficient function of the metabolic machinery begin to be affected, the metabolism of the living system is believed to become compromised leading to an acceleration of the production of reactive species and the accumulation of damage to the genetic material as metabolism continues to proceed. At this point in the life span of the organism, the slow accumulation of damage accelerates in an exponential fashion due to the increasing inefficiency that is introduced into the metabolic processes as damage to the metabolic machinery accumulates. The living system then begins to deteriorate in its ability to function normally as an exponential function of the age of the organism. This process of deterioration of the biological structure and function then manifests itself as degenerative diseases associated with aging and increased cancer induction. This biochemical scenario forms the basis for the free radical theories of aging and cancer that have been proposed for the last 60 years (Harman, 2003, 2006; Jorgenson et al., 2013; Oberley & Buettner, 1979; Spitz et al., 2000, 2004).

As the deterioration of oxidative metabolism begins to affect the stem cell compartments of living tissues, the stem cells are believed to either undergo senescence or lose their ability to regenerate normally functioning tissues or undergo neoplastic transformation where they become immortalized and begin the progression towards malignant metastatic phenotypes. Since during the transformation process CSCs presumably carry the same faulty oxidative metabolic process forward during their progression to malignancy, oxidative metabolism may represent a significant target for selectively inactivating cancer versus normal stem cells. From this theoretical construct, it is logical to propose that if the pathways (such as glucose, Trx, and GSH metabolism) that the CSCs must upregulate to survive their defect in oxidative metabolism could be simultaneously inhibited, then they would be selectively sensitized to cell killing by agents capable of inducing oxidative stress. Metabolic oxidative stress in CSCs could therefore represent an "Achilles heel" for developing novel combined-modality cancer therapies based on fundamental defects in cancer cell oxidative metabolism.

3. ROS AND NORMAL STEM CELLS

3.1. Types of major normal stem cells

As concluded by Leydig "*Omne vivum ex vivo* (all life [is] from life)," life neither ends nor begins but continues (Sell, 2004). A new life is formed by the union of an oocyte and a sperm after fertilization, which generates the first stem cell, for example, the zygote, in an organism. The zygote is a totipotent stem cell. It undergoes cleavage, proliferation, and differentiation to produce pluripotent ESCs and multipotent trophoblast stem cells. ESCs continue proliferating and differentiating to generate various multipotent tissue stem cells. These tissue stem cells have the ability to self-renew and produce various specialized cell types in a tissue or an organ during embryogenesis in an embryo and organogenesis in a fetus after they differentiate into transit-amplifying cells (TACs) or progenitor cells. In addition, they persist in various tissues in an adult, termed ASCs. ASCs are responsible for maintaining tissue homeostasis and repairing tissue injury by replenishing senescent and damaged cells for the entire life of an organism. Somatic cells can become pluripotent stem cells after their nuclei are transferred into enucleated oocytes (Gurdon, 1962; Gurdon, Elsdale, & Fischberg, 1958) or reprogrammed by ectopic expression of a panel of defined transcription factors (such as Oct4, Sox2, Klf4, and Myc) (Takahashi et al., 2007; Takahashi & Yamanaka, 2006), incubation with selective small molecules (Hou et al., 2013), or a combination of both (Zhu et al., 2010). These induced pluripotent stem cells (iPSCs) exhibit some of the same properties as ESCs, including expressing certain ESC genes and proteins, forming embryoid body *in vitro* and teratoma *in vivo*, and differentiating into various somatic cells in culture and in a tetraploid blastocyst complementation assay. Therefore, "*Omne vivum ex vivo* (all life [is] from life)" through stem cells (Fig. 1.4).

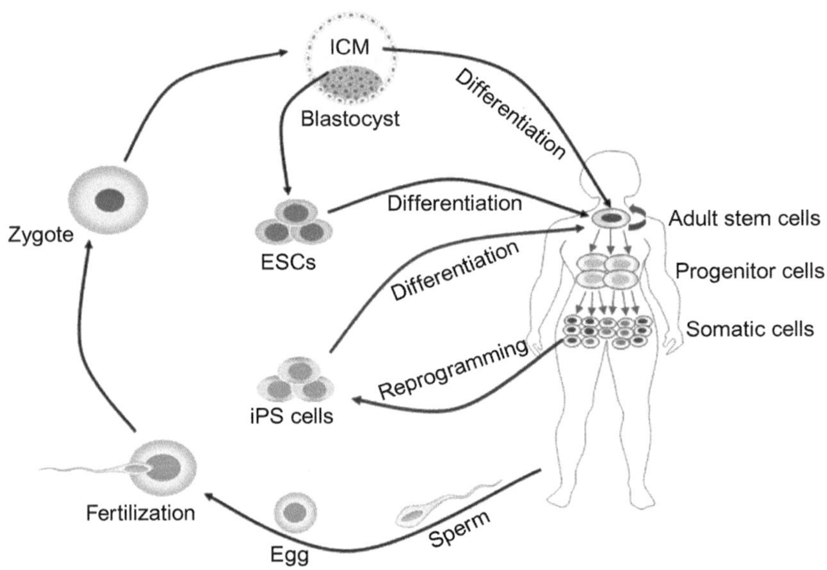

Figure 1.4 Life cycle through stem cells. ICM, inner cell mass; ESCs, embryonic stem cells; iPSCs, inducible pluripotent stem cells.

3.1.1 ESCs

ESCs can be derived from the inner cell mass (ICM) or an epiblast of an early embryo, for example, a blastocyst. The first ESC lines were derived from mouse embryos in 1981 independently by Evans and Kaufman (1981) and Martin (1981). It took another 17 years for Thomson et al. to derive the first human ESC lines from human blastocysts (Thomson et al., 1998). ESCs exhibit an almost unlimited capacity of cell proliferation *in vitro* but retain a normal karyotype. They have the potential to differentiate into all lineages of cells in an embryo body *in vitro* and in teratomas after ectopic transplantation into mice. After being injected into blastocysts, ESCs can contribute to all cell types including the germ cells in chimeric mice (Bradley, Evans, Kaufman, & Robertson, 1984) but poorly to the trophectoderm and extraembryonic endoderm (Beddington & Robertson, 1989). Therefore, ESCs are pluripotent cells. They are different from the totipotent zygote and blastomeres because the zygote and blastomeres can turn into both embryos and the placenta and other extraembryonic tissues via ESCs and trophoblast stem cells, respectively.

3.1.2 ASCs

ASCs originated from ESCs during embryogenesis with the ability to differentiate into multiple types of cells (multipotency) down to a single cell type (unipotency) in a tissue (Sell, 2004). ASCs can be found not only in fetal tissues but also in various tissues of children and adults because they can self-renew and persist through adulthood. However, unlike ESCs, ASCs cannot grow indefinitely in culture and divide infrequently *in vivo* to generate one stem cell and a TAC that then undergoes a limited number of cell divisions before terminally differentiating into mature cells in a tissue. ASCs are supposed to play an important role in the maintenance of tissue homeostasis and tissue repair after injury by replenishing aged and damaged cells in a body. Therefore, they can be found in many tissues with a high rate of cell turnover, such as those in the BM (Becker, Mc, & Till, 1963; McCulloch & Till, 1960; Till & Mc, 1961), intestine (Barker et al., 2007; Jung et al., 2011), and skin (Nowak, Polak, Pasolli, & Fuchs, 2008; Sun & Green, 1976). In addition, they can also been found in tissues undergoing a slow rate of cell turnover, such as brain (Reynolds & Weiss, 1992; Uchida et al., 2000), heart (Beltrami et al., 2003; Oh et al., 2003), and skeletal muscle (Lipton & Schultz, 1979; Mauro, 1961). Since hematopoietic stem cells (HSCs) are the first ASCs that have been isolated in high purity and the best characterized ASCs in our body, we will use HSCs as an example to discuss some of the fundamental biology of ASCs.

In the early 1960s, Till and McCulloch showed that mouse BM HSCs can self-renew and give rise to multiple lineages of progeny after transplantation into lethally irradiated animals (Becker et al., 1963; McCulloch & Till, 1960). This landmark discovery laid the foundation for modern stem cell biology and hematology research. However, as demonstrated by Till and McCulloch in their pioneering works, the cells that were originally believed to be HSCs identified in their colony-forming unit-spleen (CFU-S) assay were heterogeneous, because they had variable capacity for self-renewal (Becker et al., 1963; McCulloch & Till, 1960; Till & Mc, 1961). This finding provoked a series of investigations, aiming to prospectively identify and isolate HSCs for characterization. Through decades of research, HSCs and their progeny, including multipotent progenitors (MPPs) and various hematopoietic progenitor cells (HPCs), can now be prospectively isolated in high purity using multiparameter flow cytometry and a large array of monoclonal antibodies against various cell surface molecules not only in mice but also in humans (Fig. 1.5).

Murine HSCs, MPPs, and HPCs are lineage-negative (Lin$^-$) hematopoietic cells, because they are immature hematopoietic cells and express no

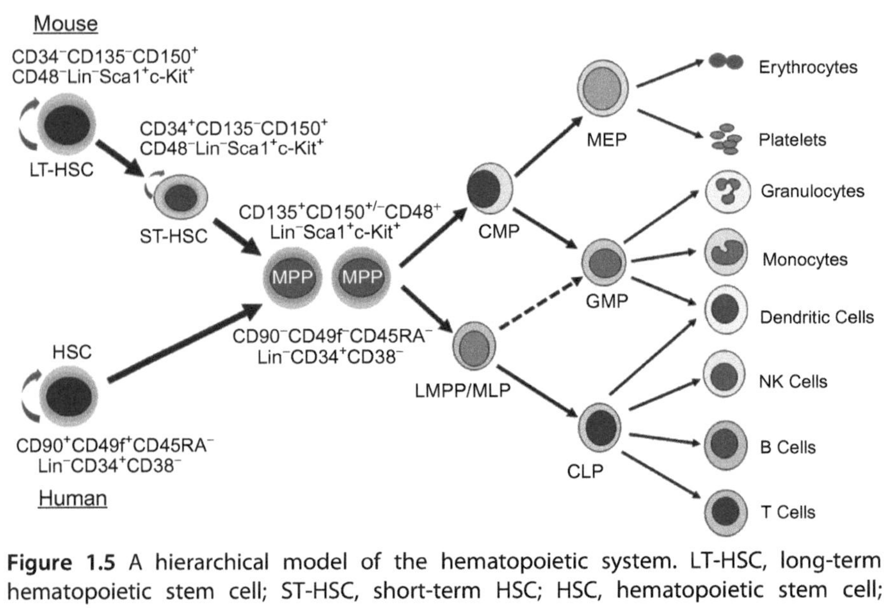

Figure 1.5 A hierarchical model of the hematopoietic system. LT-HSC, long-term hematopoietic stem cell; ST-HSC, short-term HSC; HSC, hematopoietic stem cell; MPP, multipotent progenitor; LMPP/MLP, lymphoid-primed multipotent progenitor/multilymphoid progenitor; CMP, common myeloid progenitor; CLP, common lymphoid progenitor; MEP, megakaryocyte/erythroid progenitor; GMP, granulocyte/monocyte progenitor. (See the color plate.)

detectable levels of lineage cell surface markers, such as B220, CD4, CD8, Gr-1, Mac-1, and Ter-119. Both HSCs and MPPs express c-Kit and Sca1, and thus, they are collectively called LSK ($Lin^-sca1^+c\text{-}kit^+$) cells. In contrast, HPCs express c-Kit but not Sca1 and thus are termed LS^-K^+ ($Lin^-sca1^-c\text{-}kit^+$) cells (Kondo et al., 2003). Later, Kiel et al. demonstrated that the expression of CD150 and CD48 can be used to separate HSCs and MPPs, because HSCs are $CD150^+CD48^-$LSK cells and MPPs are $CD150^{+/-}CD48^+$LSK cells (Kiel, Yilmaz, Iwashita, Terhorst, & Morrison, 2005). Alternative strategies using other cell surface markers and dye effluxing have also been used to identify and isolate mouse HSCs. These include identification of HSCs as $CD34^-$LSK cells (Osawa, Hanada, Hamada, & Nakauchi, 1996), $Thy1^{lo}Flk\text{-}2^-$LSK cells (Christensen & Weissman, 2001), and the Hoechst-effluxing side population (SP) cells (Goodell, Brose, Paradis, Conner, & Mulligan, 1996). More recently, according to the expression of CD34, CD150, and CD48, HSCs can be further differentiated into long-term or dormant HSCs ($CD34^-CD135^-CD150^+CD48^-$LSK cells) and short-term HSCs ($CD34^+CD135^-CD150^+CD48^-$LSK cells) (Wilson et al., 2008) (Fig. 1.5).

The first marker to enrich human HSCs, for example, CD34, was discovered by Civin et al. (1984). It has been widely used in clinic to harvest human HSCs from BM, mobilized peripheral blood, or cord blood for transplantation since its discovery. However, the majority of Lin^-CD34^+ cells are committed progenitors. Therefore, extensive studies have been done over the years to identify additional markers that can be used to prospectively isolate and characterize human HSCs. These studies were facilitated by the use of long-term culture-initiating cell (LTC-IC) assay and various severe combined immunodeficiency (SCID) mice as transplantation recipients. For example, Lansdorp, Sutherland, and Eaves (1990) found that $CD34^+CD45RA^-$ cells were enriched for LTC-ICs. In addition, using a humanized SCID mouse model, Baum, Weissman, Tsukamoto, Buckle, and Peault (1992) and Murray et al. (1995) showed that $Lin^-CD34^+Thy\text{-}1$ $(CD90)^+$ cell population from human fetal BM and mobilized peripheral blood contains more multipotent hematopoietic progenitors than $Lin^-CD34^+CD90^-$ cells. Subsequently, Bhatia, Wang, Kapp, Bonnet, and Dick (1997) demonstrated that CD38 can be used as an additional marker to differentiate primitive human hematopoietic cells from more committed progenitors and $Lin^-CD34^+CD38^-$ cells represent a population of highly purified SCID-repopulating cells. More recently, Notta, Doulatov, et al. (2011) found that the expression of integrin $\alpha6$ (CD49f) can be used to differentiate human HSCs and MPPs. Thus, through almost two decades of research, human HSCs can be identified as $Lin^{-}CD34^{+}CD38^{-}CD45RA^{-}CD90^{+}CD49f^{+}$ cells, whereas $Lin^{-}CD34^{+}CD38^{-}CD45RA^{-}CD90^{-}CD49f^{-}$ cells are MPPs and $Lin^{-}CD34^{+}CD38^{+}$ cells contain various committed progenitors or HPCs (Fig. 1.5).

In both rodents and humans, the hematopoietic system is organized in a hierarchical manner (Fig. 1.5). The rare HSCs reside at the top of the hierarchy and have the ability to self-renew, proliferate, and differentiate into different lineages of peripheral blood cells though MPPs and HPCs (Reya, 2003; Weissman, Anderson, & Gage, 2001). HSCs are quiescent under steady-state conditions and serve as a reserve that protects the hematopoietic system from exhaustion under various stress conditions (Wilson, Laurenti, & Trumpp, 2009). In contrast, MPPs and HPCs are proliferating cells with limited and no self-renewal ability, respectively. The proliferation and differentiation of MPPs and HPCs meet the needs of normal hematopoiesis and also allow the hematopoietic system to react swiftly and effectively to meet demands for increased production of mature cells during hematopoietic crises, such as loss of blood, hemolysis, and infection. Quiescent HSCs can be activated in response to severe hematopoietic damage when MPPs and HPCs are depleted by an exogenous stressor. Under such circumstances, activated HSCs can undergo self-renewing proliferation and differentiation to repopulate MPPs and HPCs and restore homeostasis. However, if HSCs are injured or their self-renewing ability is impaired, HSCs could be exhausted, leading to BM failure and death of the organism (Wang, Schulte, & Zhou, 2006).

The majority of HSCs are assumed to reside in the osteoblastic niche adjacent to the endosteal bone surface (Calvi et al., 2003; Zhang et al., 2003). The osteoblastic niche provides HSCs with a special environment that supports their self-renewal. This is likely achieved in part by extensive interactions between HSCs and the niche via a variety of soluble factors, such as Wnt (Sugimura et al., 2012), bone morphogenetic proteins (BMPs) (Goldman et al., 2009), angiopoietin-1 (Arai et al., 2004), osteopontin (Nilsson et al., 2005; Stier et al., 2005), thrombopoietin (TPO) (Yoshihara et al., 2007), interleukin-3 (IL-3), and IL-6 (Barria, Mikels, & Haas, 2004); various adhesion molecules, including CXCL12–CXCR4 (Sugiyama, Kohara, Noda, & Nagasawa, 2006) and N-cadherin (Haug et al., 2008); and different signaling pathways, for example, stem cell factor (SCF)/c-Kit (Broudy, 1997) and Jagged/Notch (Mercher et al., 2008). These intricate interactions promote HSC self-renewal not only by increasing HSC survival but also by keeping them quiescent in a hypoxic environment to prevent HSCs from exhaustion. In addition, sinusoidal endothelial cells (SECs) in BM have been revealed to function as an alternative HSC niche called the vascular niche (Kiel et al., 2005). The vascular niche plays an important role in hematopoietic development during embryonic and fetal development. It is also involved in regulation of HSC/HPC mobilization, proliferation, and differentiation in response to hematopoietic stress (Hattori et al., 2002). Therefore, HSCs may use either osteoblasts or endothelial cells as their niche under different circumstances to maintain a fine balance between quiescence and proliferation or self-renewal and differentiation and to respond to stress (Trumpp, Essers, & Wilson, 2010). Furthermore, recent studies have identified additional types of cells that contribute to the compositions of the HSC niche, which include arteriolar pericytes (Kunisaki et al., 2013), perivascular cells, CXCL12-abundant reticular (CAR) cells, mesenchymal stem cells, macrophages, regulatory T cells, and Schwann cells along with sympathetic nerve fibers (Chow et al., 2011; Fujisaki et al., 2011; Mendez-Ferrer et al., 2010; Omatsu et al., 2010; Yamazaki et al., 2011). These cells can also regulate various HSC activities directly and/or indirectly via affecting the function of the osteoblastic and vascular niche.

3.2. Role of ROS in stem cell physiology

3.2.1 ROS and stem cells

In the beginning of the twentieth century, Warburg (1908) observed a respiratory burst immediately after the sea urchin eggs were fertilized by the sperms to generate the zygotes. However, the biological significance of the increased O_2 consumption by the fertilized eggs was unknown until nearly 100 years after the initial observation by Warburg (Wong, Creton, & Wessel, 2004; Wong & Wessel, 2005). Wong and her colleagues revealed that the oxygen burst of the fertilized eggs is attributed to the activation of Udx1, a dual NOX, which converts O_2 to H_2O_2. H_2O_2 produced by Udx1 not only functions as a substrate of ovoperoxidase to participate in fertilization envelop cross-linking to prevent polyspermy but also acts as signaling molecule to regulate the zygote cleavage. Since then, accumulating evidence demonstrates that ROS plays a very important role in determination of the fate of various stem cells in a concentration-dependent manner. For example, Ezashi, Das, and Roberts (2005) showed that human ESCs (hESCs) underwent spontaneous differentiation when they were cultured under a normoxic condition (e.g., 21% O_2). When the cells were cultured under a physiological O_2 tension (3–5% O_2), the amount of spontaneous cell differentiation was significantly reduced without significant change in the rate of cell proliferation. However, further reduction in O_2 concentration (to 1%) inhibited the cell proliferation. The effects of O_2 on hESC differentiation and proliferation are likely in part attributable to the production of ROS, because the treatment of hESCs with BSO that increases ROS production by reducing intracellular levels of GSH promoted hESC differentiation via mitogen-activated protein kinase (MAPK) signaling (Ji et al., 2010). Similar findings were also observed in mouse ESCs (mESCs) in which ROS also acts as a signaling molecule to promote mESC cardiovascular differentiation (Sauer & Wartenberg, 2005).

The effects of ROS on ASCs, particularly on HSCs, have also been extensively studied. Low levels of ROS are permissive for HSCs to maintain their normal functions, including proliferation, differentiation, and mobilization (Juntilla et al., 2010; Kinder et al., 2010; Lewandowski, Sheehan, Bennett, & Boswell, 2010; Owusu-Ansah & Banerjee, 2009). For example, it was reported recently that HSCs from *Akt1/2* double-knockout mice exhibit a defect in long-term hematopoietic reconstitution after transplantation (Juntilla et al., 2010). The defect is attributable to the reduced production of ROS, as moderate elevation of ROS in HSCs by incubation of the cells from the knockout mice with low doses of the glutathione depleting agent, BSO, increased their clonogenicity. This is in agreement with another recent observation that ROS-dependent proliferation of HSCs also plays an important role in the early steps of hematopoietic reconstitution after HSC transplantation (Lewandowski et al., 2010). However, increased production of ROS can be detrimental to HSCs. Jang and Sharkis (2007) showed that mouse BM HSCs could be separated into ROS^{high} and ROS^{low} populations based on the intensity of dichlorodihydrofluorescein (DCF) staining. ROS^{high} HSCs exhibited a reduced ability to form the colony-forming unit granulocyte–erythrocyte–monocyte–megakaryocyte (CFU-GEMM) and LTC-ICs *in vitro* and to produce long-term engraftment after transplantation in comparison with ROS^{low} HSCs. ROS^{high} HSCs, but not ROS^{low} HSCs, became exhausted after the third transplantation. These defects in ROS^{high} HSCs were associated with an increased activation of p38 MAPK (p38) and mammalian target of rapamycin (mTOR), which could be attenuated by the treatment with an antioxidant, a p38 inhibitor, or rapamycin. Furthermore, incubation of mouse BM HSCs with high concentrations of BSO resulted in a dramatic reduction in HSC clonogenicity (Juntilla et al., 2010; Rodriguez et al.,

2011). In addition, increased production of ROS has been found to have a cause and effect relationship with HSC defects in various pathological conditions. For example, deletion of the ataxia-telangiectasia mutated (*Atm*) gene in mice increases ROS production, which comprises the ability of HSCs to self-renewal and leads to HSC premature exhaustion by disrupting HSC quiescence and stimulating HSC cycling (Ito et al., 2004, 2007). Treatment of $Atm^{-/-}$ mice with N-acetylcysteine (NAC) can restore the function of HSCs and prevent the development of BM failure (Ito et al., 2004). Subsequently, it was shown that the number of HSCs and their long-term repopulating activity were markedly reduced in association with an increased production of ROS in HSCs after the deletion of the genes encoding the O subclass of the forkhead family of transcription factors, for example, *Foxos* (*Foxo1*, *Foxo3*, and *Foxo4*) in mice (Tothova et al., 2007). These defects were associated with an increased production of ROS in HSCs and ameliorated by the treatment with NAC. In addition, increased production of ROS is also associated with HSC defect in several other pathological conditions, including deletion of *Bmi1* (Park et al., 2003; Schuringa & Vellenga, 2010), the mouse double minute 2 homologue gene (*Mdm2*) (Abbas et al., 2010), and the tuberous sclerosis 1 (*Tsc1*) gene (Chen et al., 2008); Fanconi anemia mutation (Du, Adam, Rani, Zhang, & Pang, 2008); aging (Ito et al., 2006); and post-ionizing radiation (IR) exposure (Shao et al., 2013; Wang et al., 2010).

3.2.2 Hypoxia and metabolism of stem cells

In general, stem cells are more sensitive to the adverse effects of ROS than their progeny. Increased production of ROS can inhibit stem cell self-renewal not only by promoting stem cell differentiation but also via induction of stem cell senescence and/or apoptosis as discussed in the next section. To maintain stem cell stemness, both ESCs and ASCs have developed various abilities to minimize the impact of ROS on the cells. Since O_2 is an important ingredient for the production of ROS, ESCs and various ASCs preferentially reside in a hypoxic environment to reduce ROS production (Eliasson & Jonsson, 2010; Ezashi et al., 2005; Mohyeldin, Garzon-Muvdi, & Quinones-Hinojosa, 2010; Sauer & Wartenberg, 2005; Suda, Takubo, & Semenza, 2011). ESCs are derived from ICM or epiblast of a blastocyst before implantation. They are adapted to the low levels of O_2 in the uterus where these cells reside *in vivo* (Fischer & Bavister, 1993) and undergo less differentiation when they are maintained in a hypoxic condition *in vitro* (Ezashi et al., 2005). In addition, ESCs preferentially use glycolysis as a main source of energy, which also reduces ROS production through mitochondrial OXPHOS. This is because undifferentiated ESCs possess small and immature mitochondria with lower levels of mitochondrial mass, membrane potential, and OXPHOS activity compared with the mitochondria in mature or differentiated cells (Rafalski, Mancini, & Brunet, 2012; Shyh-Chang, Daley, et al., 2013). As ESCs differentiate and commit to a specific cell fate, the cells gain more mitochondria with a more mature phenotype (such as an elongated morphology with swollen cristae and dense matrices) and exhibit increased copy numbers of mtDNA per cell and elevated levels of O_2 consumption and ATP production (Facucho-Oliveira & St John, 2009). In contrast, a switch from OXPHOS back to glycolysis occurs when differentiated cells are converted to iPSCs during reprogramming (Folmes et al., 2011; Shyh-Chang, Locasale, et al., 2013). In addition, elevation of glycolysis promotes the induction of iPSCs, whereas stimulation of OXPHOS or inhibition of glycolysis reduces the induction (Shyh-Chang, Daley, et al., 2013). These findings suggest that increased glycolysis and reduced OXPHOS are the common metabolic characteristics of pluripotent stem cells, which can help stem cells to maintain their stemness by reducing ROS production.

The endosteal osteoblastic niches of HSCs are also considered hypoxic, as they are relatively remote from blood flow (Winkler et al., 2010). It is estimated that the concentration of O_2 in these niches is below 1%. In the mouse BM, HSCs show lower blood perfusion as determined by low Hoechst 33342 (Hoe) staining after the dye injection (Parmar, Mauch, Vergilio, Sackstein, & Down, 2007). In addition, HSCs can be enriched in the BM cell populations with the most pimonidazole staining after administration of pimonidazole, a chemical marker for hypoxia (Parmar et al., 2007). Furthermore, human cord blood $CD34^+$ cells became hypoxic within a few weeks after transplantation into NOD/SCID interleukin-2 receptor γ chain knockout mice (Shima et al., 2010). These findings suggest that hypoxia is an important component of the HSC niche, which can protect HSCs from oxidative damage by reducing the production of ROS in HSCs. Similarly, other ASCs are also believed to reside in a hypoxic environment to be preserved in an undifferentiated state (Mohyeldin et al., 2010). In addition, hypoxia increases the expression and activity of HIF-1α in HSCs (Simsek et al., 2010; Takubo et al., 2010). Increased expression of HIF-1α and activity of HIF-1 alters the metabolism of HSCs by upregulating glycolysis while downregulating OXPHOS, leading to reduced production of ROS (Simsek et al., 2010; Takubo et al., 2010).

In addition, HIF-1 also plays a critical role in the maintenance of HSC quiescence not only via inhibiting ROS production but probably also via antagonizing the effect of c-Myc to upregulate the expression of p21 (Koshiji et al., 2004). Similar to ESCs, quiescent HSCs possess low mitochondria mass and immature mitochondrial morphology and thus primarily utilize glycolysis rather than OXPHOS for ATP production (Piccoli et al., 2005; Simsek et al., 2010). Therefore, HSCs are presumably better protected from oxidative stress to maintain their ability to self-renew by residing in a hypoxic environment and being in a quiescent state in the endosteal osteoblastic niches.

3.2.3 Regulation of ROS production in stem cells

Since ROS plays an important role in the determination of the fate of stem cells, its production in stem cells has to be tightly regulated. This is achieved in part via a network of transcription factors and signaling transduction pathways as discussed in the succeeding text (Fig. 1.6).

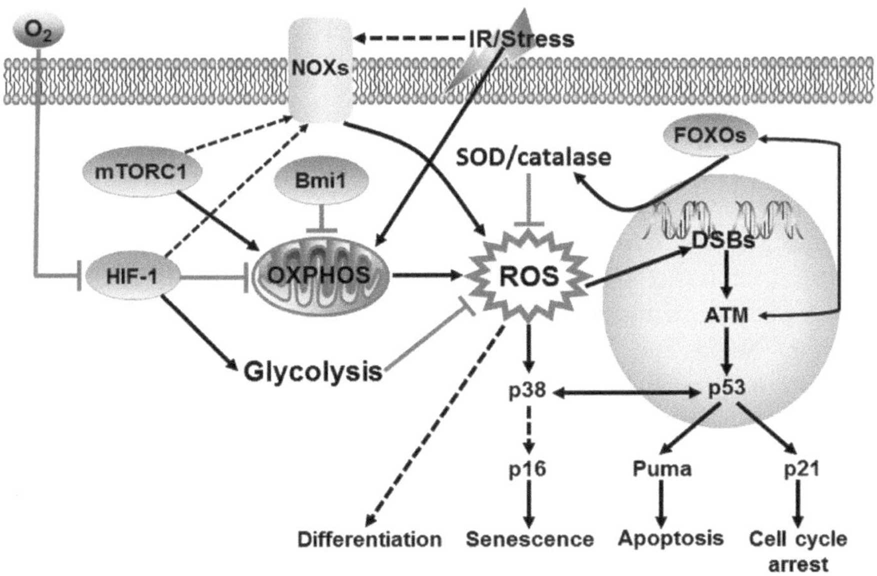

Figure 1.6 Theoretical model illustrating the regulation of ROS production in HSCs and the effects of ROS on HSCs. HSCs, hematopoietic stem cells; IR, ionizing radiation; NOXs, NADPH oxidases; mTORC1, mammalian target of rapamycin complex 1; HIF-1, hypoxia-inducible factor-1; OXPHOS, oxidative phosphorylation; SOD, superoxide dismutase; ROS, reactive oxygen species; FOXOs, forkhead box O transcription factors; DSBs, DNA double-strand breaks; ATM, ataxia-telangiectasia mutated. (See the color plate.)

HIF-1: HIF-1 is a heterodimeric transcription factor. It consists of two basic helix-loop-helix proteins: an oxygen-sensitive HIF-1α subunit and a constitutively expressed HIF-1β subunit (Semenza, 1999, 2004; Wang & Semenza, 1995). HIF-1 activity is regulated by posttranslational modification of the oxygen-dependent degradation (ODD) domain in the α subunit by a prolyl hydroxylase (PHD) at the proline residues 402 and/or 564 (Kaelin & Ratcliffe, 2008). Prolyl-hydroxylated HIF-1α is ubiquitinated by the E3-ubiquitin-ligase von Hippel–Lindau (VHL) tumor suppressor protein, leading to its subsequent degradation by the 26*S* proteasome. Under hypoxic conditions, HIF-1α proteins are stabilized due to suppression of PHD and reduced HIF-1α hydroxylation (Kaelin & Ratcliffe, 2008). The stabilized HIF-1α dimerizes with HIF-1β to form heterodimers that can bind to hypoxia response elements (HREs) to activate the transcription of numerous target genes (Semenza, 2004). The downstream target genes of HIF-1 include glucose transporter 1 (GLUT1), lactate dehydrogenase A (LDHA), and pyruvate dehydrogenase kinase 1 (PDK1) (Semenza, 1999, 2004). LDHA converts pyruvate into lactate, whereas PDK1 can phosphorylate and inactivate pyruvate dehydrogenase (PDH) that catalyzes the conversion of pyruvate into acetyl-CoA. Increase in LDHA and PDK1 expression shunts pyruvate away from OXPHOS. Therefore, HIF-1 activation induces the switch of cellular metabolism from OXPHOS to glycolysis, resulting in reduced O_2 consumption and ROS production.

Knockout of HIF-1α or HIF-1β in mice is embryonically lethal (Keith & Simon, 2007). Coculture of hESCs with human fetal liver stromal cells overexpressing HIF-1α inhibits hESC differentiation in association with an increased expression of Oct4 and Nanog (Ji et al., 2009). In addition, HIF-1 can potentially affect ESC self-renewal and differentiation via regulation of Notch and Wnt signaling (Keith & Simon, 2007; Mazumdar, Dondeti, & Simon, 2009). However, the roles of HIF-1 in the regulation of ESC hypoxic response, metabolism, and ROS production have yet to be investigated. In contrast, the effect of HIF-1 on HSCs has been extensively studied. It was shown that HSCs express higher levels of HIF-1α mRNA and protein than their progeny (Simsek et al., 2010; Takubo et al., 2010). The expression of HIF-1α in HSCs is regulated by the transcriptional factor Meis 1 (Kocabas et al., 2012; Simsek et al., 2010; Unnisa et al., 2012). Conditional knockout of Meis 1 or HIF-1α in HSCs induced a shift of HSC metabolism from glycolysis to OXPHOS. This resulted in increases in HSC ROS production, cell cycle entry and proliferation, and apoptosis and decreases in the ability of HSCs to reconstitute the hematopoietic system during serial BM transplantation and to tolerate stress such as 5-fluorouracil administration. These HSC defects could be attenuated by treatment of the knockout mice with NAC, knockdown of VHL, or conditional monoallelic knockout of VHL (Kocabas et al., 2012; Simsek et al., 2010; Takubo et al., 2010; Unnisa et al., 2012). However, overactivation of HIF-1 in HSCs induced by conditional biallelic knockout of VHL or inhibition of PHD is detrimental to HSCs and can lead to HSC premature exhaustion (Eliasson et al., 2010; Takubo et al., 2010). These findings demonstrate that HIF-1α plays an essential role in the maintenance of HSCs. Moderate levels of HIF-1 activity are beneficial for HSCs by restricting OXPHOS and ROS production and keeping HSCs in quiescence, whereas excessive levels of HIF-1 activity are harmful to HSCs. The mechanism by which excessive levels of HIF-1 activity aversively affect HSCs remains to be investigated.

FOXOs: The forkhead box O (FOXO) family of transcription factors consists of four members: FOXO1, FOXO3, FOXO4, and FOXO6 (Eijkelenboom & Burgering, 2013). They can regulate the expression of a large array of genes that affect many cellular processes, including metabolism, cell cycle regulation, differentiation, survival, and response to stress. The activities of FOXOs are normally suppressed in proliferating cells by insulin and growth factor signaling that activates phosphatidylinositol 3-kinases (PI3Ks) to increase the production of phosphatidylinositol-3-phosphate (PIP3). PIP3 functions as a second messenger to recruit and activate phosphoinositide-dependent kinase 1 (PDK1) and protein kinase B (PKB or Akt). Active Akt translocates to the nucleus and phosphorylates FOXO proteins at three consensus phosphorylation sites. Phosphorylated FOXOs are then exported from the nucleus to the cytoplasm by 14-3-3 proteins, resulting in reduced expression of target genes. In addition, the activities of FOXOs can be regulated by diverse signaling pathways and a variety of stresses. Accumulating evidence suggests that FOXOs play an important role in stem cell maintenance. For example, FOXO1 is essential for the maintenance of ESC pluripotency by regulating the expression of OCT4 and SOX2 (Zhang et al., 2011). More importantly, it was found that FOXOs maintain HSC self-renewal because the numbers of BM HSCs in $Foxo1^{-/-}/Foxo3a^{-/-}/Foxo4^{-/-}$ triple-knockout mice were dramatically reduced and their ability to produce long-term engraftment after transplantation was significantly lower than the cells from wild-type mice (Tothova et al., 2007). These HSC defects were associated with reduced expression of SOD2, catalase, and ATM, increased expression of $p16^{Ink4a}$ (p16) and production of ROS, accelerated cell cycling, and decreased cell survival. Similar findings were also observed in *Foxo3a* knockout mice, as BM HSCs from this knockout mice exhibited high levels of ROS, activation of p38, increased expression of p16 and Arf, and loss of self-renewal (Miyamoto et al., 2007). Administration of the antioxidant NAC to the knockout mice could rescue the defects of HSCs. These findings demonstrate that FOXOs maintain HSC homeostasis primarily via transcriptional regulation of antioxidants in addition to regulation of cell quiescence and DNA damage repair (Tothova & Gilliland, 2007).

Bmi1: Bmi1 is a member of the Polycomb family of transcriptional repressors that assemble in multimeric complexes and transcriptionally repress target genes through histone modifications (Sauvageau & Sauvageau, 2010). It is essential for the maintenance and self-renewal of not only normal ASCs such as HSCs but also LSCs (Lessard & Sauvageau, 2003; Molofsky et al., 2003; Park et al., 2003). *Bmi1*$^{-/-}$ mice develop progressive BM hypoplasia and die from BM failure shortly after birth (about 2 months), even though they have normal levels of HSCs in their fetal liver. However, *Bmi1*$^{-/-}$ fetal liver HSCs can only transiently reconstitute the hematopoietic system after transplantation into lethally irradiated recipients (Lessard & Sauvageau, 2003; Park et al., 2003). This is because *Bmi1*$^{-/-}$ HSCs undergo premature senescence and apoptosis. The induction of HSC premature senescence and apoptosis by *Bmi1* knockout was initially attributed to the derepression of the *Ink4a/Arf* locus that encodes two important tumor suppressors, p16 and Arf, because ectopic expression of p16 and Arf in HSCs induces cell cycle arrest and apoptosis, respectively, whereas deletion of both *Ink4a* and *Arf* partially rescues the ability of *Bmi1*$^{-/-}$ HSCs to self-renew (Oguro et al., 2006). However, a more recent study suggests that Bmi1 also regulates mitochondrial function and redox homeostasis, as the cells from *Bmi1*$^{-/-}$ mice produce increased levels of ROS and DNA damage and exhibit activation of the DNA damage response (DDR) pathway (Liu et al., 2009). The treatment of the knockout mice with the antioxidant NAC or the interruption of the DDR by *Chk2* knockout improves the function of HSCs and prolongs the life span of *Bmi1*$^{-/-}$ mice. These findings demonstrate that Bmi1 promotes HSC self-renewal at least in part by regulating mitochondrial function to inhibit ROS production and p16 expression to prevent HSCs from undergoing premature senescence. In addition, Nakamura et al. showed that overexpression of Bmi1 in HSCs confers resistance to oxidative stress (Nakamura et al., 2012). However, the mechanism by which Bmi1 regulates mitochondrial function and ROS production and confers resistance to oxidative stress has yet to be elucidated.

ATM: ATM is a serine/threonine kinase that plays a central role in coordinating the repair of DNA double-strand breaks (DSBs) and the cellular response to DNA damage (Shiloh & Ziv, 2013). Although ATM is primarily located in the nucleus, a faction of ATM is localized in the cytoplasm, particularly in the mitochondria (Ditch & Paull, 2012; Valentin-Vega et al., 2012). Recent studies suggest that the cytoplasmic ATM also functions as a redox sensor and regulator that controls the levels of ROS in a cell. ROS can directly oxidize ATM; and oxidized ATM then can form an active dimer via intermolecular disulfide bonds (Guo, Kozlov, Lavin, Person, & Paull, 2010). Activated ATM can phosphorylate BID to inhibit mitochondrial production of ROS (Maryanovich et al., 2012; Valentin-Vega et al., 2012). Alternatively, ATM may regulate ROS production via phosphorylation of HIF-1α and/or liver kinase B1 (LKB1) (Ditch & Paull, 2012). Therefore, ATM-deficient cells produce increased levels of ROS and have reduced levels of GSH biosynthesis, and $Atm^{-/-}$ mice exhibit signs of oxidative stress in different cells and multiple tissues (Ito et al., 2004; Kamsler et al., 2001; Meredith & Dodson, 1987). Particularly, it was found that $Atm^{-/-}$ mice exhibit a progressive decline in hematopoiesis with age because their HSCs show a defect in HSC self-renewal (Ito et al., 2004). This defect is associated with increased production of ROS, disruption of HSC quiescence, activation of p38, and elevated expression of p16. Treatment of $Atm^{-/-}$ mice with NAC or a p38 inhibitor can restore the function of HSCs and prevent the development of BM failure (Ito et al., 2004, 2006).

mTOR: mTOR is a serine/threonine kinase belonging to the PI3K-related kinase family that has an important role in regulating cell growth and metabolism (Laplante & Sabatini, 2012). It interacts with several proteins to form two distinct multiprotein complexes: mTOR complex 1 (mTORC1) and mTOR complex 2 (mTORC2). mTORC1 consists of mTOR, regulatory-associated protein of mTOR (Raptor), mammalian lethal with Sec13 protein 8 (mLST8, also known as GβL), proline-rich Akt substrate 40 kDa (PRAS40), and DEP-domain-containing mTOR-interacting protein (Deptor), whereas mTORC2 is composed of mTOR, rapamycin-insensitive companion of mTOR (Rictor), mammalian stress-activated protein kinase-interacting protein (mSin1), protein observed with Rictor-1 (Protor-1), mLST8, and Deptor. Rapamycin inhibits mTORC1 but not mTORC2. mTORC1 plays an important role in sensing environmental cues, such as growth signals, nutrient and energy status, and O_2 supply, to regulate cell metabolism, growth, proliferation, autophagy, and other functions. mTORC2, which is less investigated than mTORC1, regulates cell survival, cell metabolism, and cytoskeletal organization. Multiple upstream signaling pathways including the PI3K–Akt pathway can regulate the activity of mTORC1 but converge at the tuberous sclerosis 1 (TSC1) and TSC2 complex that functions as a GTPase-activating protein (GAP) for the Ras homologue enriched in brain (Rheb) GTPase. As a Rheb GAP, TSC1/2 negatively regulates mTORC1 activity by converting GTP-bound Rheb into its inactive GDP-bound state. Initially, it was reported that deletion of phosphatase and tensin homologue (*Pten*), a negative regulator of the PI3K–Akt pathway, from HSCs in adult mice promotes HSC proliferation, resulting in HSC depletion and leukemic transformation (Yilmaz et al., 2006; Zhang et al., 2006). The effect of *Pten* deletion on HSCs could be ameliorated by the treatment with rapamycin, indicating that mTORC1 is involved in regulation of HSCs. Subsequently, it was shown that overactivation of mTORC1 by conditional deletion of *Tsc1* stimulates HSC proliferation, impairs HSC self-renewal, and eventually causes HSC exhaustion in association with an increased production of ROS (Chen et al., 2008; Gan et al., 2008). Treatment with rapamycin or NAC can rescue HSC defects in *Tsc1* knockout mice, demonstrating that ROS are primarily responsible for mediating mTOR overactivation-induced HSC dysfunction.

Although each of these transcription factors and signaling pathways described in the preceding text can regulate HSC maintenance individually via regulating the production of ROS, there are extensive cross talks among them (Fig. 1.6). For example, ATM can regulate mTORC1 indirectly via phosphorylating Akt and LKB1/AMP-activated protein kinase (AMPK) (Ditch & Paull, 2012). In addition, ATM-mediated phosphorylation of HIF-1α can not only directly modulate the activity of HIF-1 but also indirectly downregulate mTORC1 (Cam, Easton, High, & Houghton, 2010). FOXO1 is a potential target of ATM (Matsuoka et al., 2007), whereas FOXO3a can increase ATM expression in HSCs (Yalcin et al., 2008) and directly interact with ATM to promote ATM activation in response to DNA damage (Tsai, Chung, Takahashi, Xu, & Hu, 2008). In contrast, FOXO1 can inhibit mTORC1 via TSC2-dependent and TSC2-independent mechanisms (Chen et al., 2010). Furthermore, there are numerous other molecules and pathways that also participate in the maintenance of HSCs via regulating ROS production (Liu, Cao, & Finkel, 2011; Suda et al., 2011). Due to space limitations, we are regretfully unable to discuss all of them here in this chapter.

3.2.4 Cellular sources of ROS in stem cells

As discussed in Section 2, cellular sources of ROS include the mitochondrial respiratory chain, NOXs, other oxidases, and various oxygenases. Different stem cells can use ROS derived from diverse cellular origins to perform unique functions. For example, the zygote of the sea urchin utilizes Udx1 (a dual NOX) to convert O_2 to H_2O_2 to prevent polyspermy and regulate cleavage (Wong et al., 2004; Wong & Wessel, 2005). hESC differentiation is associated with mitochondrial biogenesis, which leads to increases in mitochondrial mass and production of ATP and ROS (Cho et al., 2006; Saretzki et al., 2008). However, cardiovascular differentiation of ESCs depends on various NOXs for increased production of ROS (Ateghang, Wartenberg, Gassmann, & Sauer, 2006; Li et al., 2006; Sauer & Wartenberg, 2005; Schmelter, Ateghang, Helmig, Wartenberg, & Sauer, 2006). The mitochondria have frequently been considered to be the main source of cellular-derived ROS in HSCs. It has been shown that cells, including HSCs, from *Bmi1*$^{-/-}$ mice exhibit abnormal mitochondrial function, resulting in increased production of ROS (Liu et al., 2009). In addition, increased production of ROS in HSCs from *Tsc1*$^{-/-}$ mice has been attributed to the elevation of mitochondrial biogenesis and oxidative activities (Chen et al., 2008). However, compared to their progeny, HSCs are dormant and have fewer mitochondria and primarily utilize glycolysis rather than OXPHOS for ATP production (Piccoli et al., 2005; Simsek et al., 2010). Thus, it has yet to be determined whether the mitochondria are the major cellular source of ROS in HSCs. In contrast, the expression of NOX1, 2 and 4 and various regulatory subunits of NOXs has been detected in human HSCs (Piccoli et al., 2005, 2007). It was estimated that NOX-mediated extramitochondrial O_2 consumption accounts for about half of the endogenous cell respiration in human HSCs (Piccoli et al., 2005). Interestingly, our recent studies showed that NOX1, 2, and 4 are also expressed in mouse BM HSC-enriched LSK cells, whereas HPCs, Lin$^-$ cells, and mononuclear cells from mouse BM express NOX1 and 2, but not NOX4, suggesting that the expression of NOX4 is downregulated upon HSC differentiation and that NOX4 may play an important role in regulation of HSC function (Wang, Liu, et al., 2010). Moreover, exposure to IR induces persistent increases in ROS production in both human and murine HSCs (Wang, Liu, et al., 2010; Yamaguchi & Kashiwakura, 2013). However, the increases in ROS production in HSCs is not associated with mitochondrial dysfunction nor elevated mitochondrial $O_2^{\bullet-}$ production (Yamaguchi & Kashiwakura, 2013), but is related to upregulation of

NOX4 expression (Wang, Liu, et al., 2010). Inhibition of NOX4 activity with diphenyliodonium inhibits IR-induced elevation of ROS production in HSCs, suggesting that NOX4 is likely one of the main cellular sources of ROS in HSCs after radiation injury (Wang, Liu, et al., 2010).

3.3. Role of ROS in stem cell pathology

Although low levels of ROS production are required for stem cells to undergo self-renewing proliferation and proper differentiation (Ezashi et al., 2005; Juntilla et al., 2010; Kinder et al., 2010; Lewandowski et al., 2010; Owusu-Ansah & Banerjee, 2009; Sauer & Wartenberg, 2005), increased production of ROS is detrimental to stem cells and has been implicated in the pathogenesis of many pathological conditions and diseases by causing damage to stem cells. The pathological effects of ROS on stem cells are dose-dependent. A moderate increase in ROS production can impair stem cell self-renewal by promoting stem cell proliferation and differentiation, resulting in premature exhaustion of stem cells (Chen et al., 2008; Gan et al., 2008; Miyamoto et al., 2007; Tothova et al., 2007). Further increase in ROS production can induce stem cell senescence through the redox-dependent activation of the p38–p16 pathway (Shao, Wu, & Zhou, 2012). Finally, an acute and excessive increase in ROS production can induce stem cell apoptosis by activating the DNA damage response and p53 pathway (Shao et al., 2010; Yu et al., 2010) (Figs. 1.6 and 1.7).

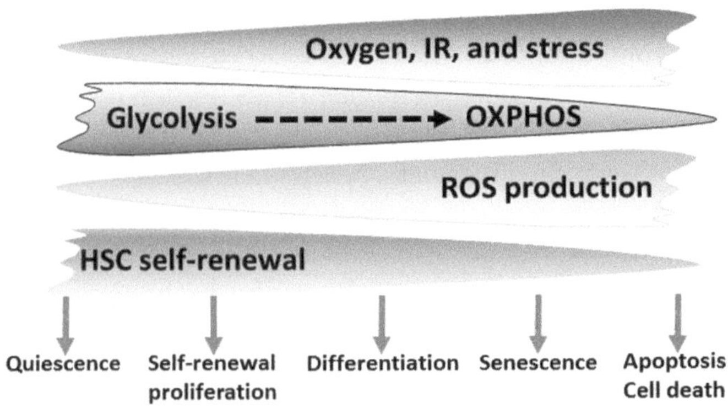

Figure 1.7 A diagram illustrating the relationship of cell metabolism, ROS production, and HSC self-renewal under increasing concentrations of oxygen and after exposure to IR and stress. HSC, hematopoietic stem cell; IR, ionizing radiation; OXPHOS, oxidative phosphorylation; ROS, reactive oxygen species.

3.3.1 ROS and stem cell differentiation

One of the fundamental characteristics of stem cells is their ability to self-renew and differentiate into different lineages of cells. However, these two functions have to be tightly regulated in order to maintain a proper balance between stem cell self-renewal and differentiation to prevent stem cells from premature exhaustion. Furthermore, stem cell differentiation to different lineages of cells has to be fine-tuned to prevent lineage skewing. It has been well documented that increased production of ROS can promote stem cell differentiation. However, the mechanisms by which ROS promotes stem cell differentiation have not been well established. It was shown that ROS enhances hESC differentiation to mesendodermal lineage by the activation of the members of the MAPK family and downregulation of the expression of pluripotent transcription factors, such as Oct4, Nanog, and Sox2 (Ji et al., 2010). The activation of selective members of the MAPK family was implicated in mediating ROS-induced cardiovascular differentiation of ESCs (Sauer & Wartenberg, 2005; Schmelter et al., 2006). In addition, Xiao et al. reported that NOX4-derived H_2O_2 promoted mESC differentiation to smooth muscle cells by the upregulation and activation of serum response factor (SRF) via phosphorylation (Xiao et al., 2009). However, the mechanisms by which H_2O_2 upregulates and activates SRF have yet to be investigated. Increased production of ROS not only promotes HSC differentiation but also causes lineage skewing (Pervaiz, Taneja, & Ghaffari, 2009; Shao, Lou, & Zhou, 2014; Suda et al., 2011). Jang and Sharkis showed that ROS^{high} HSCs exhibited accelerated exhaustion and myeloid skewed differentiation after serial transplantation (Jang & Sharkis, 2007). Similar findings were also observed in HSCs from sublethally irradiated mice (Li, Wang, Pazhanisamy, et al., 2011; Wang et al., 2012). The myeloid skewing of HSCs induced by IR was associated with increases in ROS production and DNA damage, which could be corrected after treatment with a SOD mimetic antioxidant (Li, Wang, Pazhanisamy, et al., 2011). In addition, Wang et al. showed recently that DNA damage induced by IR and telomere dysfunction promoted lymphoid differentiation of HSCs, resulting in depletion of HSCs with the capacity of lymphoid differentiation and accumulation of myeloid-biased HSCs (Wang et al., 2012). The enhanced lymphoid differentiation was attributed to the activation of the granulocyte colony-stimulating factor (G-CSF)-/Stat3-/BATF-dependent differentiation checkpoint in HSCs. *Gcsf* knockout, *Stat3* knockdown, or *Baft* deletion improved HSC self-renewal and function in response to IR or telomere shortening. However, it has yet to be determined whether ROS mediates IR-induced activation of this differentiation checkpoint in HSCs.

3.3.2 ROS and stem cell senescence

Most somatic cells have a finite growth potential, which was discovered by Hayflick and Moorhead in the early 1960s (Hayflick & Moorhead, 1961). The intrinsic replicative life span of a cell is determined by telomere length (Campisi, Kim, Lim, & Rubio, 2001). Without the expression of telomerase, telomeric sequences shorten after each time DNA replicates. When the telomeres reach a critically short length (~4 kb) after a certain number of cell doublings, known as the Hayflick limit, cells stop dividing and are irreversibly arrested at the G_1 phase, entering replicative senescence (Campisi et al., 2001; Marcotte & Wang, 2002). ESCs and the majority of tumor cells do not undergo replicative senescence because they express high levels of telomerase to prevent telomere erosion (Campisi et al., 2001; Marcotte & Wang, 2002; Zeng, 2007). Expression of telomerase is also important for ASCs such as HSCs to maintain their function (Allsopp, Morin, DePinho, Harley, & Weissman, 2003; Goytisolo et al., 2000; Greenwood & Lansdorp, 2003; Samper et al., 2002), because a deficiency in telomerase activity can lead to telomere shortening and reduction in HSC transplantation ability as seen in the late generations of telomerase RNA component (*Terc*)-null mice (Samper et al., 2002). In addition, the development of aplastic anemia or BM failure has been observed in patients with telomerase deficiency, due to mutations in telomerase reverse transcriptase (*Tert*) or *Terc* (Yamaguchi et al., 2005). In addition, many human and animal cells undergo premature senescence after exposure to oxidative and genotoxic stress or subjected to oncogenic mutations and/or aberrant activation of the p38 pathway (Serrano & Blasco, 2001). Premature senescent cells have a shortened intrinsic replicative life span without significant erosion in telomeres but are morphologically indistinguishable from replicatively senescent cells and exhibit many of the characteristics ascribed to replicatively senescent cells (Campisi et al., 2001; Marcotte & Wang, 2002; Serrano & Blasco, 2001). These changes include an enlarged and flattened appearance, increased senescence-associated β-galactosidase (SA-β-gal) activity, and elevated expression of p16 (Campisi et al., 2001; Marcotte & Wang, 2002; Serrano & Blasco, 2001). Moreover, premature and replicative senescence share common induction pathways (Campisi et al., 2001; Marcotte & Wang, 2002; Serrano & Blasco, 2001). Although ESCs are resistant to oxidative stress-induced senescence (Guo,

Chakraborty, Rajan, Wang, & Huang, 2010), HSCs are highly sensitive to the induction of premature senescence by ROS. For example, it was shown that *Atm*$^{-/-}$ mice exhibit progressive failure of hematopoietic function with aging (Ito et al., 2004). The failure is attributed primarily to HSC premature senescence resulting from an increased production of ROS, as treatment of *Atm*$^{-/-}$ mice with NAC can restore the function of HSCs and prevent the development of BM failure. Increased production of ROS also contributes to the induction of HSC senescence in mice with genetic deletion of *Bmi1* (Lessard & Sauvageau, 2003; Park et al., 2003), *Foxos* (Miyamoto et al., 2007; Tothova et al., 2007; Yalcin et al., 2008), *Mdm2* (Abbas et al., 2010), and *Tsc1* (Chen et al., 2008); Fanconi anemia mutation (Du et al., 2008); aging (Ito et al., 2006); and post-IR exposure (Li, Wang, Pazhanisamy, et al., 2011; Wang, Liu, et al., 2010; Zhang, Zhai, et al., 2013). Human HSCs exhibit increase in ROS production, accumulation of oxidative DNA damage, and impairment in self-renewal and long-term repopulation after serial transplantation into immunodeficient mice (Yahata et al., 2011). These defects can be attenuated by the treatment with NAC. Interestingly, the effects of ROS on HSCs appear not to be a nonspecific oxidative effect as previously hypothesized, but at least in part mediated by the redox-dependent activation of the p38–p16 pathway as described in the succeeding text (Wang, Liu, & Zhou, 2011).

p38: p38 is a member of the MAPK family of signal transduction kinases (Kyriakis & Avruch, 2001). It can be activated by ROS via apoptosis signal-regulating kinase 1 (ASK1) (Matsuzawa & Ichijo, 2008) and/or inactivation of protein tyrosine phosphatases (PTPs) such as MAPK phosphatases (Keyse, 2008; Patterson, Brummer, O'Brien, & Daly, 2009). Normally, ASK-1 forms an inactive complex with the repressor protein Trx in a cell. The formation of this complex is dependent on the presence of a reduced form of an intramolecular disulfide bridge between two cysteine residues of Trx. Oxidation of Trx by ROS causes dissociation of ASK-1 from Trx, resulting in the activation of ASK1 by oligomerization, interaction with TNF receptor-associated factor-2/TNF receptor-associated factor-6, and threonine autophosphorylation (Matsuzawa & Ichijo, 2008). It has been shown that ROS production from NOX4 can activate p38 via activation of ASK-1 (Chiang et al., 2006). In addition, oxidation of the catalytic cysteine of PTPs by ROS can reversibly inactivate PTPs (Shouse, Warren, & Whipple, 1931), which in turn can increase p38 activity. It remains to be determined whether ROS can activate p38 in HSCs through any of these mechanisms.

Activation of p38 regulates a variety of cellular processes such as inflammation, cell cycle arrest, and apoptosis in a cell type-specific manner. There is an increasing body of evidence demonstrating that p38 plays a critical role in the induction of senescence in response to a variety of stimuli via upregulating p16 (Antonchuk, Sauvageau, & Humphries, 2002; Geest & Coffer, 2009). For example, it was shown that a high level of Ras or Raf activation in human normal fibroblasts induced senescence by stimulating a sustained activation of p38, which in turn upregulated the expression of p16 (Antonchuk et al., 2002). Activation of the p38 pathway also contributes to the induction of p16 and cellular senescence after DNA damage resulting from exposure to genotoxic and oxidative stress and telomere shortening due to extensive replication (Kirito, Fox, & Kaushansky, 2003; Madlambayan et al., 2005; Waegell, Higley, Kincade, & Dasch, 1994). Furthermore, activation of p38 by ectopic transfection of mitogen-activated protein kinase kinase 3 (MKK3) and/or MKK6 increases p16 expression and induces senescence. In contrast, inhibition of p38 activity or downregulation of p38 expression attenuates the induction of p16 and cellular senescence by oncogenic stress, DNA damage, and telomere shortening (Kirito et al., 2003; Madlambayan et al., 2005; Waegell et al., 1994).

In addition, activation of p38 has been implicated in BM suppression in various pathological conditions associated with HSC dysfunction, including aplastic anemia and myelodysplastic syndromes (Navas et al., 2006; Zhou, Opalinska, & Verma, 2007). Furthermore, recently, it was shown that mutation of the *Atm* gene and knockout of the *Foxo3* gene induced premature senescence/exhaustion of HSCs (Ito et al., 2007; Miyamoto et al., 2007; Tothova et al., 2007). The induction of HSC senescence/exhaustion was associated with an elevated production of ROS, a selective activation of p38, and an upregulation of p16 in HSCs. Pharmacological inhibition of p38 activity rescued the defects of HSCs from *Atm* mutants and *Foxo3* knockout mice (Ito et al., 2007; Miyamoto et al., 2007; Tothova et al., 2007). These findings indicate that p38 plays an important role in regulation of HSC self-renewal and its activation by oxidative stress can mediate the induction of HSC senescence via upregulation of p16 (Ito et al., 2006). Similar findings were also observed in mice after exposure to a sublethal dose of TBI (Li, Wang, Wu, et al., 2011; Wang et al., 2011). In these irradiated mice, we found that p38 was selectively activated in irradiated hematopoietic cells and this activation sustained up to 5 weeks after IR in a long-term BM cell culture assay. Inhibition of p38 activity with a specific inhibitor attenuated IR-induced suppression of BM hematopoietic cell function in association with a significant reduction in p16 expression and SA-β-gal activity. Moreover, our *in vivo* data show that inhibition of p38 attenuated IR-induced residual BM suppression. These results suggest that p38 activation plays a role in mediating IR-induced hematopoietic cell senescence and BM suppression and that pharmacological inhibition of the p38 pathway with a specific inhibitor can be further exploited for amelioration of IR-induced residual BM injury (Li, Wang, Wu, et al., 2011; Wang et al., 2011).

p16 and Arf: The *Ink4a/Arf* locus encodes two tumor suppressors, p16 and Arf (Lowe & Sherr, 2003; Sharpless & DePinho, 1999). The transcripts for these proteins have different first exons (α for p16 and β for Arf) but share exons 2 and 3. However, there is no amino acid sequence similarity between these two proteins due to the use of alternative reading frames for their translation. p16 is a potent cyclin-dependent kinase (CDK) 4/6 inhibitor. By inhibiting CDK4/6 activity, p16 causes retinoblastoma protein (Rb) hypophosphorylation and suppresses the expression of E2F-dependent genes, resulting in restriction of G1/S cell cycle progression and formation of senescence-associated heterochromatic foci (SAHF) (Lowe & Sherr, 2003; Narita et al., 2003; Sharpless & DePinho, 1999). Once SAHF are formed after the engagement of the p16–Rb pathway, the cells become permanently growth-arrested and senescent. It has therefore been suggested that diverse stimuli can induce cellular senescence via various upstream signal transduction cascades, including the p38 and p53–p21 pathways, but converge on the p16–Rb pathway, whose activation provides an inescapable barrier preventing senescent cells from reentering the cell cycle. This suggestion is supported by the finding that activation of p53 and induction of p21 in cells undergoing senescence are transient events that occur during the onset of senescence and then subside when the expression of p16 starts rising (Campisi, 2005; Robles & Adami, 1998; te Poele, Okorokov, Jardine, Cummings, & Joel, 2002). Inactivation of p53 prior to upregulation of p16 can prevent senescence induction. However, once p16 is highly expressed, cell cycle arrest becomes irreversible by downregulation of p53, indicating that activation of the p53–p21 pathway plays an important role in the initiation of senescence, but induction of p16 is required for the maintenance of senescence (Beausejour et al., 2003; Campisi, 2005). In agreement with this suggestion, we found that IR induced p53 activation and p21 expression in HSCs prior to the induction of p16 (Li, Wang, Pazhanisamy, et al., 2011; Wang, Schulte, LaRue, Ogawa, & Zhou, 2006). While p53 activation and p21 upregulation gradually declined within a few weeks after IR, p16 expression in irradiated HSCs remained elevated and the cells subsequently became senescent, exhibiting positive SA-β-gal staining. In contrast, the biological action of Arf relies on the p53 pathway. This is because Arf can directly bind to MDM2 and cause the accumulation of p53 by segregating MDM2 from p53 and by inhibiting MDM2's E3 ubiquitin protein ligase activity for p53 (Lowe & Sherr, 2003; Sharpless & DePinho, 1999; Sherr & Weber, 2000). Therefore, activation of p53 by Arf can induce not only cellular senescence but also apoptosis, depending on which gene downstream of p53 is induced following its activation.

Upregulation of p16 and Arf has been implicated in mediating the induction of cellular senescence in HSCs. For example, increased expression of p16 and Arf was found in HSCs from *Bmi1*$^{-/-}$ mice (Park et al., 2003). However, it appears that p16 but not Arf plays an important role in mediating the induction of *Bmi1*$^{-/-}$ HSC senescence (Park et al., 2003). In addition, it has been found that knockout of both the *p16* and *Arf* genes in mice significantly increases the clonal expansion of HSCs *in vitro* but modestly promotes HSC self-renewal *in vivo* (Lessard & Sauvageau, 2003; Molofsky et al., 2003). However, knockout of the *Arf* gene alone does not provide any advantage for HSC/HPC expansion and self-renewal (Molofsky et al., 2003). In contrast, knockout p16 increases the life span of HSCs by promoting HSC self-renewal (Janzen et al., 2006; Stepanova & Sorrentino, 2005). Furthermore, mutation of the *Atm* gene also results in the upregulation of p16 and Arf in HSCs (Ito et al., 2007; Molofsky et al., 2003). Inactivation of the p16–Rb pathway by retroviral transfection of HPV E7 proteins restores the reproductive function of *Atm*$^{-/-}$ HSCs, while inhibition of the Arf–p53 pathway by E6 transfection has no such effect (Ito et al., 2004). These findings suggest that p16 plays a more significant role than Arf in the regulation of HSC self-renewal and induction of HSC senescence, even though both proteins are overexpressed in senescent HSCs. Increased expression of p16 and Arf has been found in IR-induced senescent LSK cells (Li, Wang, Pazhanisamy, et al., 2011; Wang, Schulte, LaRue, et al., 2006). However, their roles in mediating IR-induced HSC senescence and long-term BM suppression remain to be investigated.

3.3.3 ROS and stem cell apoptosis

Apoptosis is an orderly and regulated form of cell death via a genetically controlled process (Kerr, Wyllie, & Currie, 1972; Majno & Joris, 1995). The characteristics of an apoptotic cell include externalization of phosphatidylserine on the outer leaflet of the plasma membrane, cell shrinkage, condensation of the nuclear chromatin, fragmentation of the nucleus and DNA, and cellular membrane blebbing (Kerr et al., 1972; Majno & Joris, 1995). It is well established that exposure of cells to oxidative stress induces DNA damage, particularly DSBs, which activates ATM (Shiloh & Ziv, 2013). Alternatively, ROS can directly activate ATM (Ditch & Paull, 2012). Activation of ATM causes accumulation and activation of p53. Activated p53 then translocates into nucleus to transcriptionally activate the expression of various proapoptotic factors such as Fas, DR5, Puma, and Bax (Riley, Sontag, Chen, & Levine, 2008; Villunger et al., 2003) or into mitochondrial to directly interact with Bax (Chipuk et al., 2004; Chipuk, Maurer, Green, & Schuler, 2003). The interaction of Fas or DR5 with their respective ligands can directly activate the initiator caspase-8 to trigger apoptosis through mitochondria-independent and mitochondria-dependent mechanisms, whereas Puma and Bax can cause the mitochondria to release caspase-activating factors, such as cytochrome c and Apaf-1 that, in turn, activate another initiator caspase, for example, caspase-9. All these diverse upstream apoptotic pathways converge at the effector caspases (e.g., caspase-3, caspase-6, and caspase-7) whose activation leads to the final stage of cell self-destruction (Budihardjo, Oliver, Lutter, Luo, & Wang, 1999).

Although ESCs are highly resistant to senescence, they are extremely sensitive to the induction of apoptosis by ROS and other genotoxic insults (Filion et al., 2009; Guo, Chakraborty, et al., 2010; Qin et al., 2007). This is because ESCs either express low levels of p53 or are incapable of expressing cell cycle inhibitors and proapoptotic proteins in response to p53 activation (Qin et al., 2007; Zeng, 2007). However, activated p53 can induce rapid ESC apoptosis by translocating into mitochondria to directly interact with Bax (Han et al., 2008). The high sensitivity of ESCs to p53-mediated mitochondrial apoptosis may be attributable to ESC mitochondrial priming or the expression of constitutive active form Bax at the Golgi (Dumitru et al., 2012; Liu et al., 2013). These differential sensitivities to the induction of senescence and apoptosis by ROS and other insults may be an important mechanism for ESCs to maintain their genomic integrity to prevent the propagation of undesirable mutations to the resulting somatic and germ cell lineages.

In coordination with cell proliferation and differentiation, apoptosis contributes to the maintenance of hematopoietic homeostasis by regulating the size of hematopoietic lineages (Wickremasinghe & Hoffbrand, 1999).

Dysregulation of apoptosis in HSCs and their progeny can result in many pathological conditions (Wickremasinghe & Hoffbrand, 1999). For example, it has been well established that induction of apoptosis in HPCs and HSCs is primarily responsible for the induction of acute radiation syndrome after exposure to a high or moderate dose of total body irradiation that induces acute and intense oxidative stress (Domen, Gandy, & Weissman, 1998; Mauch et al., 1995). Overexpression of an antiapoptotic protein, bcl-2, throughout the hematopoietic compartment protects mice against IR-induced hematopoietic failure and death, because HSCs isolated from *bcl-2* transgenic mice are more resistant to IR-induced apoptosis (Domen et al., 1998). In contrast, bcl-2 deficiency sensitizes murine HSCs to IR (Hoyes, Cai, Potten, & Hendry, 2000). The induction of HSC apoptosis by IR is mediated by the activation of the p53 pathway, because HSCs from p53-deficient mice are less sensitive to IR than are those from wild-type mice (Cui et al., 1995; Hirabayashi et al., 1997; Lee & Bernstein, 1993) and treatment with a p53 inhibitor protected mice from IR-induced lethal damage by suppression of p53-dependent apoptosis (Komarov et al., 1999). More recently, several groups of investigators reported that Puma, a downstream target of p53 and a proapoptotic BH3-only protein, plays a critical role in mediating IR-induced HSC apoptosis (Shao et al., 2010; Yu et al., 2010). They showed that Puma was selectively induced by IR in LSK cells and LSK cells from *Puma* knockout mice were insensitive to IR-induced apoptosis. As such, Puma deficiency in mice confers resistance to high-dose IR in a hematopoietic cell-autonomous manner. In contrast, other p53 targets, such as Bim and Noxa, play a minor or moderate role in IR-induced apoptosis in hematopoietic cells (Erlacher et al., 2005; Labi et al., 2010). These findings indicate that targeting the p53–Puma pathway may represent a novel strategy to protecting HSCs from IR injury, particularly considering that transient inhibition of p53 activity with an inhibitor did not increase IR-induced carcinogenesis, while Puma knockout actually reduced IR-induced tumorigenesis in mice (Christophorou, Ringshausen, Finch, Swigart, & Evan, 2006; Labi et al., 2010; Michalak et al., 2010).

4. ROS AND TSCS

4.1. TSCs

Like their normal tissue counterparts, tumors contain phenotypically and functionally heterogeneous populations of tumor cells (Beck & Blanpain, 2013; Nguyen, Vanner, Dirks, & Eaves, 2012). These cells are structured in a hierarchical manner in some, but not all, tumors (Magee et al., 2012). Undifferentiated tumor cells that are at the apex of the hierarchy have the ability to propagate the tumor by generating all of the cells in a tumor, including undifferentiated and differentiated tumor cells and other cells in the tumor, including endothelial cells (Ricci-Vitiani et al., 2010; Wang, Chadalavada, et al., 2010) and pericytes (Cheng et al., 2013). Therefore, these cells have been named TSCs or tumor-initiating cells. Furthermore, TSCs in leukemia are called LSCs and those in solid tumors CSCs. The first definitive evidence to demonstrate the existence of TSCs was observed by Furth and Kahn (1937). They showed that leukemia could be transmitted to a recipient mouse after receiving injection of a single leukemia cell. Later, Pierce and his colleagues showed that teratocarcinomas contain undifferentiated cells that are highly tumorigenic and can differentiate into multiple types of differentiated and nontumorigenic cells (Pierce & Speers, 1988). However, the true identity of TSCs was not known until the landmark studies by Dick and his colleagues in the 1990s. They showed that the peripheral blood of acute myeloid leukemia (AML) patient contained the leukemia-initiating cells or LSCs in the $CD34^{+}CD38^{-}$ population (Bonnet & Dick, 1997; Lapidot et al., 1994). These cells were rare but capable of differentiating into leukemic blasts to initiate human AML in NOD/SCID mice. It took almost another 10 years for Clarke and his colleagues to identify the first CSCs (Al-Hajj, Wicha, Benito-Hernandez, Morrison, & Clarke, 2003). They prospectively identified and isolated $CD44^{+}CD22^{-/low}$/lineage^{-} CSCs from nine breast cancer patients. They showed that as few as 100 $CD44^{+}CD22^{-/low}$/lineage^{-} tumorigenic cells could form tumors in a xenograft transplantation mouse model and the tumors contained both newly generated $CD44^{+}CD22^{-/low}$/lineage^{-} CSCs and the phenotypically diverse mixed populations of nontumorigenic cells present in the initial tumor after serial passages. Since then, CSCs have been identified and prospectively isolated from many other cancers, including brain tumors and colorectal, lung, pancreatic, prostate, and ovarian cancers (Beck & Blanpain, 2013; Nguyen et al., 2012).

ASCs may represent an ideal cellular origin of TSCs because they have the ability to self-renew and are long-lived cells to allow the accumulation of mutations for transformation. Therefore, it has been suggested that some TSCs may be derived from ASCs because TSCs share some of the same characteristics of normal ASCs, such as self-renewal. This suggestion is supported by the studies on chronic myelogenous (or myeloid) leukemia (CML). CML is caused by the oncogenic Bcl–Abl fusion gene generated by a reciprocal translocation between chromosomes 9 and 22 in HSCs. The Bcr–Abl fusion gene only initiates CML in HSCs but not in committed murine hematopoietic progenitors (Huntly et al., 2004). However, TSCs can also originate from more differentiated progenitors that acquire the ability to self-renew by accumulation of genetic mutations and epigenetic abnormalities. For example, the oncogenic fusion genes MLL–ENL, MLL–AF9, and MOZ–TIF2 found in human AML patients can also transform mouse committed myeloid progenitors for the induction of AML (Cozzio et al., 2003; Huntly et al., 2004; Krivtsov et al., 2006). More importantly, even when a tumor-initiating mutation occurs first in ASCs, the functional TSCs become detectable in their progeny (Jamieson et al., 2004; Miyamoto, Weissman, & Akashi, 2000).

TSCs may play an important role in tumor progression, metastasis, resistance to therapy, and relapse after treatment (Baccelli & Trumpp, 2012; Clevers, 2011). Therefore, strategies that can specifically target TSCs may offer a cure or a better treatment for tumors. However, TSCs appear a moving target that is difficult to kill. This is because unlike normal ASCs, TSCs are genetically unstable. They can constantly generate new subclones that are different from the initial "parental" clones of TSCs by obtaining additional mutations as revealed by the studies on LSCs from human acute lymphoblastic leukemia (ALL) (Anderson et al., 2011; Notta, Mullighan, et al., 2011). These different clones of LSCs may compete, resulting in clonal evolution during tumor progression, metastasis, and treatment. Furthermore, TSCs and some other tumor cells have high plasticity. TSCs may change their phenotype, and nontumorigenic cells can be reprogrammed or "dedifferentiate" into TSCs in response to various environmental cues (Tang, 2012). Therefore, we need to have a better understanding of the fundamental differences between TSCs and normal ASCs to develop more effective treatments for tumors. Since ROS plays an important role in the determination of the fate of normal ASCs, it would be of a great interest to determine whether regulation of ROS production in TSCs can be exploited for therapeutic gains against leukemia and cancer as discussed in the succeeding text.

4.2. ROS and LSCs

A large body of evidence demonstrates that leukemia and solid tumor cells produce increased levels of ROS compared with normal cells (Ogasawara & Zhang, 2009; Shi, Zhang, Zheng, & Pan, 2012; Trachootham, Alexandre, &

Huang, 2009; Zhou, Shen, & Claret, 2013). The increased production of ROS is attributable to the activation of oncogenes (such as Ras, Bcr–Abl, and c-Myc) and/or inactivation of tumor suppressor genes (such as p53), resulting in mitochondrial dysfunction, aberrant metabolism, and alteration in antioxidant production. In addition, inflammation and exposure to IR and genotoxic stress can also increase ROS production. Increased production of ROS has been implicated in tumorigenesis because ROS is a potent carcinogen that causes DNA damage and gene mutations, which may eventually lead to induction of genomic instability and cell transformation. Therefore, antioxidant therapy has been extensively exploited as a preventive strategy to reduce carcinogenesis and tumorigenesis. So far, this strategy has yielded mixed results, indicating that more studies are needed to gain a better understanding of redox biology and cancer biology. In addition, because tumor cells produce higher levels of ROS, thus they are assumed to be more sensitive to oxidative stress than their normal counterparts. Therefore, the promotion of ROS production has the potential to selectively kill tumor cells without causing significantly collateral damage to normal tissues. Since tumors consist of undifferentiated tumorigenic TSCs and various differentiated nontumorigenic tumor cells and TSCs are primarily responsible for the initiation and propagation of a tumor, recent studies have started focusing on ROS in TSCs (Ogasawara & Zhang, 2009; Shi et al., 2012; Zhou et al., 2013).

It has been well established that Bcr–Abl oncoprotein stimulates CML cell production of ROS in a kinase-dependent manner (Koptyra et al., 2006; Nowicki et al., 2004). The increase production of ROS may result from the activation of Rac2 GTPase that alters mitochondrial membrane potential and electron flow through respiratory chain complex III (Nieborowska-Skorska et al., 2012). The increased production of ROS induces genomic instability by causing chronic DNA damage and inhibiting DNA damage repair in chronic-phase CML LSCs, which may contribute to the progression, drug resistance, and relapse of the disease (Slupianek et al., 2013). To our surprise, a study reported very recently by Lagadinou et al. (2013) showed that the majority of LSCs from AML patients are defined as ROS^{low} cells. However, this finding is in agreement with an earlier observation that the expression of GPx3, an enzymatic antioxidant scavenger of ROS, correlates with the abundance of LSCs in Hoxa9+Meis1-induced AML (Herault et al., 2012). These findings suggest that LSCs from AML and CML in chronic phase have different capacity of producing ROS. It remains to be determined whether the difference is related to the different phases of leukemia studied and/or underlying causes by the different oncogene mutations. In addition, it is also not known whether the production of ROS by LSCs is comparable to that of normal HSCs, ideally from the same individuals. However, AML LSCs appear more sensitive than normal HSCs to drugs that elevate the production of ROS. For example, Guzman et al. (2005) and Guzman et al. (2007) showed that parthenolide and its analogs can selectively eradicate AML and CML stem and progenitor cells. Similar findings were also observed when AML LSCs were treated with niclosamide and fenretinide (Jin et al., 2010; Zhang, Mi, et al., 2013). Interestingly, all these compounds also inhibit NF-κB, suggesting that increased production of ROS may induce synthetic lethality selectively in LSCs with NF-κB inhibition (Guzman et al., 2005, 2007; Jin et al., 2010; Zhang, Mi, et al., 2013). This is because NF-κB can protect cells against oxidative stress by regulating the expression of SOD2, several cell cycle inhibitors, and various anti-apoptotic proteins (Ahmed & Li, 2008).

4.3. ROS and CSCs

Diehn et al. (2009) were the first to report that $CD44^{+}/CD22^{-/low}$ CSCs from some human and murine breast tumors produce lower levels of ROS than corresponding nontumorigenic cancer cells. The decreased production of ROS by breast CSCs may be attributed to their increased expression of antioxidant genes, including GCLM and FOXO1, because GCLM encodes the regulatory subunit of glutamate–cysteine ligase that catalyzes the rate-limiting step of GSH synthesis and FOXOs have been shown to regulate the expression of SOD2 and catalase. In addition, high levels of expression of CD44 in CSCs may also contribute to the lower production of ROS, because CD44 can promote the uptake of cysteine for the synthesis of GSH via interaction with xCT, a glutamate–cystine transporter (Ishimoto et al., 2011). More recently, Dong et al. (2013) showed that Snail can mediate metabolic reprogramming of basal-like breast cancer cells to CSC-like cells. These CSC-like cells express decreased levels of fructose-1,6-biphosphatase (FPB1) due to the Snail–G9a–Dnmt1 complex-mediated epigenetic silencing of FPB1 via promoter methylation. Decreased expression of FPB1 increases glycolysis and NADPH production by PPC while reducing OXPHOS and ROS production. The lower levels of ROS production in CSCs not only may be important for the maintenance of CSCs but also can confer resistance to therapy, because lower ROS levels in $CD44^{+}/CD22^{-/low}$ breast CSCs are associated with less DNA damage and a higher rate of cell survival after exposure to IR and pharmacological depletion of GSH with BSO sensitizes CSCs to IR (Diehn et al., 2009). Similar findings were also observed in $CD13^+$ human liver CSCs (Haraguchi et al., 2010; Kim et al., 2012). Therefore, pharmacologically increasing ROS production has the potential to be used as an adjuvant therapy to more effectively eradicate CSCs. However, this strategy requires further investigation, because of concerns that increased production of ROS may also cause damage to normal ASCs.

5. CONCLUSION

Oxidative stress resulting from an increase in ROS production and/or a reduction in antioxidant capacity has been implicated in the pathogenesis of many diseases and aging. Although all cells in an organism can be affected by oxidative stress, the effects of ROS on stem cells have the greatest impact on the body, because they have the ability to self-renew and generate/replenish all other cells for the life span of the organism. Unfortunately, stem cells are more sensitive to oxidative stress than their differentiated progeny. Increased production of ROS can lead to stem cells exhaustion by induction of stem cell proliferation/differentiation, senescence, and/or apoptosis, which can be a major contributor to aging, degenerative diseases, and normal tissue injury induced by conventional cancer treatment and also limits the use of stem cells for regenerative medicine. Recent studies have provided a better understanding of the mechanisms whereby the production of ROS is regulated in stem cells and those by which the fate of the cells is affected by oxidative stress as discussed in this chapter, which may lead to the development of new interventions for aging-associated diseases and cancer therapy-induced normal tissue injury and novel strategies to harness the healing power of stem cells to treat diseases. In addition, an accumulating body of evidence demonstrates that many tumors contain TSCs that play an important role in tumor progression, metastasis, resistance to therapy, and relapse after treatment. The production of ROS in some TSCs is dysregulated, resulting in aberrant production of ROS, which can be exploited for the design of a better therapy for cancer and leukemia in the future.

ACKNOWLEDGMENTS

We apologize to the authors whose contributions were not directly cited owing to space limitations. The authors thank the previous and current members of Dr. Zhou's Laboratory for their work and support and Gareth Smith and Shawn Roach for their graphic design assistance. The research conducted in Dr. Zhou's Laboratory was supported in part by grants from the National Institutes of Health (R01-CA122023 and AI080421) and a grant from the Edward P. Evans Foundation and the Arkansas Research Alliance Scholarship from the Arkansas Science & Technology Authority. Dr. Spitz was supported by R01CA182804, R01CA133114, and R21CA161182.

REFERENCES

Abbas, H. A., Maccio, D. R., Coskun, S., Jackson, J. G., Hazen, A. L., Sills, T. M., et al. (2010). Mdm2 is required for survival of hematopoietic stem cells/progenitors via dampening of ROS-induced p53 activity. *Cell Stem Cell, 7*, 606–617.

Ahmad, I. M., Aykin-Burns, N., Sim, J. E., Walsh, S. A., Higashikubo, R., Buettner, G. R., et al. (2005). Mitochondrial $O_2{}^{*}$- and H_2O_2 mediate glucose deprivation-induced stress in human cancer cells. *Journal of Biological Chemistry, 280*, 4254–4263.

Ahmed, K. M., & Li, J. J. (2008). NF-kappa B-mediated adaptive resistance to ionizing radiation. *Free Radical Biology & Medicine, 44*, 1–13.

Al-Hajj, M., Wicha, M. S., Benito-Hernandez, A., Morrison, S. J., & Clarke, M. F. (2003). Prospective identification of tumorigenic breast cancer cells. *Proceedings of the National Academy of Sciences of the United States of America, 100*, 3983–3988.

Allsopp, R. C., Morin, G. B., DePinho, R., Harley, C. B., & Weissman, I. L. (2003). Telomerase is required to slow telomere shortening and extend replicative lifespan of HSCs during serial transplantation. *Blood, 102*, 517–520.

Anderson, K., Lutz, C., van Delft, F. W., Bateman, C. M., Guo, Y., Colman, S. M., et al. (2011). Genetic variegation of clonal architecture and propagating cells in leukaemia. *Nature, 469*, 356–361.

Antonchuk, J., Sauvageau, G., & Humphries, R. K. (2002). HOXB4-induced expansion of adult hematopoietic stem cells ex vivo. *Cell, 109*, 39–45.

Antonenkov, V. D., Grunau, S., Ohlmeier, S., & Hiltunen, J. K. (2010). Peroxisomes are oxidative organelles. *Antioxidants & Redox Signaling, 13*, 525–537.

Arai, F., Hirao, A., Ohmura, M., Sato, H., Matsuoka, S., Takubo, K., et al. (2004). Tie2/angiopoietin-1 signaling regulates hematopoietic stem cell quiescence in the bone marrow niche. *Cell, 118*, 149–161.

Archer, V. E. (1979). Carcinogenicity of fibers and films: A theory. *Medical Hypotheses, 5*, 1257–1262.

Arner, E. S., & Holmgren, A. (2006). The thioredoxin system in cancer. *Seminars in Cancer Biology, 16*, 420–426.

Ateghang, B., Wartenberg, M., Gassmann, M., & Sauer, H. (2006). Regulation of cardiotrophin-1 expression in mouse embryonic stem cells by HIF-1alpha and intracellular reactive oxygen species. *Journal of Cell Science, 119*, 1043–1052.

Aykin-Burns, N., Ahmad, I. M., Zhu, Y., Oberley, L. W., & Spitz, D. R. (2009). Increased levels of superoxide and H2O2 mediate the differential susceptibility of cancer cells versus normal cells to glucose deprivation. *Biochemistry Journal, 418*, 29–37.

Aykin-Burns, N., Slane, B. G., Liu, A. T., Owens, K. M., O'Malley, M. S., Smith, B. J., et al. (2011). Sensitivity to low-dose/low-LET ionizing radiation in mammalian cells harboring mutations in succinate dehydrogenase subunit C is governed by mitochondria-derived reactive oxygen species. *Radiation Research, 175*, 150–158.

Baccelli, I., & Trumpp, A. (2012). The evolving concept of cancer and metastasis stem cells. *Journal of Cell Biology, 198*, 281–293.

Balaban, R. S., Nemoto, S., & Finkel, T. (2005). Mitochondria, oxidants, and aging. *Cell, 120*, 483–495.

Barker, H. E., Cox, T. R., & Erler, J. T. (2012). The rationale for targeting the LOX family in cancer. *Nature Reviews. Cancer, 12*, 540–552.

Barker, N., van Es, J. H., Kuipers, J., Kujala, P., van den Born, M., Cozijnsen, M., et al. (2007). Identification of stem cells in small intestine and colon by marker gene Lgr5. *Nature, 449*, 1003–1007.

Barria, E., Mikels, A., & Haas, M. (2004). Maintenance and self-renewal of long-term reconstituting hematopoietic stem cells supported by amniotic fluid. *Stem Cells and Development, 13*, 548–562.

Baum, C. M., Weissman, I. L., Tsukamoto, A. S., Buckle, A. M., & Peault, B. (1992). Isolation of a candidate human hematopoietic stem-cell population. *Proceedings of the National Academy of Sciences of the United States of America, 89*, 2804–2808.

Beausejour, C. M., Krtolica, A., Galimi, F., Narita, M., Lowe, S. W., Yaswen, P., et al. (2003). Reversal of human cellular senescence: Roles of the p53 and p16 pathways. *EMBO Journal, 22*, 4212–4222.

Beck, B., & Blanpain, C. (2013). Unravelling cancer stem cell potential. *Nature Reviews. Cancer, 13*, 727–738.

Becker, A. J., Mc, C. E., & Till, J. E. (1963). Cytological demonstration of the clonal nature of spleen colonies derived from transplanted mouse marrow cells. *Nature, 197*, 452–454.

Beckman, J. S., & Koppenol, W. H. (1996). Nitric oxide, superoxide, and peroxynitrite: The good, the bad, and ugly. *American Journal of Physiology, 271*, C1424–C1437.

Bedard, K., & Krause, K. H. (2007). The NOX family of ROS-generating NADPH oxidases: Physiology and pathophysiology. *Physiological Reviews, 87*, 245–313.

Beddington, R. S., & Robertson, E. J. J. (1989). An assessment of the developmental potential of embryonic stem cells in the midgestation mouse embryo. *Development, 105*, 733–737.

Beltrami, A. P., Barlucchi, L., Torella, D., Baker, M., Limana, F., Chimenti, S., et al. (2003). Adult cardiac stem cells are multipotent and support myocardial regeneration. *Cell, 114*, 763–776.

Bhatia, M., Wang, J. C., Kapp, U., Bonnet, D., & Dick, J. E. (1997). Purification of primitive human hematopoietic cells capable of repopulating immune-deficient mice. *Proceedings of the National Academy of Sciences of the United States of America, 94*, 5320–5325.

Bize, I. B., Oberley, L. W., & Morris, H. P. (1980). Superoxide dismutase and superoxide radical in Morris hepatomas. *Cancer Research, 40*, 3686–3693.

Blackburn, R. V., Spitz, D. R., Liu, X., Galoforo, S. S., Sim, J. E., Ridnour, L. A., et al. (1999). Metabolic oxidative stress activates signal transduction and gene expression during glucose deprivation in human tumor cells. *Free Radical Biology & Medicine, 26*, 419–430.

Bonner, M. Y., & Arbiser, J. L. (2012). Targeting NADPH oxidases for the treatment of cancer and inflammation. *Cellular and Molecular Life Sciences, 69*, 2435–2442.

Bonnet, D., & Dick, J. E. (1997). Human acute myeloid leukemia is organized as a hierarchy that originates from a primitive hematopoietic cell. *Nature Medicine, 3*, 730–737.

Bosgra, S., Mennes, W., & Seinen, W. (2005). Proceedings in uncovering the mechanism behind peroxisome proliferator-induced hepatocarcinogenesis. *Toxicology, 206*, 309–323.

Boveris, A. (1977). Mitochondrial production of superoxide radical and hydrogen peroxide. *Advances in Experimental Medicine and Biology, 78*, 67–82.

Boveris, A., & Cadenas, E. (1982). Production of superoxide radicals and hydrogen peroxide in mitochondria. In L. W. Oberley (Ed.), *Superoxide Dismutase: Vol. 1.* (pp. 15–30)Boca Raton, FL: CRC Press.

Bradley, A., Evans, M., Kaufman, M. H., & Robertson, E. (1984). Formation of germ-line chimaeras from embryo-derived teratocarcinoma cell lines. *Nature, 309*, 255–256.

Broudy, V. C. (1997). Stem cell factor and hematopoiesis. *Blood, 90*, 1345–1364.

Budihardjo, I., Oliver, H., Lutter, M., Luo, X., & Wang, X. (1999). Biochemical pathways of caspase activation during apoptosis. *Annual Review of Cell and Developmental Biology, 15*, 269–290.

Buettner, G. R. (1993). The pecking order of free radicals and antioxidants: Lipid peroxidation, alpha-tocopherol, and ascorbate. *Archives of Biochemistry and Biophysics, 300*, 535–543.
Buettner, G. R. (2011). Superoxide dismutase in redox biology: The roles of superoxide and hydrogen peroxide. *Anti-Cancer Agents in Medicinal Chemistry, 11*, 341–346.
Buettner, G. R., Ng, C. F., Wang, M., Rodgers, V. G., & Schafer, F. Q. (2006). A new paradigm: Manganese superoxide dismutase influences the production of H_2O_2 in cells and thereby their biological state. *Free Radical Biology & Medicine, 41*, 1338–1350.
Calvi, L. M., Adams, G. B., Weibrecht, K. W., Weber, J. M., Olson, D. P., Knight, M. C., et al. (2003). Osteoblastic cells regulate the haematopoietic stem cell niche. *Nature, 425*, 841–846.
Cam, H., Easton, J. B., High, A., & Houghton, P. J. (2010). mTORC1 signaling under hypoxic conditions is controlled by ATM-dependent phosphorylation of HIF-1alpha. *Molecular Cell, 40*, 509–520.
Campisi, J. (2005). Senescent cells, tumor suppression, and organismal aging: Good citizens, bad neighbors. *Cell, 120*, 513–522.
Campisi, J., Kim, S. H., Lim, C. S., & Rubio, M. (2001). Cellular senescence, cancer and aging: The telomere connection. *Experimental Gerontology, 36*, 1619–1637.
Cannizzo, E. S., Clement, C. C., Sahu, R., Follo, C., & Santambrogio, L. (2011). Oxidative stress, inflamm-aging and immunosenescence. *Journal of Proteomics, 74*, 2313–2323.
Chance, B., Sies, H., & Boveris, A. (1979). Hydroperoxide metabolism in mammalian organs. *Physiological Reviews, 59*, 527–605.
Chen, C. C., Jeon, S. M., Bhaskar, P. T., Nogueira, V., Sundararajan, D., Tonic, I., et al. (2010). FoxOs inhibit mTORC1 and activate Akt by inducing the expression of Sestrin3 and Rictor. *Developmental Cell, 18*, 592–604.
Chen, C., Liu, Y., Liu, R., Ikenoue, T., Guan, K. L., & Zheng, P. (2008). TSC-mTOR maintains quiescence and function of hematopoietic stem cells by repressing mitochondrial biogenesis and reactive oxygen species. *Journal of Experimental Medicine, 205*, 2397–2408.
Cheng, L., Huang, Z., Zhou, W., Wu, Q., Donnola, S., Liu, J. K., et al. (2013). Glioblastoma stem cells generate vascular pericytes to support vessel function and tumor growth. *Cell, 153*, 139–152.
Chiang, E., Dang, O., Anderson, K., Matsuzawa, A., Ichijo, H., & David, M. (2006). Cutting edge: Apoptosis-regulating signal kinase 1 is required for reactive oxygen species-mediated activation of IFN regulatory factor 3 by lipopolysaccharide. *Journal of Immunology, 176*, 5720–5724.
Chipuk, J. E., Kuwana, T., Bouchier-Hayes, L., Droin, N. M., Newmeyer, D. D., Schuler, M., et al. (2004). Direct activation of Bax by p53 mediates mitochondrial membrane permeabilization and apoptosis. *Science, 303*, 1010–1014.
Chipuk, J. E., Maurer, U., Green, D. R., & Schuler, M. (2003). Pharmacologic activation of p53 elicits Bax-dependent apoptosis in the absence of transcription. *Cancer Cell, 4*, 371–381.
Cho, Y. M., Kwon, S., Pak, Y. K., Seol, H. W., Choi, Y. M., Park do, J., et al. (2006). Dynamic changes in mitochondrial biogenesis and antioxidant enzymes during the spontaneous differentiation of human embryonic stem cells. *Biochemical and Biophysical Research Communications, 348*, 1472–1478.
Chow, A., Lucas, D., Hidalgo, A., Mendez-Ferrer, S., Hashimoto, D., Scheiermann, C., et al. (2011). Bone marrow CD169+ macrophages promote the retention of hematopoietic stem and progenitor cells in the mesenchymal stem cell niche. *Journal of Experimental Medicine, 208*, 261–271.
Christensen, J. L., & Weissman, I. L. (2001). Flk-2 is a marker in hematopoietic stem cell differentiation: A simple method to isolate long-term stem cells. *Proceedings of the National Academy of Sciences of the United States of America, 98*, 14541–14546.

Christophorou, M. A., Ringshausen, I., Finch, A. J., Swigart, L. B., & Evan, G. I. (2006). The pathological response to DNA damage does not contribute to p53-mediated tumour suppression. *Nature, 443*, 214–217.

Civin, C. I., Strauss, L. C., Brovall, C., Fackler, M. J., Schwartz, J. F., & Shaper, J. H. (1984). Antigenic analysis of hematopoiesis. III. A hematopoietic progenitor cell surface antigen defined by a monoclonal antibody raised against KG-1a cells. *Journal of Immunology, 133*, 157–165.

Clevers, H. (2011). The cancer stem cell: Premises, promises and challenges. *Nature Medicine, 17*, 313–319.

Coso, S., Harrison, I., Harrison, C. B., Vinh, A., Sobey, C. G., Drummond, G. R., et al. (2012). NADPH oxidases as regulators of tumor angiogenesis: Current and emerging concepts. *Antioxidants & Redox Signaling, 16*, 1229–1247.

Cozzio, A., Passegue, E., Ayton, P. M., Karsunky, H., Cleary, M. L., & Weissman, I. L. (2003). Similar MLL-associated leukemias arising from self-renewing stem cells and short-lived myeloid progenitors. *Genes & Development, 17*, 3029–3035.

Cui, Y. F., Zhou, P. K., Woolford, L. B., Lord, B. I., Hendry, J. H., & Wang, D. W. (1995). Apoptosis in bone marrow cells of mice with different p53 genotypes after gamma-rays irradiation in vitro. *Journal of Environmental Pathology, Toxicology and Oncology, 14*, 159–163.

Cyr, A. R., & Domann, F. E. (2011). The redox basis of epigenetic modifications: From mechanisms to functional consequences. *Antioxidants & Redox Signaling, 15*, 551–589.

Cyr, A. R., Hitchler, M. J., & Domann, F. E. (2013). Regulation of SOD2 in cancer by histone modifications and CpG methylation: Closing the loop between redox biology and epigenetics. *Antioxidants & Redox Signaling, 18*, 1946–1955.

De Flora, S., & Ferguson, L. R. (2005). Overview of mechanisms of cancer chemopreventive agents. *Mutation Research, 591*, 8–15.

De Marco, F. (2013). Oxidative stress and HPV carcinogenesis. *Viruses, 5*, 708–731.

Dhar, S. K., & St Clair, D. K. (2012). Manganese superoxide dismutase regulation and cancer. *Free Radical Biology & Medicine, 52*, 2209–2222.

Diehn, M., Cho, R. W., Lobo, N. A., Kalisky, T., Dorie, M. J., Kulp, A. N., et al. (2009). Association of reactive oxygen species levels and radioresistance in cancer stem cells. *Nature, 458*, 780–783.

Dionisi, O., Galeotti, T., Terranova, T., & Azzi, A. (1975). Superoxide radicals and hydrogen peroxide formation in mitochondria from normal and neoplastic tissues. *Biochimica et Biophysica Acta, 403*, 292–300.

Ditch, S., & Paull, T. T. (2012). The ATM protein kinase and cellular redox signaling: Beyond the DNA damage response. *Trends in Biochemical Sciences, 37*, 15–22.

Domen, J., Gandy, K. L., & Weissman, I. L. (1998). Systemic overexpression of BCL-2 in the hematopoietic system protects transgenic mice from the consequences of lethal irradiation. *Blood, 91*, 2272–2282.

Dong, C., Yuan, T., Wu, Y., Wang, Y., Fan, T. W., Miriyala, S., et al. (2013). Loss of FBP1 by Snail-mediated repression provides metabolic advantages in basal-like breast cancer. *Cancer Cell, 23*, 316–331.

Droge, W. (2002). Free radicals in the physiological control of cell function. *Physiological Reviews, 82*, 47–95.

Du, W., Adam, Z., Rani, R., Zhang, X., & Pang, Q. (2008). Oxidative stress in Fanconi anemia hematopoiesis and disease progression. *Antioxidants & Redox Signaling, 10*, 1909–1921.

Du, Y., Zhang, H., Lu, J., & Holmgren, A. (2012). Glutathione and glutaredoxin act as a backup of human thioredoxin reductase 1 to reduce thioredoxin 1 preventing cell death by aurothioglucose. *Journal of Biological Chemistry, 287*, 38210–38219.

Du, Y., Zhang, H., Zhang, X., Lu, J., & Holmgren, A. (2013). Thioredoxin 1 is inactivated due to oxidation induced by peroxiredoxin under oxidative stress and reactivated by the glutaredoxin system. *Journal of Biological Chemistry, 288*, 32241–32247.

Dumitru, R., Gama, V., Fagan, B. M., Bower, J. J., Swahari, V., Pevny, L. H., et al. (2012). Human embryonic stem cells have constitutively active Bax at the Golgi and are primed to undergo rapid apoptosis. *Molecular Cell, 46*, 573–583.

Eijkelenboom, A., & Burgering, B. M. (2013). FOXOs: Signalling integrators for homeostasis maintenance. *Nature Reviews. Molecular Cell Biology, 14*, 83–97.

Eliasson, P., & Jonsson, J. I. (2010). The hematopoietic stem cell niche: Low in oxygen but a nice place to be. *Journal of Cellular Physiology, 222*, 17–22.

Eliasson, P., Rehn, M., Hammar, P., Larsson, P., Sirenko, O., Flippin, L. A., et al. (2010). Hypoxia mediates low cell-cycle activity and increases the proportion of long-term-reconstituting hematopoietic stem cells during in vitro culture. *Experimental Hematology, 38*, 301–310.e2.

Erlacher, M., Michalak, E. M., Kelly, P. N., Labi, V., Niederegger, H., Coultas, L., et al. (2005). BH3-only proteins Puma and Bim are rate-limiting for gamma-radiation- and glucocorticoid-induced apoptosis of lymphoid cells in vivo. *Blood, 106*, 4131–4138.

Evans, M. J., & Kaufman, M. H. (1981). Establishment in culture of pluripotential cells from mouse embryos. *Nature, 292*, 154–156.

Ezashi, T., Das, P., & Roberts, R. M. (2005). Low O2 tensions and the prevention of differentiation of hES cells. *Proceedings of the National Academy of Sciences of the United States of America, 102*, 4783–4788.

Facucho-Oliveira, J. M., & St John, J. C. (2009). The relationship between pluripotency and mitochondrial DNA proliferation during early embryo development and embryonic stem cell differentiation. *Stem Cell Reviews, 5*, 140–158.

Fath, M. A., Ahmad, I. M., Smith, C. J., Spence, J., & Spitz, D. R. (2011). Enhancement of carboplatin-mediated lung cancer cell killing by simultaneous disruption of glutathione and thioredoxin metabolism. *Clinical Cancer Research, 17*, 6206–6217.

Filion, T. M., Qiao, M., Ghule, P. N., Mandeville, M., van Wijnen, A. J., Stein, J. L., et al. (2009). Survival responses of human embryonic stem cells to DNA damage. *Journal of Cellular Physiology, 220*, 586–592.

Finkel, T. (2005). Radical medicine: Treating ageing to cure disease. *Nature Reviews. Molecular Cell Biology, 6*, 971–976.

Finkel, T., & Holbrook, N. J. (2000). Oxidants, oxidative stress and the biology of ageing. *Nature, 408*, 239–247.

Fischer, B., & Bavister, B. D. (1993). Oxygen tension in the oviduct and uterus of rhesus monkeys, hamsters and rabbits. *Journal of Reproduction and Fertility, 99*, 673–679.

Fliedner, S. M., Kaludercic, N., Jiang, X. S., Hansikova, H., Hajkova, Z., Sladkova, J., et al. (2012). Warburg effect's manifestation in aggressive pheochromocytomas and paragangliomas: Insights from a mouse cell model applied to human tumor tissue. *PLoS One, 7*, e40949.

Folmes, C. D., Nelson, T. J., Martinez-Fernandez, A., Arrell, D. K., Lindor, J. Z., Dzeja, P. P., et al. (2011). Somatic oxidative bioenergetics transitions into pluripotency-dependent glycolysis to facilitate nuclear reprogramming. *Cell Metabolism, 14*, 264–271.

Forman, H. J., & Kennedy, J. A. (1974). Role of superoxide radical in mitochondrial dehydrogenase reactions. *Biochemical and Biophysical Research Communications, 60*, 1044–1050.

Forman, H. J., & Kennedy, J. (1975). Superoxide production and electron transport in mitochondrial oxidation of dihydroorotic acid. *Journal of Biological Chemistry, 250*, 4322–4326.

Fridovich, I. (1978). The biology of oxygen radicals. *Science, 201*, 875–880.

Fujisaki, J., Wu, J., Carlson, A. L., Silberstein, L., Putheti, P., Larocca, R., et al. (2011). In vivo imaging of Treg cells providing immune privilege to the haematopoietic stem-cell niche. *Nature, 474*, 216–219.

Furth, J., & Kahn, M. (1937). The transmission of leukaemia of mice with a single cell. *American Journal of Cancer, 31*, 276–282.

Gan, B., Sahin, E., Jiang, S., Sanchez-Aguilera, A., Scott, K. L., Chin, L., et al. (2008). mTORC1-dependent and -independent regulation of stem cell renewal, differentiation, and mobilization. *Proceedings of the National Academy of Sciences of the United States of America, 105*, 19384–19389.

Geest, C. R., & Coffer, P. J. (2009). MAPK signaling pathways in the regulation of hematopoiesis. *Journal of Leukocyte Biology, 86*, 237–250.

Gimm, O., Armanios, M., Dziema, H., Neumann, H. P., & Eng, C. (2000). Somatic and occult germ-line mutations in SDHD, a mitochondrial complex II gene, in nonfamilial pheochromocytoma. *Cancer Research, 60*, 6822–6825.

Gius, D., & Spitz, D. R. (2006). Redox signaling in cancer biology. *Antioxidants & Redox Signaling, 8*, 1249–1252.

Goldman, D. C., Bailey, A. S., Pfaffle, D. L., Al Masri, A., Christian, J. L., & Fleming, W. H. (2009). MP4 regulates the hematopoietic stem cell niche. *Blood, 114*, 4393–4401.

Gonda, T. A., Tu, S., & Wang, T. C. (2009). Chronic inflammation, the tumor microenvironment and carcinogenesis. *Cell Cycle, 8*, 2005–2013.

Goodell, M. A., Brose, K., Paradis, G., Conner, A. S., & Mulligan, R. C. (1996). Isolation and functional properties of murine hematopoietic stem cells that are replicating in vivo. *Journal of Experimental Medicine, 183*, 1797–1806.

Goytisolo, F. A., Samper, E., Martin-Caballero, J., Finnon, P., Herrera, E., Flores, J. M., et al. (2000). Short telomeres result in organismal hypersensitivity to ionizing radiation in mammals. *Journal of Experimental Medicine, 192*, 1625–1636.

Greenwood, M. J., & Lansdorp, P. M. (2003). Telomeres, telomerase, and hematopoietic stem cell biology. *Archives of Medical Research, 34*, 489–495.

Gukovsky, I., Li, N., Todoric, J., Gukovskaya, A., & Karin, M. (2013). Inflammation, autophagy, and obesity: Common features in the pathogenesis of pancreatitis and pancreatic cancer. *Gastroenterology, 144*, 1199–1209.e4.

Guo, Y. L., Chakraborty, S., Rajan, S. S., Wang, R., & Huang, F. (2010). Effects of oxidative stress on mouse embryonic stem cell proliferation, apoptosis, senescence, and self-renewal. *Stem Cells and Development, 19*, 1321–1331.

Guo, Z., Kozlov, S., Lavin, M. F., Person, M. D., & Paull, T. T. (2010). ATM activation by oxidative stress. *Science, 330*, 517–521.

Gurdon, J. B. (1962). The developmental capacity of nuclei taken from intestinal epithelium cells of feeding tadpoles. *Journal of Embryology and Experimental Morphology, 10*, 622–640.

Gurdon, J. B., Elsdale, T. R., & Fischberg, M. (1958). Sexually mature individuals of Xenopus laevis from the transplantation of single somatic nuclei. *Nature, 182*, 64–65.

Guzman, M. L., Rossi, R. M., Karnischky, L., Li, X., Peterson, D. R., Howard, D. S., et al. (2005). The sesquiterpene lactone parthenolide induces apoptosis of human acute myelogenous leukemia stem and progenitor cells. *Blood, 105*, 4163–4169.

Guzman, M. L., Rossi, R. M., Neelakantan, S., Li, X., Corbett, C. A., Hassane, D. C., et al. (2007). An orally bioavailable parthenolide analog selectively eradicates acute myelogenous leukemia stem and progenitor cells. *Blood, 110*, 4427–4435.

Hadzic, T., Aykin-Burns, N., Zhu, Y., Coleman, M. C., Leick, K., Jacobson, G. M., et al. (2010). Paclitaxel combined with inhibitors of glucose and hydroperoxide metabolism enhances breast cancer cell killing via H_2O_2-mediated oxidative stress. *Free Radical Biology & Medicine, 48*, 1024–1033.

Halliwell, B., & Gutteridge, J. M. (2007). *Free radicals in biology and medicine* (4th ed.). New York, NY: Oxford University Press Inc.

Han, M. K., Song, E. K., Guo, Y., Ou, X., Mantel, C., & Broxmeyer, H. E. (2008). SIRT1 regulates apoptosis and Nanog expression in mouse embryonic stem cells by controlling p53 subcellular localization. *Cell Stem Cell, 2*, 241–251.

Haraguchi, N., Ishii, H., Mimori, K., Tanaka, F., Ohkuma, M., Kim, H. M., et al. (2010). CD13 is a therapeutic target in human liver cancer stem cells. *Journal of Clinical Investigation, 120*, 3326–3339.

Harman, D. (2003). The free radical theory of aging. *Antioxidants & Redox Signaling, 5*, 557–561.

Harman, D. (2006). Free radical theory of aging: An update: Increasing the functional life span. *Annals of the New York Academy of Sciences, 1067*, 10–21.

Hattori, K., Heissig, B., Wu, Y., Dias, S., Tejada, R., Ferris, B., et al. (2002). Placental growth factor reconstitutes hematopoiesis by recruiting VEGFR1(+) stem cells from bone-marrow microenvironment. *Nature Medicine, 8*, 841–849.

Haug, J. S., He, X. C., Grindley, J. C., Wunderlich, J. P., Gaudenz, K., Ross, J. T., et al. (2008). N-cadherin expression level distinguishes reserved versus primed states of hematopoietic stem cells. *Cell Stem Cell, 2*, 367–379.

Hayflick, L., & Moorhead, P. S. (1961). The serial cultivation of human diploid cell strains. *Experimental Cell Research, 25*, 585–621.

Herault, O., Hope, K. J., Deneault, E., Mayotte, N., Chagraoui, J., Wilhelm, B. T., et al. (2012). A role for GPx3 in activity of normal and leukemia stem cells. *Journal of Experimental Medicine, 209*, 895–901.

Hirabayashi, Y., Matsuda, M., Matumura, T., Mitsui, H., Sasaki, H., Tukada, T., et al. (1997). The p53-deficient hemopoietic stem cells: Their resistance to radiation-apoptosis, but lasted transiently. *Leukemia, 11*(Suppl 3), 489–492.

Hitchler, M. J., & Domann, F. E. (2009). Metabolic defects provide a spark for the epigenetic switch in cancer. *Free Radical Biology & Medicine, 47*, 115–127.

Hitchler, M. J., & Domann, F. E. (2012). Redox regulation of the epigenetic landscape in cancer: A role for metabolic reprogramming in remodeling the epigenome. *Free Radical Biology & Medicine, 53*, 2178–2187.

Hou, P., Li, Y., Zhang, X., Liu, C., Guan, J., Li, H., et al. (2013). Pluripotent stem cells induced from mouse somatic cells by small-molecule compounds. *Science, 341*, 651–654.

Hoyes, K. P., Cai, W. B., Potten, C. S., & Hendry, J. H. (2000). Effect of bcl-2 deficiency on the radiation response of clonogenic cells in small and large intestine, bone marrow and testis. *International Journal of Radiation Biology, 76*, 1435–1442.

Huntly, B. J., Shigematsu, H., Deguchi, K., Lee, B. H., Mizuno, S., Duclos, N., et al. (2004). MOZ-TIF2, but not BCR-ABL, confers properties of leukemic stem cells to committed murine hematopoietic progenitors. *Cancer Cell, 6*, 587–596.

Imaoka, S., Osada, M., Minamiyama, Y., Yukimura, T., Toyokuni, S., Takemura, S., et al. (2004). Role of phenobarbital-inducible cytochrome P450s as a source of active oxygen species in DNA-oxidation. *Cancer Letters, 203*, 117–125.

Irwin, M. E., Rivera-Del Valle, N., & Chandra, J. (2013). Redox control of leukemia: From molecular mechanisms to therapeutic opportunities. *Antioxidants & Redox Signaling, 18*, 1349–1383.

Ishii, N., Fujii, M., Hartman, P. S., Tsuda, M., Yasuda, K., Senoo-Matsuda, N., et al. (1998). A mutation in succinate dehydrogenase cytochrome b causes oxidative stress and ageing in nematodes. *Nature, 394*, 694–697.

Ishimoto, T., Nagano, O., Yae, T., Tamada, M., Motohara, T., Oshima, H., et al. (2011). CD44 variant regulates redox status in cancer cells by stabilizing the xCT subunit of system xc(−) and thereby promotes tumor growth. *Cancer Cell, 19*, 387–400.

Ito, K., Hirao, A., Arai, F., Matsuoka, S., Takubo, K., Hamaguchi, I., et al. (2004). Regulation of oxidative stress by ATM is required for self-renewal of haematopoietic stem cells. *Nature, 431*, 997–1002.

Ito, K., Hirao, A., Arai, F., Takubo, K., Matsuoka, S., Miyamoto, K., et al. (2006). Reactive oxygen species act through p38 MAPK to limit the lifespan of hematopoietic stem cells. *Nature Medicine*, *12*, 446–451.

Ito, K., Takubo, K., Arai, F., Satoh, H., Matsuoka, S., Ohmura, M., et al. (2007). Regulation of reactive oxygen species by Atm is essential for proper response to DNA double-strand breaks in lymphocytes. *Journal of Immunology*, *178*, 103–110.

Jamieson, C. H., Ailles, L. E., Dylla, S. J., Muijtjens, M., Jones, C., Zehnder, J. L., et al. (2004). Granulocyte-macrophage progenitors as candidate leukemic stem cells in blast-crisis CML. *New England Journal of Medicine*, *351*, 657–667.

Jang, Y. Y., & Sharkis, S. J. (2007). A low level of reactive oxygen species selects for primitive hematopoietic stem cells that may reside in the low-oxygenic niche. *Blood*, *110*, 3056–3063.

Janzen, V., Forkert, R., Fleming, H. E., Saito, Y., Waring, M. T., Dombkowski, D. M., et al. (2006). Stem-cell ageing modified by the cyclin-dependent kinase inhibitor p16INK4a. *Nature*, *443*, 421–426.

Ji, A. R., Ku, S. Y., Cho, M. S., Kim, Y. Y., Kim, Y. J., Oh, S. K., et al. (2010). Reactive oxygen species enhance differentiation of human embryonic stem cells into mesendodermal lineage. *Experimental & Molecular Medicine*, *42*, 175–186.

Ji, L., Liu, Y. X., Yang, C., Yue, W., Shi, S. S., Bai, C. X., et al. (2009). Self-renewal and pluripotency is maintained in human embryonic stem cells by co-culture with human fetal liver stromal cells expressing hypoxia inducible factor 1alpha. *Journal of Cellular Physiology*, *221*, 54–66.

Jin, Y., Lu, Z., Ding, K., Li, J., Du, X., Chen, C., et al. (2010). Antineoplastic mechanisms of niclosamide in acute myelogenous leukemia stem cells: Inactivation of the NF-kappaB pathway and generation of reactive oxygen species. *Cancer Research*, *70*, 2516–2527.

Jorgenson, T. C., Zhong, W., & Oberley, T. D. (2013). Redox imbalance and biochemical changes in cancer. *Cancer Research*, *73*, 6118–6123.

Jung, P., Sato, T., Merlos-Suarez, A., Barriga, F. M., Iglesias, M., Rossell, D., et al. (2011). Isolation and in vitro expansion of human colonic stem cells. *Nature Medicine*, *17*, 1225–1227.

Juntilla, M. M., Patil, V. D., Calamito, M., Joshi, R. P., Birnbaum, M. J., & Koretzky, G. A. (2010). AKT1 and AKT2 maintain hematopoietic stem cell function by regulating reactive oxygen species. *Blood*, *115*, 4030–4038.

Kaelin, W. G., Jr., & Ratcliffe, P. J. (2008). Oxygen sensing by metazoans: The central role of the HIF hydroxylase pathway. *Molecular Cell*, *30*, 393–402.

Kaewpila, S., Venkataraman, S., Buettner, G. R., & Oberley, L. W. (2008). Manganese superoxide dismutase modulates hypoxia-inducible factor-1 alpha induction via superoxide. *Cancer Research*, *68*, 2781–2788.

Kamata, T. (2009). Roles of Nox1 and other Nox isoforms in cancer development. *Cancer Science*, *100*, 1382–1388.

Kamsler, A., Daily, D., Hochman, A., Stern, N., Shiloh, Y., Rotman, G., et al. (2001). Increased oxidative stress in ataxia telangiectasia evidenced by alterations in redox state of brains from Atm-deficient mice. *Cancer Research*, *61*, 1849–1854.

Keith, B., & Simon, M. C. (2007). Hypoxia-inducible factors, stem cells, and cancer. *Cell*, *129*, 465–472.

Kerr, J. F., Wyllie, A. H., & Currie, A. R. (1972). Apoptosis: A basic biological phenomenon with wide-ranging implications in tissue kinetics. *British Journal of Cancer*, *26*, 239–257.

Kessenbrock, K., Plaks, V., & Werb, Z. (2010). Matrix metalloproteinases: Regulators of the tumor microenvironment. *Cell*, *141*, 52–67.

Keyse, S. M. (2008). Dual-specificity MAP kinase phosphatases (MKPs) and cancer. *Cancer Metastasis Reviews*, *27*, 253–261.

Kiel, M. J., Yilmaz, O. H., Iwashita, T., Terhorst, C., & Morrison, S. J. (2005). SLAM family receptors distinguish hematopoietic stem and progenitor cells and reveal endothelial niches for stem cells. *Cell, 121*, 1109–1121.

Kim, H. M., Haraguchi, N., Ishii, H., Ohkuma, M., Okano, M., Mimori, K., et al. (2012). Increased CD13 expression reduces reactive oxygen species, promoting survival of liver cancer stem cells via an epithelial-mesenchymal transition-like phenomenon. *Annals of Surgical Oncology, 19*(Suppl 3), S539–S548.

Kinder, M., Wei, C., Shelat, S. G., Kundu, M., Zhao, L., Blair, I. A., et al. (2010). Hematopoietic stem cell function requires 12/15-lipoxygenase-dependent fatty acid metabolism. *Blood, 115*, 5012–5022.

Kirito, K., Fox, N., & Kaushansky, K. (2003). Thrombopoietin stimulates Hoxb4 expression: An explanation for the favorable effects of TPO on hematopoietic stem cells. *Blood, 102*, 3172–3178.

Kobayashi, C. I., & Suda, T. (2012). Regulation of reactive oxygen species in stem cells and cancer stem cells. *Journal of Cellular Physiology, 27*, 421–430.

Kocabas, F., Zheng, J., Thet, S., Copeland, N. G., Jenkins, N. A., DeBerardinis, R. J., et al. (2012). Meis1 regulates the metabolic phenotype and oxidant defense of hematopoietic stem cells. *Blood, 120*, 4963–4972.

Komarov, P. G., Komarova, E. A., Kondratov, R. V., Christov-Tselkov, K., Coon, J. S., Chernov, M. V., et al. (1999). A chemical inhibitor of p53 that protects mice from the side effects of cancer therapy. *Science, 285*, 1733–1737.

Kondo, M., Wagers, A. J., Manz, M. G., Prohaska, S. S., Scherer, D. C., Beilhack, G. F., et al. (2003). Biology of hematopoietic stem cells and progenitors: Implications for clinical application. *Annual Review of Immunology, 21*, 759–806.

Koppenol, W. H. (2001). The Haber-Weiss cycle—70 years later. *Redox Report, 6*, 229–234.

Koptyra, M., Falinski, R., Nowicki, M. O., Stoklosa, T., Majsterek, I., Nieborowska-Skorska, M., et al. (2006). BCR/ABL kinase induces self-mutagenesis via reactive oxygen species to encode imatinib resistance. *Blood, 108*, 319–327.

Koshiji, M., Kageyama, Y., Pete, E. A., Horikawa, I., Barrett, J. C., & Huang, L. E. (2004). HIF-1alpha induces cell cycle arrest by functionally counteracting Myc. *EMBO Journal, 23*, 1949–1956.

Krivtsov, A. V., Twomey, D., Feng, Z., Stubbs, M. C., Wang, Y., Faber, J., et al. (2006). Transformation from committed progenitor to leukaemia stem cell initiated by MLL-AF9. *Nature, 442*, 818–822.

Kunisaki, Y., Bruns, I., Scheiermann, C., Ahmed, J., Pinho, S., Zhang, D., et al. (2013). Arteriolar niches maintain haematopoietic stem cell quiescence. *Nature, 502*, 637–643.

Kyriakis, J. M., & Avruch, J. (2001). Mammalian mitogen-activated protein kinase signal transduction pathways activated by stress and inflammation. *Physiological Reviews, 81*, 807–869.

Labi, V., Erlacher, M., Krumschnabel, G., Manzl, C., Tzankov, A., Pinon, J., et al. (2010). Apoptosis of leukocytes triggered by acute DNA damage promotes lymphoma formation. *Genes & Development, 24*, 1602–1607.

Lagadinou, E. D., Sach, A., Callahan, K., Rossi, R. M., Neering, S. J., Minhajuddin, M., et al. (2013). BCL-2 inhibition targets oxidative phosphorylation and selectively eradicates quiescent human leukemia stem cells. *Cell Stem Cell, 12*, 329–341.

Lambeth, J. D. (2007). Nox enzymes, ROS, and chronic disease: An example of antagonistic pleiotropy. *Free Radical Biology & Medicine, 43*, 332–347.

Lansdorp, P. M., Sutherland, H. J., & Eaves, C. J. (1990). Selective expression of CD45 isoforms on functional subpopulations of CD34+ hemopoietic cells from human bone marrow. *Journal of Experimental Medicine, 172*, 363–366.

Lapidot, T., Sirard, C., Vormoor, J., Murdoch, B., Hoang, T., Caceres-Cortes, J., et al. (1994). A cell initiating human acute myeloid leukaemia after transplantation into SCID mice. *Nature, 367*, 645–648.
Laplante, M., & Sabatini, D. M. (2012). mTOR signaling in growth control and disease. *Cell, 149*, 274–293.
Lee, J. M., & Bernstein, A. (1993). p53 Mutations increase resistance to ionizing radiation. *Proceedings of the National Academy of Sciences of the United States of America, 90*, 5742–5746.
Lee, S., Kim, S. M., & Lee, R. T. (2013). Thioredoxin and thioredoxin target proteins: From molecular mechanisms to functional significance. *Antioxidants & Redox Signaling, 18*, 1165–1207.
Lehninger, A. L. (1976). *Biochemistry*. New York, NY: Worth Publishers Inc, pp. 245–441, 467–471, 849–850.
Lessard, J., & Sauvageau, G. (2003). Bmi-1 determines the proliferative capacity of normal and leukaemic stem cells. *Nature, 423*, 255–260.
Lewandowski, J. P., Sheehan, K. B., Bennett, P. E., Jr., & Boswell, R. E. (2010). Mago Nashi, Tsunagi/Y14, and Ranshi form a complex that influences oocyte differentiation in Drosophila melanogaster. *Developmental Biology, 339*, 307–319.
Li, J., Stouffs, M., Serrander, L., Banfi, B., Bettiol, E., Charnay, Y., et al. (2006). The NADPH oxidase NOX4 drives cardiac differentiation: Role in regulating cardiac transcription factors and MAP kinase activation. *Molecular Biology of the Cell, 17*, 3978–3988.
Li, H., Wang, Y., Pazhanisamy, S. K., Shao, L., Batinic-Haberle, I., Meng, A., et al. (2011). Mn(III) meso-tetrakis-(N-ethylpyridinium-2-yl) porphyrin mitigates total body irradiation-induced long-term bone marrow suppression. *Free Radical Biology & Medicine, 51*, 30–37.
Li, D., Wang, Y., Wu, H., Lu, L., Zhang, H., Chang, J., et al. (2011). Mitigation of ionizing radiation-induced bone marrow suppression by p38 inhibition and G-CSF administration. *Journal of Radiation Research, 52*, 712–716.
Li, Y., Wicha, M. S., Schwartz, S. J., & Sun, D. (2011). Implications of cancer stem cell theory for cancer chemoprevention by natural dietary compounds. *Journal of Nutritional Biochemistry, 22*, 799–806.
Li, S., Yan, T., Yang, J. Q., Oberley, T. D., & Oberley, L. W. (2000). The role of cellular glutathione peroxidase redox regulation in the suppression of tumor cell growth by manganese superoxide dismutase. *Cancer Research, 60*, 3927–3939.
Lipton, B. H., & Schultz, E. (1979). Developmental fate of skeletal muscle satellite cells. *Science, 205*, 1292–1294.
Liu, J., Cao, L., Chen, J., Song, S., Lee, I. H., Quijano, C., et al. (2009). Bmi1 regulates mitochondrial function and the DNA damage response pathway. *Nature, 459*, 387–392.
Liu, J., Cao, L., & Finkel, T. (2011). Oxidants, metabolism, and stem cell biology. *Free Radical Biology & Medicine, 51*, 2158–2162.
Liu, J. C., Guan, X., Ryan, J. A., Rivera, A. G., Mock, C., Agarwal, V., et al. (2013). High mitochondrial priming sensitizes hESCs to DNA-damage-induced apoptosis. *Cell Stem Cell, 13*, 483–491.
Lowe, S. W., & Sherr, C. J. (2003). Tumor suppression by Ink4a-Arf: Progress and puzzles. *Current Opinion in Genetics & Development, 13*, 77–83.
Lu, J., & Holmgren, A. (2014). The thioredoxin antioxidant system. *Free Radical Biology & Medicine, 66*, 75–87.
Madlambayan, G. J., Rogers, I., Kirouac, D. C., Yamanaka, N., Mazurier, F., Doedens, M., et al. (2005). Dynamic changes in cellular and microenvironmental composition can be controlled to elicit in vitro human hematopoietic stem cell expansion. *Experimental Hematology, 33*, 1229–1239.

Magee, J. A., Ikenoue, T., Nakada, D., Lee, J. Y., Guan, K. L., & Morrison, S. J. (2012). Temporal changes in PTEN and mTORC2 regulation of hematopoietic stem cell self-renewal and leukemia suppression. *Cell Stem Cell, 11*, 415–428.

Majno, G., & Joris, I. (1995). Apoptosis, oncosis, and necrosis. An overview of cell death. *American Journal of Pathology, 146*, 3–15.

Marcotte, R., & Wang, E. (2002). Replicative senescence revisited. *Journals of Gerontology. Series A, Biological Sciences and Medical Sciences, 57*, B257–B269.

Martin, G. R. (1981). Isolation of a pluripotent cell line from early mouse embryos cultured in medium conditioned by teratocarcinoma stem cells. *Proceedings of the National Academy of Sciences of the United States of America, 78*, 7634–7638.

Maryanovich, M., Oberkovitz, G., Niv, H., Vorobiyov, L., Zaltsman, Y., Brenner, O., et al. (2012). The ATM-BID pathway regulates quiescence and survival of haematopoietic stem cells. *Nature Cell Biology, 14*, 535–541.

Matsuoka, S., Ballif, B. A., Smogorzewska, A., McDonald, E. R., 3rd., Hurov, K. E., Luo, J., et al. (2007). ATM and ATR substrate analysis reveals extensive protein networks responsive to DNA damage. *Science, 316*, 1160–1166.

Matsuzawa, A., & Ichijo, H. (2008). Redox control of cell fate by MAP kinase: Physiological roles of ASK1-MAP kinase pathway in stress signaling. *Biochimica et Biophysica Acta, 1780*, 1325–1336.

Mauch, P., Constine, L., Greenberger, J., Knospe, W., Sullivan, J., Liesveld, J. L., et al. (1995). Hematopoietic stem cell compartment: Acute and late effects of radiation therapy and chemotherapy. *International Journal of Radiation Oncology, Biology, Physics, 31*, 1319–1339.

Mauro, A. (1961). Satellite cell of skeletal muscle fibers. *Journal of Biophysical and Biochemical Cytology, 9*, 493–495.

Mazumdar, J., Dondeti, V., & Simon, M. C. (2009). Hypoxia-inducible factors in stem cells and cancer. *Journal of Cellular and Molecular Medicine, 13*, 4319–4328.

McCulloch, E. A., & Till, J. E. (1960). The radiation sensitivity of normal mouse bone marrow cells, determined by quantitative marrow transplantation into irradiated mice. *Radiation Research, 13*, 115–125.

Mendez-Ferrer, S., Michurina, T. V., Ferraro, F., Mazloom, A. R., Macarthur, B. D., Lira, S. A., et al. (2010). Mesenchymal and haematopoietic stem cells form a unique bone marrow niche. *Nature, 466*, 829–834.

Menon, S. G., & Goswami, P. C. (2007). A redox cycle within the cell cycle: Ring in the old with the new. *Oncogene, 26*, 1101–1109.

Mercher, T., Cornejo, M. G., Sears, C., Kindler, T., Moore, S. A., Maillard, I., et al. (2008). Notch signaling specifies megakaryocyte development from hematopoietic stem cells. *Cell Stem Cell, 3*, 314–326.

Meredith, M. J., & Dodson, M. L. (1987). Impaired glutathione biosynthesis in cultured human ataxia-telangiectasia cells. *Cancer Research, 47*, 4576–4581.

Michalak, E. M., Vandenberg, C. J., Delbridge, A. R., Wu, L., Scott, C. L., Adams, J. M., et al. (2010). Apoptosis-promoted tumorigenesis: Gamma-irradiation-induced thymic lymphomagenesis requires Puma-driven leukocyte death. *Genes & Development, 24*, 1608–1613.

Miyamoto, K., Araki, K. Y., Naka, K., Arai, F., Takubo, K., Yamazaki, S., et al. (2007). Foxo3a is essential for maintenance of the hematopoietic stem cell pool. *Cell Stem Cell, 1*, 101–112.

Miyamoto, T., Weissman, I. L., & Akashi, K. (2000). AML1/ETO-expressing nonleukemic stem cells in acute myelogenous leukemia with 8;21 chromosomal translocation. *Proceedings of the National Academy of Sciences of the United States of America, 97*, 7521–7526.

Mohyeldin, A., Garzon-Muvdi, T., & Quinones-Hinojosa, A. (2010). Oxygen in stem cell biology: A critical component of the stem cell niche. *Cell Stem Cell, 7*, 150–161.

Molofsky, A. V., Pardal, R., Iwashita, T., Park, I. K., Clarke, M. F., & Morrison, S. J. (2003). Bmi-1 dependence distinguishes neural stem cell self-renewal from progenitor proliferation. *Nature*, *425*, 962–967.

Monfort, A., & Wutz, A. (2013). Breathing-in epigenetic change with vitamin C. *EMBO Reports*, *14*, 337–346.

Murray, L., Chen, B., Galy, A., Chen, S., Tushinski, R., Uchida, N., et al. (1995). Enrichment of human hematopoietic stem cell activity in the CD34+Thy-1+Lin- subpopulation from mobilized peripheral blood. *Blood*, *85*, 368–378.

Nagano, O., Okazaki, S., & Saya, H. (2013). Redox regulation in stem-like cancer cells by CD44 variant isoforms. *Oncogene*, *32*, 5191–5198.

Nakamura, H., Nakamura, K., & Yodoi, J. (1997). Redox regulation of cellular activation. *Annual Review of Immunology*, *15*, 51–69.

Nakamura, S., Oshima, M., Yuan, J., Saraya, A., Miyagi, S., Konuma, T., et al. (2012). Bmi1 confers resistance to oxidative stress on hematopoietic stem cells. *PLoS One*, *7*, e36209.

Narita, M., Nunez, S., Heard, E., Lin, A. W., Hearn, S. A., Spector, D. L., et al. (2003). Rb-mediated heterochromatin formation and silencing of E2F target genes during cellular senescence. *Cell*, *113*, 703–716.

Nathan, C., & Cunningham-Bussel, A. (2013). Beyond oxidative stress: An immunologist's guide to reactive oxygen species. *Nature Reviews. Immunology*, *13*, 349–361.

Navas, T. A., Mohindru, M., Estes, M., Ma, J. Y., Sokol, L., Pahanish, P., et al. (2006). Inhibition of overactivated p38 MAPK can restore hematopoiesis in myelodysplastic syndrome progenitors. *Blood*, *108*, 4170–4177.

Nguyen, P., Awwad, R. T., Smart, D. D., Spitz, D. R., & Gius, D. (2006). Thioredoxin reductase as a novel molecular target for cancer therapy. *Cancer Letters*, *236*, 164–174.

Nguyen, L. V., Vanner, R., Dirks, P., & Eaves, C. J. (2012). Cancer stem cells: An evolving concept. *Nature Reviews. Cancer*, *12*, 133–143.

Nieborowska-Skorska, M., Kopinski, P. K., Ray, R., Hoser, G., Ngaba, D., Flis, S., et al. (2012). Rac2-MRC-cIII-generated ROS cause genomic instability in chronic myeloid leukemia stem cells and primitive progenitors. *Blood*, *119*, 4253–4263.

Nilsson, S. K., Johnston, H. M., Whitty, G. A., Williams, B., Webb, R. J., Denhardt, D. T., et al. (2005). Osteopontin, a key component of the hematopoietic stem cell niche and regulator of primitive hematopoietic progenitor cells. *Blood*, *106*, 1232–1239.

Nordgren, M., & Fransen, M. (2014). Peroxisomal metabolism and oxidative stress. *Biochimie*, *98*, 56–62.

Notta, F., Doulatov, S., Laurenti, E., Poeppl, A., Jurisica, I., & Dick, J. E. (2011). Isolation of single human hematopoietic stem cells capable of long-term multilineage engraftment. *Science*, *333*, 218–221.

Notta, F., Mullighan, C. G., Wang, J. C., Poeppl, A., Doulatov, S., Phillips, L. A., et al. (2011). Evolution of human BCR-ABL1 lymphoblastic leukaemia-initiating cells. *Nature*, *469*, 362–367.

Nowak, J. A., Polak, L., Pasolli, H. A., & Fuchs, E. (2008). Hair follicle stem cells are specified and function in early skin morphogenesis. *Cell Stem Cell*, *3*, 33–43.

Nowicki, M. O., Falinski, R., Koptyra, M., Slupianek, A., Stoklosa, T., Gloc, E., et al. (2004). BCR/ABL oncogenic kinase promotes unfaithful repair of the reactive oxygen species-dependent DNA double-strand breaks. *Blood*, *104*, 3746–3753.

Oberley, L. W. (2005). Mechanism of the tumor suppressive effect of MnSOD overexpression. *Biomedicine and Pharmacotherapy*, *59*, 143–148.

Oberley, L. W., Bize, I. B., Sahu, S. K., Leuthauser, S. W., & Gruber, H. E. (1978). Superoxide dismutase activity of normal murine liver, regenerating liver, and H6 hepatoma. *Journal of the National Cancer Institute*, *61*, 375–379.

Oberley, L. W., & Buettner, G. R. (1979). Role of superoxide dismutase in cancer: A review. *Cancer Research*, *39*, 1141–1149.

Oberley, L. W., Oberley, T. D., & Buettner, G. R. (1980). Cell differentiation, aging and cancer: The possible roles of superoxide and superoxide dismutases. *Medical Hypotheses, 6*, 249–268.
Oberley, L. W., Oberley, T. D., & Buettner, G. R. (1981). Cell division in normal and transformed cells: The possible role of superoxide and hydrogen peroxide. *Medical Hypotheses, 7*, 21–42.
Ogasawara, M. A., & Zhang, H. (2009). Redox regulation and its emerging roles in stem cells and stem-like cancer cells. *Antioxidants & Redox Signaling, 11*, 1107–1122.
Oguro, H., Iwama, A., Morita, Y., Kamijo, T., van Lohuizen, M., & Nakauchi, H. (2006). Differential impact of Ink4a and Arf on hematopoietic stem cells and their bone marrow microenvironment in Bmi1-deficient mice. *Journal of Experimental Medicine, 203*, 2247–2253.
Oh, H., Bradfute, S. B., Gallardo, T. D., Nakamura, T., Gaussin, V., Mishina, Y., et al. (2003). Cardiac progenitor cells from adult myocardium: Homing, differentiation, and fusion after infarction. *Proceedings of the National Academy of Sciences of the United States of America, 100*, 12313–12318.
Ohshima, H., Tatemichi, M., & Sawa, T. (2003). Chemical basis of inflammation-induced carcinogenesis. *Archives of Biochemistry and Biophysics, 417*, 3–11.
Omatsu, Y., Sugiyama, T., Kohara, H., Kondoh, G., Fujii, N., Kohno, K., et al. (2010). The essential functions of adipo-osteogenic progenitors as the hematopoietic stem and progenitor cell niche. *Immunity, 33*, 387–399.
Osawa, M., Hanada, K., Hamada, H., & Nakauchi, H. (1996). Long-term lymphohematopoietic reconstitution by a single CD34-low/negative hematopoietic stem cell. *Science, 273*, 242–245.
Owens, K. M., Aykin-Burns, N., Dayal, D., Coleman, M. C., Domann, F. E., & Spitz, D. R. (2012). Genomic instability induced by mutant succinate dehydrogenase subunit D (SDHD) is mediated by O2(−·) and H2O2. *Free Radical Biology & Medicine, 52*, 160–166.
Owusu-Ansah, E., & Banerjee, U. (2009). Reactive oxygen species prime Drosophila haematopoietic progenitors for differentiation. *Nature, 461*, 537–541.
Pani, G., Galeotti, T., & Chiarugi, P. (2010). Metastasis: Cancer cell's escape from oxidative stress. *Cancer Metastasis Reviews, 29*, 351–378.
Park, I. K., Qian, D., Kiel, M., Becker, M. W., Pihalja, M., Weissman, I. L., et al. (2003). Bmi-1 is required for maintenance of adult self-renewing haematopoietic stem cells. *Nature, 423*, 302–305.
Parmar, K., Mauch, P., Vergilio, J. A., Sackstein, R., & Down, J. D. (2007). Distribution of hematopoietic stem cells in the bone marrow according to regional hypoxia. *Proceedings of the National Academy of Sciences of the United States of America, 104*, 5431–5436.
Patterson, K. I., Brummer, T., O'Brien, P. M., & Daly, R. J. (2009). Dual-specificity phosphatases: Critical regulators with diverse cellular targets. *Biochemistry Journal, 418*, 475–489.
Payne, S. L., Fogelgren, B., Hess, A. R., Seftor, E. A., Wiley, E. L., Fong, S. F., et al. (2005). Lysyl oxidase regulates breast cancer cell migration and adhesion through a hydrogen peroxide-mediated mechanism. *Cancer Research, 65*, 11429–11436.
Pei, S., Minhajuddin, M., Callahan, K. P., Balys, M., Ashton, J. M., Neering, S. J., et al. (2013). Targeting aberrant glutathione metabolism to eradicate human acute myelogenous leukemia cells. *Journal of Biological Chemistry, 288*, 33542–33558.
Pennington, J. D., Jacobs, K. M., Sun, L., Bar-Sela, G., Mishra, M., & Gius, D. (2007). Thioredoxin and thioredoxin reductase as redox-sensitive molecular targets for cancer therapy. *Current Pharmaceutical Design, 13*, 3368–3377.
Penta, J. S., Johnson, F. M., Wachsman, J. T., & Copeland, W. C. (2001). Mitochondrial DNA in human malignancy. *Mutation Research, 488*, 119–133.

Pervaiz, S., Taneja, R., & Ghaffari, S. (2009). Oxidative stress regulation of stem and progenitor cells. *Antioxidants and Redox Signaling, 11*, 2777–2789.
Peskin, A. V., Koen, Y. M., Zbarsky, I. B., & Konstantinov, A. A. (1977). Superoxide dismutase and glutathione peroxidase activities in tumors. *FEBS Letters, 78*, 41–45.
Piccoli, C., D'Aprile, A., Ripoli, M., Scrima, R., Lecce, L., Boffoli, D., et al. (2007). Bone-marrow derived hematopoietic stem/progenitor cells express multiple isoforms of NADPH oxidase and produce constitutively reactive oxygen species. *Biochemical and Biophysical Research Communications, 353*, 965–972.
Piccoli, C., Ria, R., Scrima, R., Cela, O., D'Aprile, A., Boffoli, D., et al. (2005). Characterization of mitochondrial and extra-mitochondrial oxygen consuming reactions in human hematopoietic stem cells. Novel evidence of the occurrence of NAD(P)H oxidase activity. *Journal of Biological Chemistry, 280*, 26467–26476.
Pierce, G. B., & Speers, W. C. (1988). Tumors as caricatures of the process of tissue renewal: Prospects for therapy by directing differentiation. *Cancer Research, 48*, 1996–2004.
Powis, G., Mustacich, D., & Coon, A. (2000). The role of the redox protein thioredoxin in cell growth and cancer. *Free Radical Biology & Medicine, 29*, 312–322.
Qin, H., Yu, T., Qing, T., Liu, Y., Zhao, Y., Cai, J., et al. (2007). Regulation of apoptosis and differentiation by p53 in human embryonic stem cells. *Journal of Biological Chemistry, 282*, 5842–5852.
Rafalski, V. A., Mancini, E., & Brunet, A. (2012). Energy metabolism and energy-sensing pathways in mammalian embryonic and adult stem cell fate. *Journal of Cell Science, 125*, 5597–5608.
Reuter, S., Gupta, S. C., Chaturvedi, M. M., & Aggarwal, B. B. (2010). Oxidative stress, inflammation, and cancer: How are they linked? *Free Radical Biology and Medicine, 49*, 1603–1616.
Reya, T. (2003). Regulation of hematopoietic stem cell self-renewal. *Recent Progress in Hormone Research, 58*, 283–295.
Reynolds, B. A., & Weiss, S. (1992). Generation of neurons and astrocytes from isolated cells of the adult mammalian central nervous system. *Science, 255*, 1707–1710.
Rhee, S. G., Chae, H. Z., & Kim, K. (2005). Peroxiredoxins: A historical overview and speculative preview of novel mechanisms and emerging concepts in cell signaling. *Free Radical Biology & Medicine, 38*, 1543–1552.
Ricci-Vitiani, L., Pallini, R., Biffoni, M., Todaro, M., Invernici, G., Cenci, T., et al. (2010). Tumour vascularization via endothelial differentiation of glioblastoma stem-like cells. *Nature, 468*, 824–828.
Riley, T., Sontag, E., Chen, P., & Levine, A. (2008). Transcriptional control of human p53-regulated genes. *Nature Reviews. Molecular Cell Biology, 9*, 402–412.
Robles, S. J., & Adami, G. R. (1998). Agents that cause DNA double strand breaks lead to p16INK4a enrichment and the premature senescence of normal fibroblasts. *Oncogene, 16*, 1113–1123.
Rodriguez, S., Wang, L., Mumaw, C., Srour, E. F., Lo Celso, C., Nakayama, K., et al. (2011). The SKP2 E3 ligase regulates basal homeostasis and stress-induced regeneration of HSCs. *Blood, 117*, 6509–6519.
Rusyn, I., Asakura, S., Pachkowski, B., Bradford, B. U., Denissenko, M. F., Peters, J. M., et al. (2004). Expression of base excision DNA repair genes is a sensitive biomarker for in vivo detection of chemical-induced chronic oxidative stress: Identification of the molecular source of radicals responsible for DNA damage by peroxisome proliferators. *Cancer Research, 64*, 1050–1057.
Rutter, J., Reick, M., Wu, L. C., & McKnight, S. L. (2001). Regulation of clock and NPAS2 DNA binding by the redox state of NAD cofactors. *Science, 293*, 510–514.
Saeidnia, S., & Abdollahi, M. (2013). Toxicological and pharmacological concerns on oxidative stress and related diseases. *Toxicology and Applied Pharmacology, 273*, 442–455.

Sahu, S. K., Oberley, L. W., Stevens, R. H., & Riley, E. F. (1977). Superoxide dismutase activity of Ehrlich ascites tumor cells. *Journal of the National Cancer Institute, 58*, 1125–1128.

Samper, E., Fernandez, P., Eguia, R., Martin-Rivera, L., Bernad, A., Blasco, M. A., et al. (2002). Long-term repopulating ability of telomerase-deficient murine hematopoietic stem cells. *Blood, 99*, 2767–2775.

Saretzki, G., Walter, T., Atkinson, S., Passos, J. F., Bareth, B., Keith, W. N., et al. (2008). Downregulation of multiple stress defense mechanisms during differentiation of human embryonic stem cells. *Stem Cells, 26*, 455–464.

Sarsour, E. H., Kumar, M. G., Chaudhuri, L., Kalen, A. L., & Goswami, P. C. (2009). Redox control of the cell cycle in health and disease. *Antioxidants & Redox Signaling, 11*, 2985–3011.

Sato, A., Okada, M., Shibuya, K., Watanabe, E., Seino, S., Narita, Y., et al. (2013). Pivotal role for ROS activation of p38 MAPK in the control of differentiation and tumor-initiating capacity of glioma-initiating cells. *Stem Cell Research, 12*, 119–131.

Sauer, H., & Wartenberg, M. (2005). Reactive oxygen species as signaling molecules in cardiovascular differentiation of embryonic stem cells and tumor-induced angiogenesis. *Antioxidants & Redox Signaling, 7*, 1423–1434.

Sauvageau, M., & Sauvageau, G. (2010). Polycomb group proteins: Multi-faceted regulators of somatic stem cells and cancer. *Cell Stem Cell, 7*, 299–313.

Scarbrough, P. M., Mapuskar, K. A., Mattson, D. M., Gius, D., Watson, W. H., & Spitz, D. R. (2012). Simultaneous inhibition of glutathione- and thioredoxin-dependent metabolism is necessary to potentiate 17AAG-induced cancer cell killing via oxidative stress. *Free Radical Biology & Medicine, 52*, 436–443.

Schafer, F. Q., & Buettner, G. R. (2001). Redox environment of the cell as viewed through the redox state of the glutathione disulfide/glutathione couple. *Free Radical Biology & Medicine, 30*, 1191–1212.

Schmelter, M., Ateghang, B., Helmig, S., Wartenberg, M., & Sauer, H. (2006). Embryonic stem cells utilize reactive oxygen species as transducers of mechanical strain-induced cardiovascular differentiation. *FASEB Journal, 20*, 1182–1184.

Schuringa, J. J., & Vellenga, E. (2010). Role of the polycomb group gene BMI1 in normal and leukemic hematopoietic stem and progenitor cells. *Current Opinion in Hematology, 17*, 294–299.

Seitz, H. K., & Stickel, F. (2006). Risk factors and mechanisms of hepatocarcinogenesis with special emphasis on alcohol and oxidative stress. *Biological Chemistry, 387*, 349–360.

Sell, S. (2004). Stem cell handbook. In S. Sell (Ed.), *Stem cells* (pp. 1–18). Totowa, NJ: Humana Press.

Semenza, G. L. (1999). Regulation of mammalian O_2 homeostasis by hypoxia-inducible factor 1. *Annual Review of Cell and Developmental Biology, 15*, 551–578.

Semenza, G. L. (2004). Hydroxylation of HIF-1: Oxygen sensing at the molecular level. *Physiology (Bethesda), 19*, 176–182.

Semenza, G. L. (2013). Cancer-stromal cell interactions mediated by hypoxia-inducible factors promote angiogenesis, lymphangiogenesis, and metastasis. *Oncogene, 32*, 4057–4063.

Serrano, M., & Blasco, M. A. (2001). Putting the stress on senescence. *Current Opinion in Cell Biology, 13*, 748–753.

Shao, L., Lou, Y., & Zhou, D. (2014). Hematopoietic stem cell injury induced by ionizing radiation. *Antioxidants & Redox Signaling, 20*, 1447–1462.

Shao, L., Sun, Y., Zhang, Z., Feng, W., Gao, Y., Cai, Z., et al. (2010). Deletion of proapoptotic Puma selectively protects hematopoietic stem and progenitor cells against high-dose radiation. *Blood, 115*, 4707–4714.

Shao, L., Wang, Y., Chang, J., Luo, Y., Meng, A., & Zhou, D. (2013). Hematopoietic stem cell senescence and cancer therapy-induced long-term bone marrow injury. *Translational Cancer Research, 2*, 397–411.

Shao, L., Wu, L., & Zhou, D. (2012). Sensitization of tumor cells to cancer therapy by molecularly targeted inhibition of the inhibitor of nuclear factor kappaB kinase. *Translational Cancer Research*, *1*, 100–108.

Sharpless, N. E., & DePinho, R. A. (1999). The INK4A/ARF locus and its two gene products. *Current Opinion in Genetics & Development*, *9*, 22–30.

Sherr, C. J., & Weber, J. D. (2000). The ARF/p53 pathway. *Current Opinion in Genetics & Development*, *10*, 94–99.

Shi, X., Zhang, Y., Zheng, J., & Pan, J. (2012). Reactive oxygen species in cancer stem cells. *Antioxidants & Redox Signaling*, *16*, 1215–1228.

Shiloh, Y., & Ziv, Y. (2013). The ATM protein kinase: Regulating the cellular response to genotoxic stress, and more. *Nature Reviews. Molecular Cell Biology*, *14*, 197–210.

Shima, H., Takubo, K., Tago, N., Iwasaki, H., Arai, F., Takahashi, T., et al. (2010). Acquisition of G(0) state by CD34-positive cord blood cells after bone marrow transplantation. *Experimental Hematology*, *38*, 1231–1240.

Shouse, S. S., Warren, S. L., & Whipple, G. H. I. I. (1931). Aplasia of marrow and fatal intoxication in dogs produced by Roentgen radiation of all bones. *Journal of Experimental Medicine*, *53*, 421–435.

Shyh-Chang, N., Daley, G. Q., & Cantley, L. C. (2013). Stem cell metabolism in tissue development and aging. *Development*, *140*, 2535–2547.

Shyh-Chang, N., Locasale, J. W., Lyssiotis, C. A., Zheng, Y., Teo, R. Y., Ratanasirintrawoot, S., et al. (2013). Influence of threonine metabolism on S-adenosylmethionine and histone methylation. *Science*, *339*, 222–226.

Sies, H. (1991). Oxidative stress: From basic research to clinical application. *American Journal of Medicine*, *91*, 31S–38S.

Simons, A. L., Ahmad, I. M., Mattson, D. M., Dornfeld, K. J., & Spitz, D. R. (2007). 2-Deoxy-D-glucose combined with cisplatin enhances cytotoxicity via metabolic oxidative stress in human head and neck cancer cells. *Cancer Research*, *67*, 3364–3370.

Simons, A. L., Parsons, A. D., Foster, K. A., Orcutt, K. P., Fath, M. A., & Spitz, D. R. (2009). Inhibition of glutathione and thioredoxin metabolism enhances sensitivity to perifosine in head and neck cancer cells. *Journal of Oncology*, *2009*, 519563.

Simsek, T., Kocabas, F., Zheng, J., Deberardinis, R. J., Mahmoud, A. I., Olson, E. N., et al. (2010). The distinct metabolic profile of hematopoietic stem cells reflects their location in a hypoxic niche. *Cell Stem Cell*, *7*, 380–390.

Slane, B. G., Aykin-Burns, N., Smith, B. J., Kalen, A. L., Goswami, P. C., Domann, F. E., et al. (2006). Mutation of succinate dehydrogenase subunit C results in increased O2.-, oxidative stress, and genomic instability. *Cancer Research*, *66*, 7615–7620.

Slupianek, A., Falinski, R., Znojek, P., Stoklosa, T., Flis, S., Doneddu, V., et al. (2013). BCR-ABL1 kinase inhibits uracil DNA glycosylase UNG2 to enhance oxidative DNA damage and stimulate genomic instability. *Leukemia*, *27*, 629–634.

Sobhakumari, A., Love-Homan, L., Fletcher, E. V., Martin, S. M., Parsons, A. D., Spitz, D. R., et al. (2012). Susceptibility of human head and neck cancer cells to combined inhibition of glutathione and thioredoxin metabolism. *PLoS One*, *7*, e48175.

Spitz, D. R., Azzam, E. I., Li, J. J., & Gius, D. (2004). Metabolic oxidation/reduction reactions and cellular responses to ionizing radiation: A unifying concept in stress response biology. *Cancer Metastasis Reviews*, *23*, 311–322.

Spitz, D. R., Malcolm, R. R., & Roberts, R. J. (1990). Cytotoxicity and metabolism of 4-hydroxy-2-nonenal and 2-nonenal in H_2O_2-resistant cell lines. Do aldehydic by-products of lipid peroxidation contribute to oxidative stress? *Biochemical Journal*, *267*, 453–459.

Spitz, D. R., Sim, J. E., Ridnour, L. A., Galoforo, S. S., & Lee, Y. J. (2000). Glucose deprivation-induced oxidative stress in human tumor cells. A fundamental defect in metabolism? *Annals of the New York Academy of Sciences*, *899*, 349–362.

Spitz, D. R., Sullivan, S. J., Malcolm, R. R., & Roberts, R. J. (1991). Glutathione dependent metabolism and detoxification of 4-hydroxy-2-nonenal. *Free Radical Biology & Medicine*, *11*, 415–423.

Springer, E. L. (1980). Comparative study of the cytoplasmic organelles of epithelial cell lines derived from human carcinomas and nonmalignant tissues. *Cancer Research*, *40*, 803–817.

Stepanova, L., & Sorrentino, B. P. (2005). A limited role for p16Ink4a and p19Arf in the loss of hematopoietic stem cells during proliferative stress. *Blood*, *106*, 827–832.

Stier, S., Ko, Y., Forkert, R., Lutz, C., Neuhaus, T., Grunewald, E., et al. (2005). Osteopontin is a hematopoietic stem cell niche component that negatively regulates stem cell pool size. *Journal of Experimental Medicine*, *201*, 1781–1791.

Suda, T., Takubo, K., & Semenza, G. L. (2011). Metabolic regulation of hematopoietic stem cells in the hypoxic niche. *Cell Stem Cell*, *9*, 298–310.

Sugimura, R., He, X. C., Venkatraman, A., Arai, F., Box, A., Semerad, C., et al. (2012). Noncanonical Wnt signaling maintains hematopoietic stem cells in the niche. *Cell*, *150*, 351–365.

Sugiyama, T., Kohara, H., Noda, M., & Nagasawa, T. (2006). Maintenance of the hematopoietic stem cell pool by CXCL12-CXCR4 chemokine signaling in bone marrow stromal cell niches. *Immunity*, *25*, 977–988.

Sun, T. T., & Green, H. (1976). Differentiation of the epidermal keratinocyte in cell culture: Formation of the cornified envelope. *Cell*, *9*, 511–521.

Takahashi, K., Tanabe, K., Ohnuki, M., Narita, M., Ichisaka, T., Tomoda, K., et al. (2007). Induction of pluripotent stem cells from adult human fibroblasts by defined factors. *Cell*, *131*, 861–872.

Takahashi, K., & Yamanaka, S. (2006). Induction of pluripotent stem cells from mouse embryonic and adult fibroblast cultures by defined factors. *Cell*, *126*, 663–676.

Takubo, K., Goda, N., Yamada, W., Iriuchishima, H., Ikeda, E., Kubota, Y., et al. (2010). Regulation of the HIF-1alpha level is essential for hematopoietic stem cells. *Cell Stem Cell*, *7*, 391–402.

Tamada, M., Nagano, O., Tateyama, S., Ohmura, M., Yae, T., Ishimoto, T., et al. (2012). Modulation of glucose metabolism by CD44 contributes to antioxidant status and drug resistance in cancer cells. *Cancer Research*, *72*, 1438–1448.

Tang, D. G. (2012). Understanding cancer stem cell heterogeneity and plasticity. *Cell Research*, *22*, 457–472.

te Poele, R. H., Okorokov, A. L., Jardine, L., Cummings, J., & Joel, S. P. (2002). DNA damage is able to induce senescence in tumor cells in vitro and in vivo. *Cancer Research*, *62*, 1876–1883.

Thomson, J. A., Itskovitz-Eldor, J., Shapiro, S. S., Waknitz, M. A., Swiergiel, J. J., Marshall, V. S., et al. (1998). Embryonic stem cell lines derived from human blastocysts. *Science*, *282*, 1145–1147.

Till, J. E., & Mc, C. E. (1961). A direct measurement of the radiation sensitivity of normal mouse bone marrow cells. *Radiation Research*, *14*, 213–222.

Tothova, Z., & Gilliland, D. G. (2007). FoxO transcription factors and stem cell homeostasis: Insights from the hematopoietic system. *Cell Stem Cell*, *1*, 140–152.

Tothova, Z., Kollipara, R., Huntly, B. J., Lee, B. H., Castrillon, D. H., Cullen, D. E., et al. (2007). FoxOs are critical mediators of hematopoietic stem cell resistance to physiologic oxidative stress. *Cell*, *128*, 325–339.

Trachootham, D., Alexandre, J., & Huang, P. (2009). Targeting cancer cells by ROS-mediated mechanisms: A radical therapeutic approach? *Nature Reviews. Drug Discovery*, *8*, 579–591.

Trumpp, A., Essers, M., & Wilson, A. (2010). Awakening dormant haematopoietic stem cells. *Nature Reviews. Immunology*, *10*, 201–209.

Tsai, W. B., Chung, Y. M., Takahashi, Y., Xu, Z., & Hu, M. C. (2008). Functional interaction between FOXO3a and ATM regulates DNA damage response. *Nature Cell Biology, 10*, 460–467.

Uchida, N., Buck, D. W., He, D., Reitsma, M. J., Masek, M., Phan, T. V., et al. (2000). Direct isolation of human central nervous system stem cells. *Proceedings of the National Academy of Sciences of the United States of America, 97*, 14720–14725.

Ueda, S., Masutani, H., Nakamura, H., Tanaka, T., Ueno, M., & Yodoi, J. (2002). Redox control of cell death. *Antioxidants & Redox Signaling, 4*, 405–414.

Unnisa, Z., Clark, J. P., Roychoudhury, J., Thomas, E., Tessarollo, L., Copeland, N. G., et al. (2012). Meis1 preserves hematopoietic stem cells in mice by limiting oxidative stress. *Blood, 120*, 4973–4981.

Valentin-Vega, Y. A., Maclean, K. H., Tait-Mulder, J., Milasta, S., Steeves, M., Dorsey, F. C., et al. (2012). Mitochondrial dysfunction in ataxia-telangiectasia. *Blood, 119*, 1490–1500.

Villunger, A., Michalak, E. M., Coultas, L., Mullauer, F., Bock, G., Ausserlechner, M. J., et al. (2003). p53- and drug-induced apoptotic responses mediated by BH3-only proteins puma and noxa. *Science, 302*, 1036–1038.

Voet, J. G., Voet, D., & Pratt, C. (1999). *Fundamentals of biochemistry: Electron transport and oxidative phosphorylation* (pp. 492–525). New York: John Wiley & Sons, Inc.

Waegell, W. O., Higley, H. R., Kincade, P. W., & Dasch, J. R. (1994). Growth acceleration and stem cell expansion in Dexter-type cultures by neutralization of TGF-beta. *Experimental Hematology, 22*, 1051–1057.

Wang, R., Chadalavada, K., Wilshire, J., Kowalik, U., Hovinga, K. E., Geber, A., et al. (2010). Glioblastoma stem-like cells give rise to tumour endothelium. *Nature, 468*, 829–833.

Wang, M., Kirk, J. S., Venkataraman, S., Domann, F. E., Zhang, H. J., Schafer, F. Q., et al. (2005). Manganese superoxide dismutase suppresses hypoxic induction of hypoxia-inducible factor-1alpha and vascular endothelial growth factor. *Oncogene, 24*, 8154–8166.

Wang, Y., Liu, L., Pazhanisamy, S. K., Li, H., Meng, A., & Zhou, D. (2010). Total body irradiation causes residual bone marrow injury by induction of persistent oxidative stress in murine hematopoietic stem cells. *Free Radical Biology & Medicine, 48*, 348–356.

Wang, Y., Liu, L., & Zhou, D. (2011). Inhibition of p38 MAPK attenuates ionizing radiation-induced hematopoietic cell senescence and residual bone marrow injury. *Radiation Research, 176*, 743–752.

Wang, Y., Schulte, B. A., LaRue, A. C., Ogawa, M., & Zhou, D. (2006). Total body irradiation selectively induces murine hematopoietic stem cell senescence. *Blood, 107*, 358–366.

Wang, Y., Schulte, B. A., & Zhou, D. (2006). Hematopoietic stem cell senescence and long-term bone marrow injury. *Cell Cycle, 5*, 35–38.

Wang, G. L., & Semenza, G. L. (1995). Purification and characterization of hypoxia-inducible factor 1. *Journal of Biological Chemistry, 270*, 1230–1237.

Wang, J., Sun, Q., Morita, Y., Jiang, H., Gross, A., Lechel, A., et al. (2012). A differentiation checkpoint limits hematopoietic stem cell self-renewal in response to DNA damage. *Cell, 148*, 1001–1014.

Warburg, O. (1908). Beobachtungen über die Oxydationsprozesse im Seeigelei. *Zeitschrift für Physiologische Chemie, 57*, 1–6.

Wei, S. J., Botero, A., Hirota, K., Bradbury, C. M., Markovina, S., Laszlo, A., et al. (2000). Thioredoxin nuclear translocation and interaction with redox factor-1 activates the activator protein-1 transcription factor in response to ionizing radiation. *Cancer Research, 60*, 6688–6695.

Weissman, I. L., Anderson, D. J., & Gage, F. (2001). Stem and progenitor cells: Origins, phenotypes, lineage commitments, and transdifferentiations. *Annual Review of Cell and Developmental Biology, 17*, 387–403.

Weyemi, U., Redon, C. E., Parekh, P. R., Dupuy, C., & Bonner, W. M. (2013). NADPH Oxidases NOXs and DUOXs as putative targets for cancer therapy. *Anti-Cancer Agents in Medicinal Chemistry, 13*, 502–514.

Wickremasinghe, R. G., & Hoffbrand, A. V. (1999). Biochemical and genetic control of apoptosis: Relevance to normal hematopoiesis and hematological malignancies. *Blood, 93*, 3587–3600.

Wilson, A., Laurenti, E., Oser, G., van der Wath, R. C., Blanco-Bose, W., Jaworski, M., et al. (2008). Hematopoietic stem cells reversibly switch from dormancy to self-renewal during homeostasis and repair. *Cell, 135*, 1118–1129.

Wilson, A., Laurenti, E., & Trumpp, A. (2009). Balancing dormant and self-renewing hematopoietic stem cells. *Current Opinion in Genetics & Development, 19*, 461–468.

Winkler, I. G., Barbier, V., Wadley, R., Zannettino, A. C., Williams, S., & Levesque, J. P. (2010). Positioning of bone marrow hematopoietic and stromal cells relative to blood flow in vivo: Serially reconstituting hematopoietic stem cells reside in distinct nonperfused niches. *Blood, 116*, 375–385.

Wong, J. L., Creton, R., & Wessel, G. M. (2004). The oxidative burst at fertilization is dependent upon activation of the dual oxidase Udx1. *Developmental Cell, 7*, 801–814.

Wong, J. L., & Wessel, G. M. (2005). Reactive oxygen species and Udx1 during early sea urchin development. *Developmental Biology, 288*, 317–333.

Xiao, Q., Luo, Z., Pepe, A. E., Margariti, A., Zeng, L., & Xu, Q. (2009). Embryonic stem cell differentiation into smooth muscle cells is mediated by Nox4-produced H2O2. *American Journal of Physiology. Cell Physiology, 296*, C711–C723.

Yahata, T., Takanashi, T., Muguruma, Y., Ibrahim, A. A., Matsuzawa, H., Uno, T., et al. (2011). Accumulation of oxidative DNA damage restricts the self-renewal capacity of human hematopoietic stem cells. *Blood, 118*, 2941–2950.

Yalcin, S., Zhang, X., Luciano, J. P., Mungamuri, S. K., Marinkovic, D., Vercherat, C., et al. (2008). Foxo3 is essential for the regulation of ataxia telangiectasia mutated and oxidative stress-mediated homeostasis of hematopoietic stem cells. *Journal of Biological Chemistry, 283*, 25692–25705.

Yamaguchi, H., Calado, R. T., Ly, H., Kajigaya, S., Baerlocher, G. M., Chanock, S. J., et al. (2005). Mutations in TERT, the gene for telomerase reverse transcriptase, in aplastic anemia. *New England Journal of Medicine, 352*, 1413–1424.

Yamaguchi, M., & Kashiwakura, I. (2013). Role of reactive oxygen species in the radiation response of human hematopoietic stem/progenitor cells. *PLoS One, 8*, e70503.

Yamazaki, S., Ema, H., Karlsson, G., Yamaguchi, T., Miyoshi, H., Shioda, S., et al. (2011). Nonmyelinating Schwann cells maintain hematopoietic stem cell hibernation in the bone marrow niche. *Cell, 147*, 1146–1158.

Yankovskaya, V., Horsefield, R., Tornroth, S., Luna-Chavez, C., Miyoshi, H., Leger, C., et al. (2003). Architecture of succinate dehydrogenase and reactive oxygen species generation. *Science, 299*, 700–704.

Yilmaz, O. H., Valdez, R., Theisen, B. K., Guo, W., Ferguson, D. O., Wu, H., et al. (2006). Pten dependence distinguishes haematopoietic stem cells from leukaemia-initiating cells. *Nature, 441*, 475–482.

Yoshihara, H., Arai, F., Hosokawa, K., Hagiwara, T., Takubo, K., Nakamura, Y., et al. (2007). Thrombopoietin/MPL signaling regulates hematopoietic stem cell quiescence and interaction with the osteoblastic niche. *Cell Stem Cell, 1*, 685–697.

Yu, H., Shen, H., Yuan, Y., XuFeng, R., Hu, X., Garrison, S. P., et al. (2010). Deletion of Puma protects hematopoietic stem cells and confers long-term survival in response to high-dose gamma-irradiation. *Blood, 115*, 3472–3480.

Zeng, X. (2007). Human embryonic stem cells: Mechanisms to escape replicative senescence? *Stem Cell Reviews*, *3*, 270–279.
Zhang, J., Grindley, J. C., Yin, T., Jayasinghe, S., He, X. C., Ross, J. T., et al. (2006). PTEN maintains haematopoietic stem cells and acts in lineage choice and leukaemia prevention. *Nature*, *441*, 518–522.
Zhang, H., Mi, J. Q., Fang, H., Wang, Z., Wang, C., Wu, L., et al. (2013). Preferential eradication of acute myelogenous leukemia stem cells by fenretinide. *Proceedings of the National Academy of Sciences of the United States of America*, *110*, 5606–5611.
Zhang, J., Niu, C., Ye, L., Huang, H., He, X., Tong, W. G., et al. (2003). Identification of the haematopoietic stem cell niche and control of the niche size. *Nature*, *425*, 836–841.
Zhang, X., Yalcin, S., Lee, D. F., Yeh, T. Y., Lee, S. M., Su, J., et al. (2011). FOXO1 is an essential regulator of pluripotency in human embryonic stem cells. *Nature Cell Biology*, *13*, 1092–1099.
Zhang, H., Zhai, Z., Wang, Y., Zhang, J., Wu, H., Li, C., et al. (2013). Resveratrol ameliorates ionizing irradiation-induced long-term hematopoietic stem cell injury in mice. *Free Radical Biology & Medicine*, *54*, 40–50.
Zhou, L., Opalinska, J., & Verma, A. (2007). p38 MAP kinase regulates stem cell apoptosis in human hematopoietic failure. *Cell Cycle*, *6*, 534–537.
Zhou, F., Shen, Q., & Claret, F. X. (2013). Novel roles of reactive oxygen species in the pathogenesis of acute myeloid leukemia. *Journal of Leukocyte Biology*, *94*, 423–429.
Zhu, S., Li, W., Zhou, H., Wei, W., Ambasudhan, R., Lin, T., et al. (2010). Reprogramming of human primary somatic cells by OCT4 and chemical compounds. *Cell Stem Cell*, *7*, 651–655.

CHAPTER TWO

Emerging Regulatory Paradigms in Glutathione Metabolism

Yilin Liu, Annastasia S. Hyde, Melanie A. Simpson, Joseph J. Barycki[1]
Department of Biochemistry and the Redox Biology Center, University of Nebraska, Lincoln, Nebraska, USA
[1]Corresponding author: e-mail address: jbarycki2@unl.edu

Contents

Abstract

One of the hallmarks of cancer is the ability to generate and withstand unusual levels of oxidative stress. In part, this property of tumor cells is conferred by elevation of the cellular redox buffer glutathione. Though enzymes of the glutathione synthesis and salvage pathways have been characterized for several decades, we still lack a comprehensive understanding of their independent and coordinate regulatory mechanisms. Recent studies have further revealed that overall central metabolic pathways are frequently altered in various tumor types, resulting in significant increases in biosynthetic capacity and feeding into glutathione synthesis. In this review, we will discuss the enzymes and pathways affecting glutathione flux in cancer and summarize current

Advances in Cancer Research, Volume 122
ISSN 0065-230X
http://dx.doi.org/10.1016/B978-0-12-420117-0.00002-5

models for regulating cellular glutathione through both *de novo* synthesis and efficient salvage. In addition, we examine the integration of glutathione metabolism with other altered fates of intermediary metabolites and highlight remaining questions about molecular details of the accepted regulatory modes.

1. INTRODUCTION

Glutathione is a critical low-molecular-weight thiol that participates in numerous cellular functions in mammalian systems (Fig. 2.1). Intracellular glutathione concentrations range between 0.5 and 10 mM, whereas extracellular glutathione concentrations are significantly lower, with estimated values in the micromolar range (Meister & Anderson, 1983). Glutathione is predominantly found in its reduced state (GSH) or as its most commonly observed oxidation product, GSSG, which is formed from two molecules of glutathione linked by a disulfide bond. The GSH/GSSG redox system maintains an overall reducing environment in the cell, with GSH/GSSG ratios ranging from 30:1 to >100:1 in the cytosol, nucleus, and mitochondria. This ratio is considerably lower in the endoplasmic reticulum with values of 1:1 to 3:1 reported, conditions in which disulfide bonds are generated in proteins traversing the secretory pathway (Dickinson & Forman, 2002; Lu, 2009, 2013; Meister & Anderson, 1983). GSSG is converted to its reduced state by glutathione reductase, using NADPH generated primarily by the pentose phosphate pathway (Fig. 2.2).

One of the primary functions of glutathione is cellular detoxification. Glutathione transferases (GSTs) conjugate the tripeptide via its thiol group to endogenous and exogenous electrophilic compounds (Hayes, Flanagan, & Jowsey, 2005; Mannervik, 2012; Townsend & Tew, 2003),

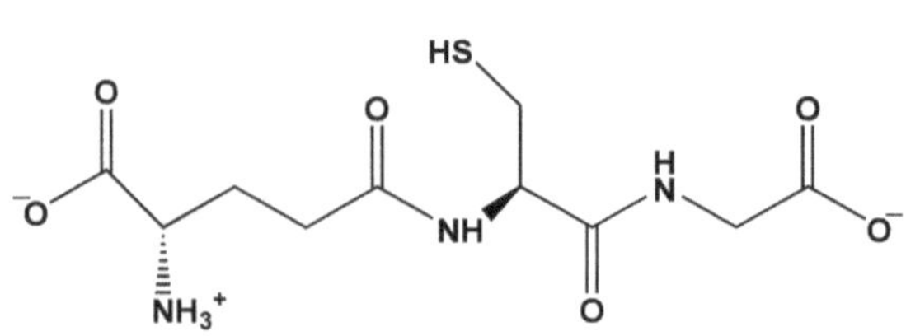

Glutathione

Figure 2.1 Structure of glutathione. GSH is a tripeptide composed of glutamate, cysteine, and glycine. Its γ-glutamyl peptide bond makes it resistant to nonspecific proteolytic cleavage.

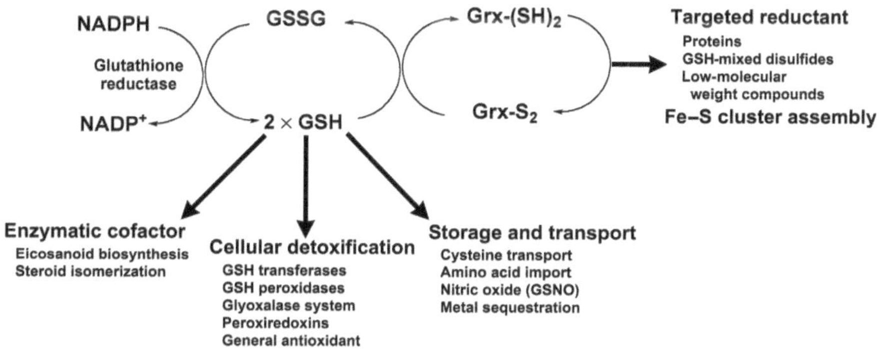

Figure 2.2 Functions of glutathione and enzymes that maintain redox homeostasis. Glutathione is maintained in a reduced state by the action of NADPH-dependent glutathione reductase. Besides its role as a general antioxidant, GSH participates in biosynthetic pathways, signaling processes, detoxification, and storage and transport of key metabolites and sustains glutaredoxin (Grx)-mediated reductive processes. Additional functions are described in the text.

reducing their reactivity and facilitating their clearance from the cell via members of the multidrug resistance-associated protein family (Bachhawat, Thakur, Kaur, & Zulkifli, 2013). A comprehensive review of this topic can be found in this volume. The selenocysteine-containing glutathione peroxidases use glutathione to reduce H_2O_2 or lipid peroxides to water or their corresponding alcohols, respectively, generating oxidized glutathione (Brigelius-Flohe & Maiorino, 2013). The glyoxalase system, composed of glyoxalase I and glyoxalase II, also requires GSH to detoxify methylglyoxal, glyoxal, and other α-oxoaldehydes (Sousa Silva, Gomes, Ferreira, Ponces Freire, & Cordeiro, 2013). However, there is no net oxidation of the cofactor. GSH has also been shown to be involved in the sequestration and transport of metals, including mercury, zinc, and copper (Wang & Ballatori, 1998).

In addition to cellular detoxification, members of membrane-associated proteins in eicosanoid and glutathione metabolism family, which are distantly related to GSTs, contribute to eicosanoid biosynthesis (Hayes et al., 2005). Prostaglandin E synthase converts prostaglandin H_2 to prostaglandin E_2, using GSH as a cofactor. Leukotriene C_4 synthase adds GSH to leukotriene A_4 to produce leukotriene C_4 (Lam, Penrose, Freeman, & Austen, 1994; Welsch et al., 1994; Yoshimoto, Soberman, Lewis, & Austen, 1985). The γ-glutamyl peptide bond of the attached glutathione is cleaved by another glutathione metabolic enzyme, γ-glutamyltranspeptidase 5

(GGT5), to generate leukotriene D_4 (Carter, Shi, Barrios, & Lieberman, 1998). Glutathione is also used in the storage and transport of cysteine (Lieberman et al., 1996). A different isozyme of γ-glutamyltranspeptidase, GGT1, initiates the enzymatic cleavage of extracellular glutathione, leading to the degradation of glutathione to its component amino acids (Hanigan & Ricketts, 1993). Cysteine derived from extracellular glutathione is then imported into the cell and used for protein and intracellular glutathione production (discussed in the succeeding text).

The glutaredoxin enzyme family (Lillig, Berndt, & Holmgren, 2008) is dependent on the GSH pools maintained by glutathione reductase. Glutaredoxins, also known as thiol transferases, catalyze the reversible glutathionylation of protein thiol groups. Dithiol glutaredoxins have an active site Cys-X-X-Cys motif and reduce selected protein disulfides at the expense of two molecules of GSH, generating GSSG. Monothiol glutaredoxins, such as Grx5 in humans, retain the N-terminal cysteine residue of this motif. Glutaredoxins have been shown to be critical in iron–sulfur cluster assembly and to catalyze the reduction of dehydroascorbate to ascorbate.

Given the diverse functions of glutathione, from involvement in signaling pathways to cellular detoxification, aberrant glutathione metabolism has been observed in multiple disease states. GSH levels are known to diminish with increased age, and these reduced GSH levels are correlated with increased oxidative damage. In the human eye, GSH levels have been shown to drop significantly, and this diminution may contribute to cataract formation (Fan et al., 2012; Harding, 1970). Similarly, brain GSH levels are lower in patients suffering from neurodegenerative diseases including Parkinson's and Alzheimer's and psychiatric disorders such as schizophrenia (Currais & Maher, 2013).

A recent comprehensive study evaluated previously reported GSH levels in numerous cancer types relative to normal controls (Gamcsik, Kasibhatla, Teeter, & Colvin, 2012). In general, GSH levels were found to be higher in many cancer types, including head and neck, breast, ovarian, colorectal, and lung. This is in agreement with the long-held hypothesis that rapidly dividing tumor cells experience increased levels of oxidative stress and require higher levels of GSH. In addition, elevated GSH in concert with higher GST expression protects these cells against administered chemotherapeutic agents (Hayes et al., 2005; Townsend & Tew, 2003). However, GSH levels tended to be lower in other cancers, including liver and brain, suggesting that the contributions of GSH to disease progression may be more complex than anticipated (Gamcsik et al., 2012).

2. GLUTATHIONE HOMEOSTASIS

GSH biosynthesis requires sufficient quantities of glutamate, cysteine, and glycine to maintain appropriate levels of the tripeptide. As discussed in the succeeding text, the availability of these precursors reflects the overall metabolic status of the cell. Meister and coworkers proposed a γ-glutamyl cycle (Fig. 2.3) to highlight the central role of GSH in amino acid uptake (Meister, 1973, 1974; Orlowski & Meister, 1970). In this proposed pathway,

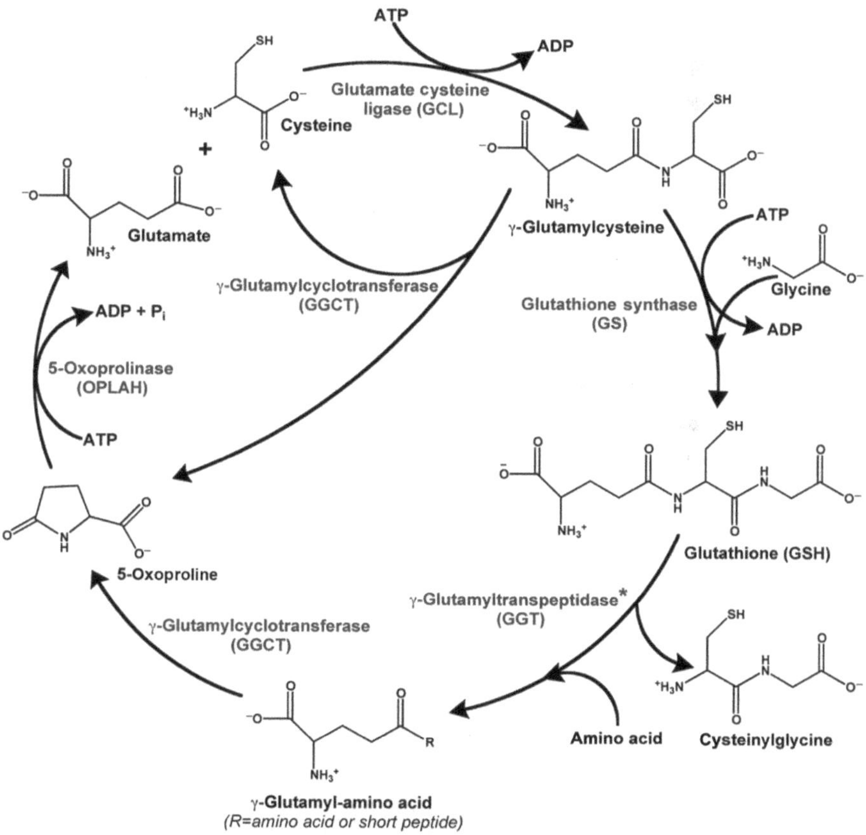

Figure 2.3 Canonical γ-glutamyl cycle. The γ-glutamyl cycle was initially proposed to describe the involvement of glutathione in transpeptidation-dependent amino acid transport, as discussed in the text (Orlowski & Meister, 1970). Since its initial description, GGCT has been suggested to use accumulated γ-glutamylcysteine as a substrate as well. A newly characterized enzyme, ChaC1, may provide a route for the intracellular degradation of glutathione, producing 5-oxoproline and cysteinyl glycine (not shown). * Please note that GGT is an ectoenzyme located at the plasma membrane.

GSH is synthesized by the sequential action of two ATP-dependent ligases, glutamate cysteine ligase (GCL) and glutathione synthetase (GS). Human GCL, also known as γ-glutamylcysteine synthetase, is composed of a 73 kDa catalytic subunit (GCLC) and a 31 kDa regulatory subunit (GCLM). GCL catalyzes the formation of a peptide bond between the γ-carboxylate of glutamate and the α-amino group of cysteine (Orlowski & Meister, 1971) and is the rate-limiting enzyme in GSH biosynthesis. GS then couples the α-carboxylate of γ-glutamylcysteine to the α-amino group of glycine to produce glutathione (Snoke & Bloch, 1955).

Although several plasma membrane transporters capable of importing intact glutathione have been identified (reviewed in Bachhawat et al. (2013)), the primary means of importing glutathione into a cell appears to be through a scavenging pathway (Lieberman et al., 1996; Lu, 2013). The salvage of extracellular glutathione is initiated by γ-glutamyltranspeptidase (GGT), a membrane-anchored glycosylated enzyme that cleaves the γ-glutamyl peptide bond of glutathione (Ikeda & Taniguchi, 2005; Keillor, Castonguay, & Lherbet, 2005; Kinlough, Poland, Bruns, & Hughey, 2005). The enzyme reaction proceeds via a γ-glutamyl enzyme intermediate, in which the γ-glutamyl moiety is covalently linked to the threonine nucleophile at the active site. The released cysteinylglycine is then cleaved to its component amino acids by a dipeptidase (Dickinson & Forman, 2002; Lu, 2013) and imported into the cell. The acyl-enzyme intermediate of γ-glutamyltranspeptidase is resolved by the nucleophilic attack of water or another amino acid/peptide to generate glutamate or a new γ-glutamyl peptide, respectively. Studies by Meister and colleagues suggested that the transpeptidation reaction was preferred and facilitated the import of amino acids into the cell (Meister, 1974; Orlowski & Meister, 1970; Thompson & Meister, 1975). γ-Glutamylcyclotransferase (GGCT) cyclizes the γ-glutamyl moiety of the imported peptides to produce 5-oxoproline, also known as pyroglutamate or pyrrolidone carboxylate, and liberates the coupled amino acid or peptide (Board, Moore, & Smith, 1978; Oakley et al., 2008; Orlowski, Richman, & Meister, 1969). To complete the proposed cycle, human 5-oxoprolinase (OPLAH), a 275 kDa homodimer, couples ATP hydrolysis to the opening of the ring structure to generate glutamate (Van der Werf, Orlowski, & Meister, 1971; Van Der Werf, Stephani, & Meister, 1974).

Now that each of the enzymes of the cycle has been identified and characterized, studies that address the concerted functionality of this pathway can be undertaken. Although the requirement of GCL and GS for glutathione biosynthesis is well documented, the contributions of the salvage pathway enzymes remain open for debate. Genome sequencing has led to the identification of close homologues of glutathione salvage enzymes in organisms that do not appear to synthesize glutathione and therefore may have alternate functions, as discussed in the succeeding text. Other studies suggest that the cycle is not as straightforward as proposed. For example, although the transpeptidase activity of GGT was proposed to be required for amino acid import (Meister, 1973; Orlowski & Meister, 1970; Sekura & Meister, 1974), GGT-null mice produced viable offspring (Lieberman et al., 1996). In addition, the observed phenotypic changes could be ameliorated by supplementation with *N*-acetylcysteine, a cysteine surrogate, suggesting that the primary function of GGT is to liberate cysteine from glutathione produced in the liver for biosynthetic purposes in peripheral tissue. Thus, the regulation of glutathione metabolism is more complex than originally suggested, and further work is needed to understand the contributions of glutathione to disease progression.

2.1. Glutathione biosynthesis

2.1.1 Glutamate cysteine ligase

Human GCL (Fig. 2.4) catalyzes the rate-limiting step in glutathione biosynthesis, and its activity is subject to multiple levels of regulation (Griffith & Mulcahy, 1999; Lu, 2013). Transcriptional regulation of the catalytic and regulatory subunits of the enzyme is mediated through several established pathways including Nrf2-Keap1 and NF-κB (Griffith & Mulcahy, 1999; Lu, 2013). GCL is also extensively regulated at the protein level. The enzyme is feedback-inhibited by glutathione (Richman & Meister, 1975) and is allosterically activated by the addition of the regulatory subunit (GCLM) to the catalytic subunit (GCLC). The resulting heterodimer is two to five-fold more efficient in the production of glutathione, due to improved catalytic efficiency of the enzyme and alleviation of feedback inhibition by glutathione (Backos, Fritz, Roede, Petersen, & Franklin, 2011; Lu, 2013). The relative ratio of GCLM/GCLC differs considerably from tissue to tissue, and this site-specific ratio may largely dictate glutathione production (Chen, Shertzer, Schneider, Nebert, & Dalton, 2005). Formation of the GCL heterodimer has been examined in human and closely related systems, but the molecular details of how complex formation improves flux through the biosynthetic pathway are not clear. This section summarizes published regulatory models and addresses associated limitations that leave the models open to further interpretation.

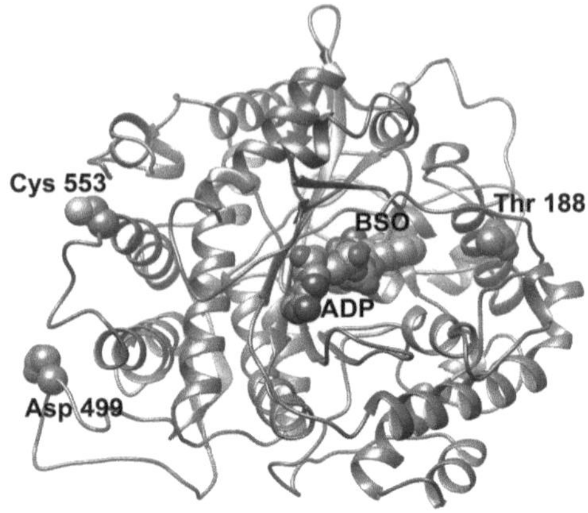

Figure 2.4 Homology model of human GCL. A homology model of human GCLC was generated from the closely related *S. cerevisiae* GCL structure (Biterova & Barycki, 2009). Potential sites of posttranslational modifications as discussed in the text are illustrated in space-filling representation. Docked into the enzyme active site are ADP and the mechanism-based inhibitor L-*S*,*R*-buthionine sulfoximine (BSO), also in space-filling representation. (See the color plate.)

2.1.2 Redox regulation

An attractive model of redox regulation of complex formation was postulated based on the observation of a disulfide-linked species in the early preparations of the enzyme purified from rat kidney (Huang, Chang, Anderson, & Meister, 1993; Seelig, Simondsen, & Meister, 1984). The model suggested that at a lower reduced glutathione concentration, a disulfide bond would form between the catalytic and modifier subunits and increase activity to produce more glutathione. Once glutathione levels were restored, the disulfide bond would be reduced, the complex would no longer be stabilized, and flux through the glutathione biosynthetic pathway would slow. However, the identities of these cysteine residues have not been established. Studies of *Drosophila* GCL indicate that an intersubunit disulfide bond can be formed but it is not required for allosteric activation of the enzyme (Fraser, Kansagra, Kotecki, Saunders, & McLellan, 2003). The disulfide-linked species is a relatively minor fraction of the total enzyme in mouse tissue, and the holoenzyme forms readily without any oxidation (Krejsa et al., 2010). Unpublished work from our group using recombinant GCLC and GCLM suggests that although a disulfide can form, it is not required for stabilization of the human GCLC/GCLM complex or activation of the enzyme. However, agents that stimulate oxidative stress (e.g., pherone and H_2O_2) have been shown to rapidly increase GCL activity independently of increased enzyme expression, suggesting that other cysteine oxidation events may indeed regulate GCL activity (Krejsa et al., 2010; Ochi, 1995, 1996). However, the precise oxidized species of GCL have not been detected.

2.1.3 Phosphorylation

Additional posttranslational modifications of human GCL have been reported. Hormone-mediated phosphorylation of rat liver GCL resulted in modest but significant reductions of enzymatic activity and correspondingly lower levels of glutathione (Lu, Kuhlenkamp, Garcia-Ruiz, & Kaplowitz, 1991). Subsequent studies indicated that protein kinase A, protein kinase C, and Ca^{2+}/calmodulin-dependent kinase II (CKII) could each phosphorylate the catalytic subunit of GCL *in vitro*, resulting in an approximately 20% reduction in V_{max} (Sun, Huang, & Lu, 1996). Although the precise sites of phosphorylation were not identified, the authors suggested that Thr 132, Thr 188, Ser 331, Thr 506, Ser 567, and Ser 591 are potential CKII target sites. The examination of a homology model of GCLC generated using the Phyre2 Web server (Kelley & Sternberg, 2009) suggests that, with the exception of Ser 591, each of these residues is relatively solvent-accessible and thus a possible candidate for phosphorylation. Of the remaining residues, Thr 188 (Fig. 2.4) is located on the periphery of the enzyme active site, and its phosphorylation could impact substrate binding and/or catalysis. However, direct identification of specific phosphorylation sites has not been done to establish the biological significance of phosphorylation given the relatively modest, yet significant, reduction in enzymatic activity.

2.1.4 Proteolysis

During apoptosis, caspase-3 cleaves the catalytic subunit of GCL into 60 and 13 kDa fragments, and this modification may contribute to reduced cellular GSH levels observed during this process (Siitonen et al., 1999). The site of proteolytic cleavage was mapped to Asp 499 by mutational analysis (Franklin et al., 2002). In a recent review, the authors indicated that *in vitro* cleavage of recombinant GCLC by caspase-3 did not impact the overall structure of GCLC or its enzymatic activity (Franklin et al., 2009). The examination of a homology model of human GCL (Willis et al., 2011) indicates that Asp 499 is predicted to be in an extended surface-exposed loop remote from the enzyme active site and that cleavage would not directly impact the core structural features (Fig. 2.4). Thus, it is unclear how significant this observation is, with respect to maintenance of glutathione levels.

2.1.5 Lipid adducts

A reactive product of lipid peroxidation, 4-hydroxy-2-nonenal, was shown to rapidly increase GCL activity and GSH levels in the lung adenocarcinoma cell line, A549 (Backos et al., 2011). A corresponding increase in isolated catalytic subunit activity was observed with 4-hydroxy-2-nonenal treatment. The modification of Cys 553 by 4-hydroxy-2-nonenal at a 100-fold molar excess resulted in an approximately two-fold increase in V_{max} and modest reductions in the K_m values for glutamate and ATP. Comparable results were observed with another cysteine-modifying agent, *N*-ethylmaleimide, suggesting that any number of modifications of this cysteine residue may activate the enzyme. Within the regulatory subunit, Cys 35 was identified as the primary site of modification by 4-hydroxy-2-nonenal. Isolated subunits when preincubated with 4-hydroxy-2-nonenal had impaired ability to form the holoenzyme. The modification of Cys 553 of GCLC impaired but did not completely block complex formation, whereas the modification of Cys 35 of GCLM precluded subunit association. A telling observation was that in the preformed complex, only Cys 35 was reactive with 4-hydroxy-2-nonenal, suggesting that Cys 553 may be inaccessible to chemical modification in the complex. Conversely, Cys 35 is solvent-accessible in either case and is considerably more reactive than any of the other 5 cysteine residues of GCLM. These studies clearly indicate that Cys 553 of GCLC and Cys 35 of GCLM are reactive towards alkylating agents and these modifications impact GCL function *in vitro*, but direct adduct formation was not demonstrated in cell lysates or tissue samples.

2.1.6 Glycation

Recently, glycation of GCLM and GCLC has also been reported, and this posttranslational modification was shown to impact heterodimer formation, with modest effects on kinetic constants (Backos et al., 2013). Pretreatment of individual subunits with 2-deoxy-D-ribose as a chemical modifier impaired heterodimer formation, but treatment of the holoenzyme had no apparent effect on oligomeric state. Extended incubations with 30 mM 2-deoxy-D-ribose (24 h) resulted in decreased activities for GCLC alone and the holoenzyme complex with no significant effects on the K_m values for each substrate. Similarly, K_i values for glutathione remained unchanged. These results suggest that glycation inactivated the enzyme directly but slowly over time. As with 4-hydroxy-2-nonenal modification, glycation products of GCLC and GCLM were not identified in biological samples. Similarly, attempts to identify the sites of modification using recombinant protein were unsuccessful, so it is difficult to assign functional significance to this observation. Cys 553 has been previously implicated in GCL complex formation (Tu & Anders, 1998), and thiol modification at this position may be an important mode of regulation. However, direct evidence for modification at Cys 553 under physiologically relevant conditions was not shown.

2.2. Glutathione synthetase

GS is largely overlooked when considering the regulation of cellular glutathione levels. Human GS is a homodimeric enzyme that has been extensively characterized using both structural (Gogos & Shapiro, 2002) and kinetic approaches (Brown et al., 2011; Dinescu, Brown, Barelier, Cundari, & Anderson, 2010; Dinescu, Cundari, Bhansali, Luo, & Anderson, 2004; Luo, Huang, Babaoglu, & Anderson, 2000; Slavens, Brown, Barakat, Cundari, & Anderson, 2011; Snoke & Bloch, 1955). It is a member of the ATP-grasp superfamily (Fawaz, Topper, & Firestine, 2011) and rapidly catalyzes the ligation of γ-glutamylcysteine and glycine to generate glutathione. Based on these kinetic studies, flux through the glutathione biosynthetic pathway is thought to be largely controlled by GCL activity (Backos et al., 2011; Lu, 2013). Unlike GCL, GS is not feedback-inhibited by GSH nor does it have an associated regulatory subunit. Furthermore, there have been no validated reports of posttranslational modification of GS having any impact on enzymatic activity. As such, GS activity appears to be primarily controlled at the level of transcription and substrate availability.

2.3. GSH in development

Targeted disruption of either GS (Winkler et al., 2011) or the catalytic subunit of GCL (Dalton, Dieter, Yang, Shertzer, & Nebert, 2000; Shi et al., 2000) results in an embryonic lethal phenotype in null homozygous mice, demonstrating the central importance of glutathione in development. Although GS-null mice were not viable, heterozygous mice had no discernible phenotype other than a 50% reduction in both protein levels and enzymatic activity. Glutathione levels were normal and γ-glutamylcysteine did not accumulate, supporting the assertion that GCL activity is typically the limiting factor in glutathione production (Winkler et al., 2011). Similarly, heterozygous GCLC-deficient mice had no overt phenotype, but had approximately two-fold reductions in GCLC levels and activity (Dalton et al., 2000; Shi et al., 2000). However, glutathione levels were only reduced by approximately 20%, suggesting other compensatory mechanisms in heterozygous mice (Dalton et al., 2000). A liver-specific targeted disruption of GCLC did not impact embryogenesis or development, but severe abnormalities of the liver manifested within 4 weeks of birth (Chen et al., 2007). Plasma glutathione levels were reduced dramatically, although glutathione levels in peripheral tissues remained relatively unchanged. In contrast, GCLM-deficient mice have less dramatic phenotypes (Yang et al., 2002). Without its regulatory subunit, GCLC activity and glutathione levels are reduced to between 10% and 40% of wild-type mice depending on the tissue, but no overt phenotype is observed in the absence of an applied stress.

2.4. GSH and cancer

Increased glutathione synthesis and the resulting lower levels of reactive oxygen species (ROS) have been suggested to confer a growth advantage to cancer cells (Traverso et al., 2013). In several human cancer types, including liver, skin, colorectal, lung, head and neck, breast, and prostate cancers, GCLC and glutathione levels are increased relative to normal tissue (Gamcsik et al., 2012). In a study involving human breast tumors, subpopulations of cancer stem cells were shown to have lower levels of ROS primarily through increased expression of GCLM and GS (Diehn et al., 2009). Pharmacological inhibition of glutathione synthesis by L-*S*,*R*-buthionine sulfoximine in these cancer stem cells reduced their ability to form colonies and increased their sensitivity to ionizing radiation. Similarly, L-*S*,*R*-buthionine sulfoximine treatment of MDA-MB231 human breast carcinoma cells enhanced 2-deoxy-D-glucose-induced cytotoxicity (Andringa et al., 2006). Experimentally induced reductions of GCLC protein levels and corresponding reductions in intracellular glutathione sensitized human colon cancer cells to cisplatin (Iida et al., 2001). Prenatal exposure to benzo [a]pyrene, a polycyclic aromatic hydrocarbon, led to increased premature ovarian failure and ovarian tumorigenesis in GCLM-null mice relative to controls, due to their diminished glutathione levels (Lim et al., 2013). This subset of studies highlights the therapeutic potential of targeting GSH homeostasis in a variety of cancer types.

3. GLUTATHIONE SALVAGE

3.1. γ-Glutamyltranspeptidase

GGT initiates the degradation of extracellular glutathione. There are seven potential full-length isoforms of the enzyme, but only GGT1 and GGT5 have been validated as active transpeptidases (Heisterkamp, Groffen, Warburton, & Sneddon, 2008). GGT1 is the canonical GGT, found at the surface of epithelial cells and involved in glutathione salvage, whereas GGT5 appears to be primarily involved in leukotriene biosynthesis and the degradation of select glutathione conjugates (Carter et al., 1998). GGT is a member of the N-terminal nucleophile superfamily, which is functionally defined by an autocatalytic cleavage reaction that generates a new N-terminal residue within the protein (Oinonen & Rouvinen, 2000). This residue then serves as the nucleophile in the enzyme-catalyzed reaction. Considerable detail is known about this autoprocessing and activation event based on studies of GGT from several model organisms (Boanca, Sand, & Barycki, 2006; Boanca et al., 2007; Castonguay, Lherbet, & Keillor, 2003; Okada, Suzuki, Wada, Kumagai, & Fukuyama, 2006, 2007; Suzuki & Kumagai, 2002). Interestingly, organisms that do not maintain an intracellular pool of glutathione (e.g., *H. pylori*) express GGT, suggesting that its main function is the salvage of glutathione to provide a growth advantage (Shibayama et al., 2007). Recently, the crystal structure of human GGT1 was determined, which will aid in detailed mechanistic and inhibitor design studies (West et al., 2013). A comprehensive review of GGT can be found within this volume.

In response to oxidative stress, glutathione and glutathione conjugates are actively transported out of the cell by multidrug resistance-associated proteins (Bachhawat et al., 2013). To recover the rapidly exported glutathione and glutathione conjugates, GGT initiates salvage, cleaving the γ-glutamyl peptide bond to generate cysteinylglycine and transferring the γ-glutamyl group to another amino acid or peptide. Intriguingly, Meister and coworkers provided evidence that the γ-glutamyl group could be transferred to cystine, imported into the cell, and reduced intracellularly to generate γ-glutamylcysteine independently of GCL (Thompson & Meister, 1975). The cysteinylglycine liberated is proposed to be cleaved by nonspecific dipeptidases and cysteine and glycine imported into the cell. PEPT2, a di- and tripeptide transporter, has also been demonstrated to contribute to uptake of the intact cysteinylglycine (Frey et al., 2007).

3.2. γ-Glutamylcyclotransferase

Imported γ-glutamyl peptides are processed in the cell by GGCT, which cyclizes glutamate, generating 5-oxoproline and the free amino acid or peptide. Although GGCT was first purified nearly 40 years ago (Board et al., 1978; Orlowski et al., 1969; Palekar, Tate, & Meister, 1974), the gene encoding this enzyme has only recently been identified (Oakley et al., 2008). It shares sequence and structural similarity to γ-glutamylamine cyclotransferase, which degrades γ-glutamyl-ε-lysine liberated from proteins covalently crosslinked by transglutaminases (Oakley, Coggan, & Board, 2010). Furthermore, this enzyme family is broadly represented in bacteria, plants, and other higher eukaryotes. Initial studies by Meister and coworkers indicated that GGCT purified from rat liver has relatively broad substrate specificity, consistent with the relatively nonspecific transfer of a γ-glutamyl moiety to acceptor substrates by GGT. These observations led to the proposal of a GGT/GGCT pathway for import of amino acids, as a component function of the γ-glutamyl cycle (Orlowski et al., 1969; Palekar et al., 1974). However, GGCT involvement in amino acid salvage has not been directly demonstrated using systems in which enzyme expression has been experimentally altered. Since the substrate specificity of human GGCT has not been extensively characterized, it remains unclear if the enzyme is involved in amino acid import, intracellular cleavage of γ-glutamylcysteine (Oakley et al., 2008), or apoptotic signaling (Masuda et al., 2006).

Several recent reports have suggested that increased expression of GGCT, initially designated as C7orf24, may have utility as a tumor marker. Using a proteomics approach, GGCT was found at higher levels in bladder urothelial carcinoma samples relative to normal controls. Manipulation of GGCT expression indicated that the enzyme provided a growth advantage to cancerous cells by an unknown mechanism (Kageyama et al., 2007). In human osteosarcoma, enzyme levels were considerably higher in both cell lines and primary tumors (Uejima et al., 2011). Knockdown of GGCT by siRNA resulted in cells with slower cell growth rates, increased clustering, and reduced invasiveness and motility as measured by standard *in vitro* methods, consistent with GGCT being involved in tumor progression (Uejima et al., 2011). Similarly, reduction of GGCT levels by siRNA inhibited the growth of human lung squamous cell carcinoma in a mouse model (Hama et al., 2012). GGCT is an effective diagnostic marker in esophageal squamous cell carcinoma, where higher levels of expression correlate with a more aggressive phenotype (Takemura et al., 2014).

Unfortunately, the impacts on GSH levels within the cell were not reported in those studies in which GGCT levels were manipulated. Promoter analysis of GGCT suggests that its expression is cell cycle-dependent, consistent with a potential role in cancer cell proliferation (Ohno et al., 2011). In contrast, GGCT was initially characterized as a novel cytochrome *c*-releasing factor, and overexpression in HeLa cells resulted in apoptosis, suggesting that GGCT may have alternate metabolic functions in different cell types (Masuda et al., 2006).

3.3. 5-Oxoprolinase

Historically, the primary functions of 5-oxoprolinase (OPLAH) have been associated with glutathione salvage, with the enzyme catalyzing the ATP-dependent cleavage of 5-oxoproline to generate glutamate to complete the pathway (Griffith, Bridges, & Meister, 1978; Orlowski & Meister, 1970; Van der Werf et al., 1971). Recently, studies of the γ-glutamyl cycle suggest that 5-oxoproline levels may reflect nutritional status, particularly with respect to glycine availability (Friesen, Novak, Hasman, & Innis, 2007). When glycine becomes limiting, increased levels of 5-oxoproline are observed, stimulating amino acid uptake. This is consistent with the observation that 5-oxoproline levels increase dramatically in GS deficiency (Dahl et al., 1997; Wellner, Sekura, Meister, & Larsson, 1974). This led to the suggestion that the conversion of γ-glutamylcysteine to 5-oxoproline and cysteine by GGCT may be the major source of 5-oxoproline (Oakley et al., 2008). However, there are examples of patients that present with 5-oxoprolinuria who have normal circulating glutathione concentrations despite increased 5-oxoproline levels (Calpena et al., 2013). These levels show glutathione homeostasis is normal in these patients. Therefore, 5-oxoproline may have additional functions, perhaps as a glutamate reservoir. In addition, the coordinate regulation of GGCT and OPLAH may provide another level of control in glutathione biosynthesis.

OPLAH protein levels and activity have been shown to decrease in some human tumor tissues, including lung and colon, although the extent and overall pattern are not conclusive (Chen et al., 1998). The authors suggest that differential expression of OPLAH between normal and cancerous tissues may be exploited by the use of the cysteine prodrug, L-2-oxothiazolidine-4-carboxylic acid (OTZ). OPLAH can hydrolyze OTZ, and the resulting product spontaneously decarboxylates to generate cysteine. In this fashion, GSH production would be bolstered in normal cells relative to cancer cells. However, OPLAH levels are elevated in ovarian cancer (Chien et al., 2009) and unchanged in colon cancer (Chen et al., 1998), illustrating potential pitfalls of this general approach. In searching for potential cancer biomarkers, several groups have identified 5-oxoproline as a biomolecule that is more abundant in tumor cells, such as nasopharyngeal carcinoma (Tang et al., 2011) and bladder cancer (Kim et al., 2010). In addition, comparison of genome-wide methylation status in normal versus colorectal cancer cells revealed OPLAH as a potential biomarker (Naumov et al., 2013). A comparable study also identified OPLAH as a potential marker in gastric cancer (Nanjo et al., 2012). In each of these studies, there was no direct association between OPLAH and glutathione metabolism reported. Given the diversity of metabolic fates for glutamate, it is likely that the originally proposed γ-glutamyl cycle is overly simplistic, and comprehensive metabolic labeling studies are needed to examine flux of intermediates through the proposed cycle.

4. PRECURSOR AVAILABILITY

An often overlooked determinant of cellular glutathione levels is precursor availability. Glutamate, cysteine, and glycine are amino acids that intersect with major metabolic hubs that reflect on nitrogen and sulfur metabolism, one-carbon metabolism, and overall energy levels within the cell. Although cysteine is generally considered the limiting component of glutathione biosynthesis, glutamate and glycine partitioning can impact glutathione levels. An emerging trend in tumor cell biology is the concept of metabolic reprogramming to meet core demands on bioenergetics, biosynthesis, and redox homeostasis (Cairns, Harris, & Mak, 2011). For example, Nrf2 is a well-studied transcriptional regulator responsible for mounting a response to xenobiotic and oxidative stress (Jaramillo & Zhang, 2013; Sporn & Liby, 2012). In addition to its involvement in the production of detoxifying enzymes, Nrf2 has recently been shown to participate in metabolic reprogramming, diverting glucose and glutamine to biosynthetic pathways (Mitsuishi et al., 2012). As discussed in the succeeding text, Nrf2 promotes purine nucleotide synthesis and glutamine metabolism. Stimulation of these pathways also provides precursors for increased glutathione synthesis and reducing equivalents, generated via the pentose phosphate pathway, to maintain GSH pools.

4.1. Glutamate/glutamine

Central to the concept of metabolic reprogramming is glutamine bioavailability, as some cancer cells have been shown to depend heavily on this amino acid for rapid proliferation and protection against ROS (Daye & Wellen, 2012; DeBerardinis et al., 2007; Le et al., 2012). Although considered a nonessential amino acid under normal metabolic conditions, glutamine becomes a conditional essential amino acid, needed for biosynthetic purposes. Glutamine can serve as a nitrogen donor in the production of amino sugars, nucleotides, and other amino acids. After donation of the γ-amino group, the remaining glutamate can be incorporated directly into glutathione or transformed into proline. Alternatively, its α-amino group can serve as a nitrogen donor via transamination to produce aspartate and alanine from oxaloacetate and pyruvate, respectively, and through more complex biosynthesis pathways, integrated into threonine, serine, glycine, and cysteine. The residual carbon skeleton, α-ketoglutarate, can then feed into the citric acid cycle. As such, glutamine/glutamate availability is directly tied to synthesis of key biomolecules.

The complex regulatory network that controls glutamine metabolism shares considerable overlap with pathways implicated in tumor progression. Glutamine is rapidly taken up in HeLa (cervical adenocarcinoma) and RT112 (urinary bladder carcinoma) cells by SLC1A5 in a Na^+-dependent manner. Accumulated intracellular glutamine drives the uptake of essential amino acids via SLC7A5/SLC3A2-mediated antiport (Nicklin et al., 2009). The resulting increase of intracellular leucine levels activates the mTORC1 pathway, and the increased availability of amino acid precursors stimulates proliferation. In contrast, mTORC1 signaling does not need to be primed by glutamine uptake in the breast cancer cell line, MCF-7, as sufficient levels of glutamine are present (Nicklin et al., 2009).

c-Myc has been shown to increase SLC1A5 and SLC7A5 mRNA levels and to stimulate glutaminolysis through suppression of miR-23a/b in human P-493 B lymphoma cells and PC3 prostate cancer cells (Gao et al., 2009). The resulting increase in mitochondrial glutaminase expression (GLS1) promoted flux through the tricarboxylic acid cycle (TCA) and more robust glutathione biosynthesis. siRNA-mediated knockdown of GLS1 in both the lymphoma and prostate cancer cell lines resulted in reduced ATP levels, diminished glutathione levels, and increased levels of ROS, firmly establishing glutamine metabolism as a target of c-Myc and as a potential target for cancer therapeutics.

Phosphate-activated mitochondrial glutaminase (GLS2) has also been shown to initiate glutaminolysis to protect against ROS through increased glutathione production (Suzuki et al., 2010). However, GLS2 has different kinetic constants and tissue distribution relative to GLS1 and is under the control of p53, a tumor suppressor. In this context, GLS2 is thought to contribute to tumor suppression as its overexpression inhibits tumor cell growth in cell culture models and GLS2 expression levels are down in liver tumors. In addition to stimulating GLS2 expression, p53 also activates TP53-inducible glycolysis and apoptosis regulator, which stimulates the pentose phosphate pathway, resulting in increased NADPH levels and, through the action of glutathione reductase, increased GSH/GSSG ratios. Although induction of GLS1 or GLS2 leads to increased antioxidant capacity as a result of increases in reduced glutathione levels, the opposing effects of GLS1 and GLS2 in tumor progression suggest that the timing of and the flux through glutaminolysis can lead to tumor promotion or suppression.

4.2. Cysteine

As discussed in the preceding text, studies of GGT1-null mice suggested that one of the primary functions of glutathione was to serve as a means of conveyance for cysteine from the liver to peripheral tissue (Lieberman et al., 1996). This assertion is supported by the observation that the addition of *N*-acetylcysteine to the diet nearly restored the growth rate of the null animals. Thus, the GGT1-initiated degradation of glutathione ultimately leads to the liberation and import of cysteine into the cell. GGT1 expression is upregulated in several cancer types (Pompella, Corti, Paolicchi, Giommarelli, & Zunino, 2007; Strasak et al., 2008), and the prevailing hypothesis is that the ability to effectively scavenge extracellular glutathione confers a growth advantage to these metabolically demanding cells (Hanigan, Gallagher, Townsend, & Gabarra, 1999).

Another source of cysteine for glutathione biosynthesis originates from methionine (Lu & Mato, 2012). Methionine adenosyltransferase catalyzes the formation of S-adenosylmethionine (AdoMet, SAM) from methionine and ATP. AdoMet serves as a potent methyl donor in hundreds of biologically significant reactions with concomitant production of *S*-adenosylhomocysteine. Hydrolysis of *S*-adenosylhomocysteine produces homocysteine that can be converted to methionine by either methionine synthase, using 5-methyltetrahydrofolate as a methyl donor, or betaine homocysteine methyltransferase, which requires the choline metabolite, betaine. Alternatively, homocysteine can be partitioned into the transsulfuration pathway in which two PLP-dependent enzymes, cystathionine β-synthase and cystathionine γ-lyase, catalyze the condensation of homocysteine and serine to produce cystathionine and its subsequent cleavage to α-ketobutyrate and cysteine.

A significant fraction of cysteine for glutathione biosynthesis can be produced via the transsulfuration pathway. For example, approximately half of the cysteine used for glutathione biosynthesis in HepG2 cells is acquired through the transsulfuration pathway (Mosharov, Cranford, & Banerjee, 2000). In contrast to the conclusions drawn from the studies of GGT1-null mice, targeted disruption of GCLC in hepatocytes did not lead to dramatic reductions in cysteine or glutathione levels in extrahepatic tissue (Chen et al., 2007). This suggests that tissues can derive significant amounts of cysteine from methionine and are not solely dependent on salvage of cysteine from circulating glutathione.

Cystine import has recently emerged as an important process in maintaining intracellular cysteine levels. System x_c^- is a cysteine/glutamate antiporter composed of two subunits, xCT and 4F2 heavy chain. xCT-null mice have higher plasma levels of cysteine and lower circulating levels of glutathione (Sato et al., 2005). Although not required for normal development, xCT is important in maintaining plasma redox balance, and there is considerable evidence that cystine is vital for cancer cell survival. Chronic lymphocytic leukemia cells do not efficiently uptake cystine and require coculture with bone marrow-derived stromal cells to maintain cellular glutathione levels (Zhang et al., 2012). The stromal cells have high levels of system x_c^- and rapidly assimilate cystine, which is reduced intracellularly and secreted back into the microenvironment. The chronic lymphocytic leukemia cells use the secreted cysteine for glutathione production and improved resistance to oxidative stress. Cancer stem cells have elevated levels of CD44 variants that have been shown to stabilize xCT, increasing cystine import and leading to increased intracellular glutathione levels (Ishimoto et al., 2011). In addition, Nrf2 signaling stimulates the expression of xCT to help combat increasing levels of oxidative stress (Sasaki et al., 2002).

4.3. Glycine

Glycine represents another critical biomolecule that connects glutathione to other metabolic processes (Locasale, 2013). Metabolic profiling studies have demonstrated that significant amounts of 3-phosphoglycerate are diverted from glycolysis into serine and ultimately glycine biosynthesis in cultured cancer cell lines (Jain et al., 2012). Phosphoglycerate dehydrogenase, which catalyzes the committed step in this biosynthetic pathway, has been found to be upregulated in many human cancer samples (Locasale et al., 2011). Additional studies have shown that glycine is used for the biosynthesis of purines in rapidly dividing cancer cells and may potentially serve as a one-carbon donor for downstream methylation events. In these studies, a significant fraction of exogenously added glycine was incorporated into glutathione. Overall, it was estimated that approximately two-thirds of the total intracellular glycine arose from biosynthesis and the remainder from exogenous sources (Jain et al., 2012). The partitioning of glycine into its various metabolic fates allows cancer cells to control proliferation rates and increases capacity to combat oxidative stress.

5. REMAINING QUESTIONS AND EMERGING PATHWAYS

5.1. 5-Oxoproline

Although there have been a number of correlative and observational studies that indicate dynamic variations in 5-oxoproline levels, the precise cellular functions of 5-oxoproline beyond GSH salvage remain unclear. The concentrations in normal human plasma and various tissues are between 20 and 50 μM (Van Der Werf, Griffith, & Meister, 1975) but can reach several millimolar concentrations in pathological conditions, leading to anion gap metabolic acidosis and ataxia (Ristoff & Larsson, 2007). Several studies have suggested that 5-oxoproline levels indicate the availability of glycine and/or cysteine (Metges et al., 2000). In normal individuals consuming a low-protein diet, elevated 5-oxoproline levels were reduced with increased protein content or supplementation with other sources of nitrogen, such as urea (Jackson, Persaud, Meakins, & Bundy, 1996). When cysteine is limiting, the γ-glutamylphosphate intermediate of GCL can spontaneously cyclize to form 5-oxoproline (Orlowski & Meister, 1971). Increased urinary excretion of 5-oxoproline has been noted in children gaining weight rapidly during recovery from severe malnutrition. Glycine supplementation restored 5-oxoproline to normal levels, suggesting that excess γ-glutamylcysteine was being converted to 5-oxoproline and cysteine (Meakins, Persaud, & Jackson, 1998; Persaud, Forrester, & Jackson, 1996). Similarly, a study of pregnant women and their newborn infants has shown an inverse correlation between 5-oxoproline and dimethylglycine, a glycine precursor (Friesen et al., 2007). Given the number of nutritional states that lead to increased

5-oxoproline production, it is surprising that OPLAH is not more effective in hydrolyzing this metabolite.

5-Oxoprolinuria has been shown to be associated with urea cycle defects, such as ornithine transcarbamylase deficiency and carbamoyl phosphate synthetase deficiency (Mayatepek, 1999). Nephropathic cystinosis mimics the effects of cysteine deficiency and results from a genetic defect in the lysosomal cystine transporter, which leads to a high level of cystine accumulation in the lysosome with corresponding low levels of cysteine in the cytosol (Kumar & Bachhawat, 2010). Similarly, defects in cystathionine synthase, which converts homocysteine to cysteine, lead to elevated 5-oxoproline levels as well. Homocysteine can replace cysteine as a substrate for GCL, forming γ-glutamylhomocysteine. However, this dipeptide is not tolerated by GS and is efficiently processed by GGCT to generate 5-oxoproline and homocysteine (Stokke, Marstein, Jellum, & Lie, 1982). Perhaps, 5-oxoproline can serve as a signaling molecule to alter metabolism and limit flux through unproductive metabolic pathways.

As with GGT1, OPLAH is a highly conserved enzyme across diverse species. For example, human and *Methanosarcina acetivorans* 5-oxoprolinase share ~45% sequence identity over the length of this nearly 1300 amino acid protein. This striking conservation is more remarkable considering that *M. acetivorans* do not appear to synthesize glutathione or have close homologues of either GGT or GGCT, suggesting that the enzyme may have additional enzymatic activities. Eukaryotic 5-oxoprolinases are homodimeric enzymes with two hydantoinase domains (HyuA and HyuB) per subunit. Hydantoinases can act on a number of five- and six-membered ring substrates, perhaps indicating that OPLAH may hydrolyze other substrates besides 5-oxoproline. Studies on the enzymatic properties of OPLAH have been relatively limited, and a large range of kinetic constants has been observed (Kumar & Bachhawat, 2010; Ohkama-Ohtsu et al., 2008; Williamson & Meister, 1982). The enzyme appears to require both monovalent and divalent cations to maintain catalytic activity and is stabilized by the presence of 5-oxoproline. Oxidation of the protein leads to inactivation (Van der Werf et al., 1971; Watanabe, Abe, Ishikawa, & Iijima, 2004). A large-scale study by Ischiropoulos et al. reported that 5-oxoprolinase in mouse kidney, liver, and lung is S-nitrosylated (Doulias, Tenopoulou, Greene, Raju, & Ischiropoulos, 2013). The impacts of nitrosylation on the enzyme activity and stability have yet to be elucidated, but it is possible that some of the 24 cysteine residues found in human OPLAH may be sites of posttranslational regulation. Additional studies are warranted to define the breadth of reactions catalyzed by OPLAH.

5.2. ChaC1

An exciting development has been the identification of a cytosolic pathway for the degradation of glutathione (Kumar et al., 2012). Human ChaC1 has the ability to cleave the γ-glutamyl group from glutathione specifically, as the enzyme had minimal activity with γ-glutamyl dipeptide substrates. The observed kinetic constants suggest that ChaC1 activity is physiologically relevant, as the measured K_m for glutathione was approximately 3 mM. Homology modeling of human ChaC1 indicated that the enzyme adopts a fold similar to GGCT, despite low sequence identity. The validity of the model was confirmed by mutational analysis of ChaC1, in which a structurally conserved glutamate, Glu 116, was demonstrated to be critical for catalysis. In an organic sulfur auxotroph strain of *S. cerevisiae*, ChaC1 expression conferred the ability to grow on glutathione as the sole sulfur source. Its ability to initiate glutathione degradation is consistent with the recent observation that amino acid starvation induced by bacterial pathogens leads to dramatic upregulation of ChaC1 expression (Tattoli et al., 2012).

ChaC1 was initially identified as part of the unfolded protein response (UPR) cascade, downstream of ATF4 (Mungrue, Pagnon, Kohannim, Gargalovic, & Lusis, 2009). Overexpression of ChaC1 led to apoptosis, whereas knockdown by siRNA attenuated apoptosis in HEK293 cells, consistent with its involvement in the UPR. Its ability to cleave the γ-glutamyl peptide bond of glutathione likely facilitates apoptosis, as lower GSH levels are a hallmark of this process. Similarly, in head and neck squamous cell carcinoma cells, overexpression of ChaC1 resulted in slower proliferation rates and induced apoptosis (Joo, Ritchie, Kamarajan, Miao, & Kapila, 2012). In contrast, neither overexpression nor reduction of ChaC1 protein levels impacted apoptosis in breast cancer and ovarian cancer cell lines. siRNA-mediated knockdown of ChaC1 resulted in reduced migration and proliferation, whereas overexpression of ChaC1 increased migration and proliferation (Goebel et al., 2012). An analysis of human breast and ovarian tumors indicated that higher ChaC1 levels correlated with more advanced stage and poorer prognosis.

Overall, these studies highlight our limited understanding of ChaC1 function. Tumor cells generally tend to have higher levels of ROS and require elevated glutathione levels for survival. However, initial observations indicated that ChaC1 participated in apoptosis and GSH degradation. Therefore, it is surprising that breast and ovarian tumors would have higher ChaC1 levels. Perhaps, variants of ChaC1 have alternate functions. There is some ambiguity as to the proper start site for translation of human ChaC1, as there is another potential initiator methionine that would result in a protein that is shorter by 43-amino acid residues. In addition, there are two isoforms of the enzyme, with variant 2 resulting in a 45-amino acid deletion near the center of the coding region. It is unclear how each of these changes would impact ChaC1 function, and perhaps, isoform distributions of ChaC1 may be an important consideration (Goebel et al., 2012). Currently, only the standard ChaC1 isoform (Kumar et al., 2012) has been characterized *in vitro*, and further studies are clearly needed.

5.3. Additional functions of γ-glutamylcysteine and glutathione

The contributions of γ-glutamylcysteine and glutathione to cellular function have been expanded in two recent studies. Mitochondrial γ-glutamylcysteine was shown to be sufficient to respond to oxidative stress irrespective of the cytosolic glutathione concentration (Quintana-Cabrera & Bolanos, 2013). Specifically, γ-glutamylcysteine was shown to be an enzymatic cofactor for glutathione peroxidase 1, and this system was able to control mitochondrial H_2O_2 concentrations to limit cellular damage. However, it is unclear how γ-glutamylcysteine is partitioned between detoxification and glutathione synthesis pathways and how oxidized γ-glutamylcysteine is reduced, as this process is not likely to be mediated by glutathione reductase (Fig. 2.5).

Glutathione has been shown to participate in a futile cycle that depletes NADPH, leading to oxidative stress (Sullivan et al., 2013). Patients with hereditary leiomyomatosis and renal cell carcinoma have reduced fumarate hydratase (fumarase) activity. As flux through the citric acid cycle is compromised, cells can accumulate fumarate, which serves as a proto-oncometabolite. Reduced glutathione adds across the double bond of fumarate to generate succinated glutathione. This recently identified metabolite is a substrate of glutathione reductase, which can regenerate reduced glutathione at the expense of NADPH. As a result, the NADPH ratio drops concomitant with an increase in oxidative stress, leading to stabilization of HIF1α. Identifying the potential contributions of succinated glutathione to disease progression in other tissue types is an exciting area of future research.

6. SUMMARY

Glutathione is a critical detoxification precursor and a vital component of redox homeostasis thought to be especially beneficial in promoting growth and survival of tumor cells. Though the enzymes of the γ-glutamyl

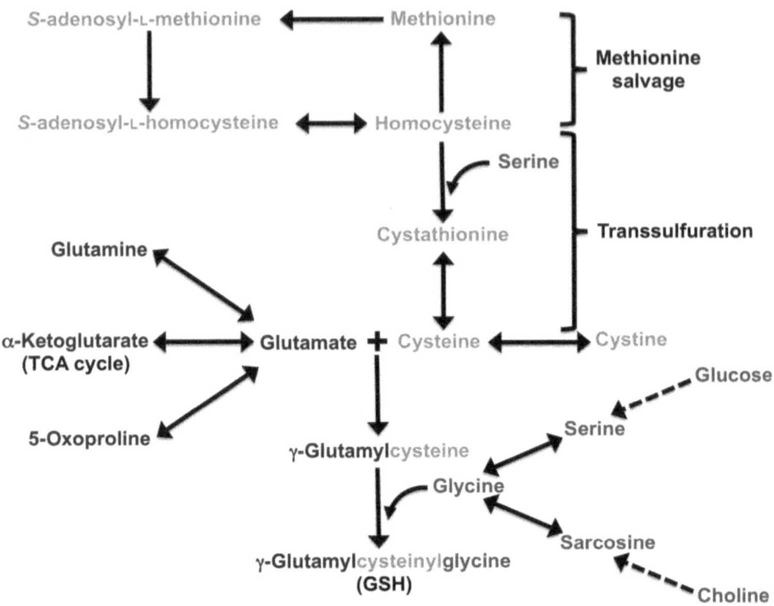

Figure 2.5 Precursors of glutathione synthesis. Metabolic routes for the three amino acids needed for GSH biosynthesis are outlined and color-coded by amino acid provision pathways: light grey (cysteine) denotes intimate cross talk with methionine metabolism (Lu & Mato, 2012; Zhang et al., 2005); medium grey (glycine) is directly linked to carbohydrate, serine, and choline levels (Locasale, 2013); and black (glutamine/glutamate) indicates coordination with proline and TCA cycle metabolites (Daye & Wellen, 2012). Although not illustrated, one-carbon metabolism links glycine metabolism and methionine salvage pathways.

cycle have been studied over a period of several decades, new complexities have arisen through the characterization of genetically engineered mouse models that perturb glutathione levels without directly impacting disease progression. Further advances in technologies such as mass spectrometry have improved detection limits for relevant metabolites and revealed points of regulated flux through glutathione synthesis and salvage pathways that intersect with other central metabolic pathways also subject to reprogramming in cancer. However, as these innovations have answered mechanistic questions about the roles of glutathione in cancer, new questions have come to the forefront. For example, increases in steady-state glutathione within blood-borne and peripheral solid tumors are not accompanied consistently by the upregulation of the biosynthetic or salvage enzymes, and liver tumors have actually shown reduced glutathione, despite the pivotal importance of glutathione to the metabolism of this tissue, in general.

To resolve these questions, it will be necessary to focus attention on the structural and functional regulation of the γ-glutamyl cycle enzymes and systematically assess glutathione with respect to each perturbation. It is also increasingly evident that the pathways of central metabolism that are altered to increase general biosynthetic potential in cancer (e.g., serine and glycine metabolism and pentose phosphate pathway) are intimately related to glutathione levels. It is tempting to speculate that as tumor cells reprogram metabolism in response to microenvironment conditions, there may be coevolution of γ-glutamyl cycle enzyme adaptations that optimally select with, or reciprocally to, the other central metabolic pathways. Detailed examination of metabolite flux in conjunction with the cross talking pathways will be important to dissect these mechanisms and further understand the role of glutathione.

ACKNOWLEDGMENT

This work was made possible by the National Institutes of Health Grants R01 GM077289 (J. J. B.), R01 CA165574 (M. A. S.), and P20 RR-17675 (National Center for Research Resources).

REFERENCES

Andringa, K. K., Coleman, M. C., Aykin-Burns, N., Hitchler, M. J., Walsh, S. A., Domann, F. E., et al. (2006). Inhibition of glutamate cysteine ligase activity sensitizes human breast cancer cells to the toxicity of 2-deoxy-D-glucose. *Cancer Research*, *66*, 1605–1610.

Bachhawat, A. K., Thakur, A., Kaur, J., & Zulkifli, M. (2013). Glutathione transporters. *Biochimica et Biophysica Acta*, *1830*, 3154–3164.

Backos, D. S., Fritz, K. S., McArthur, D. G., Kepa, J. K., Donson, A. M., Petersen, D. R., et al. (2013). Glycation of glutamate cysteine ligase by 2-deoxy-d-ribose and its potential impact on chemoresistance in glioblastoma. *Neurochemical Research*, *38*, 1838–1849.

Backos, D. S., Fritz, K. S., Roede, J. R., Petersen, D. R., & Franklin, C. C. (2011). Posttranslational modification and regulation of glutamate-cysteine ligase by the alpha, beta-unsaturated aldehyde 4-hydroxy-2-nonenal. *Free Radical Biology & Medicine*, *50*, 14–26.

Biterova, E. I., & Barycki, J. J. (2009). Mechanistic details of glutathione biosynthesis revealed by crystal structures of Saccharomyces cerevisiae glutamate cysteine ligase. *The Journal of Biological Chemistry*, *284*, 32700–32708.

Boanca, G., Sand, A., & Barycki, J. J. (2006). Uncoupling the enzymatic and autoprocessing activities of Helicobacter pylori gamma-glutamyltranspeptidase. *The Journal of Biological Chemistry*, *281*, 19029–19037.

Boanca, G., Sand, A., Okada, T., Suzuki, H., Kumagai, H., Fukuyama, K., et al. (2007). Autoprocessing of Helicobacter pylori gamma-glutamyltranspeptidase leads to the formation of a threonine-threonine catalytic dyad. *The Journal of Biological Chemistry*, *282*, 534–541.

Board, P. G., Moore, K. A., & Smith, J. E. (1978). Purification and properties of gamma-glutamylcyclotransferase from human erythrocytes. *The Biochemical Journal*, *173*, 427–431.

Brigelius-Flohe, R., & Maiorino, M. (2013). Glutathione peroxidases. *Biochimica et Biophysica Acta, 1830*, 3289–3303.

Brown, T. R., Drummond, M. L., Barelier, S., Crutchfield, A. S., Dinescu, A., Slavens, K. D., et al. (2011). Aspartate 458 of human glutathione synthetase is important for cooperativity and active site structure. *Biochemical and Biophysical Research Communications, 411*, 536–542.

Cairns, R. A., Harris, I. S., & Mak, T. W. (2011). Regulation of cancer cell metabolism. *Nature Reviews Cancer, 11*, 85–95.

Calpena, E., Casado, M., Martinez-Rubio, D., Nascimento, A., Colomer, J., Gargallo, E., et al. (2013). 5-Oxoprolinuria in heterozygous patients for 5-oxoprolinase (OPLAH) missense changes. *JIMD Reports, 7*, 123–128.

Carter, B. Z., Shi, Z. Z., Barrios, R., & Lieberman, M. W. (1998). Gamma-glutamyl leukotrienase, a gamma-glutamyl transpeptidase gene family member, is expressed primarily in spleen. *The Journal of Biological Chemistry, 273*, 28277–28285.

Castonguay, R., Lherbet, C., & Keillor, J. W. (2003). Kinetic studies of rat kidney gamma-glutamyltranspeptidase deacylation reveal a general base-catalyzed mechanism. *Biochemistry, 42*, 11504–11513.

Chen, X., Schecter, R. L., Griffith, O. W., Hayward, M. A., Alpert, L. C., & Batist, G. (1998). Characterization of 5-oxo-L-prolinase in normal and tumor tissues of humans and rats: A potential new target for biochemical modulation of glutathione. *Clinical Cancer Research, 4*, 131–138.

Chen, Y., Shertzer, H. G., Schneider, S. N., Nebert, D. W., & Dalton, T. P. (2005). Glutamate cysteine ligase catalysis: Dependence on ATP and modifier subunit for regulation of tissue glutathione levels. *The Journal of Biological Chemistry, 280*, 33766–33774.

Chen, Y., Yang, Y., Miller, M. L., Shen, D., Shertzer, H. G., Stringer, K. F., et al. (2007). Hepatocyte-specific Gclc deletion leads to rapid onset of steatosis with mitochondrial injury and liver failure. *Hepatology, 45*, 1118–1128.

Chien, J., Fan, J. B., Bell, D. A., April, C., Klotzle, B., Ota, T., et al. (2009). Analysis of gene expression in stage I serous tumors identifies critical pathways altered in ovarian cancer. *Gynecologic Oncology, 114*, 3–11.

Currais, A., & Maher, P. (2013). Functional consequences of age-dependent changes in glutathione status in the brain. *Antioxidants and Redox Signaling, 19*, 813–822.

Dahl, N., Pigg, M., Ristoff, E., Gali, R., Carlsson, B., Mannervik, B., et al. (1997). Missense mutations in the human glutathione synthetase gene result in severe metabolic acidosis, 5-oxoprolinuria, hemolytic anemia and neurological dysfunction. *Human Molecular Genetics, 6*, 1147–1152.

Dalton, T. P., Dieter, M. Z., Yang, Y., Shertzer, H. G., & Nebert, D. W. (2000). Knockout of the mouse glutamate cysteine ligase catalytic subunit (Gclc) gene: Embryonic lethal when homozygous, and proposed model for moderate glutathione deficiency when heterozygous. *Biochemical and Biophysical Research Communications, 279*, 324–329.

Daye, D., & Wellen, K. E. (2012). Metabolic reprogramming in cancer: Unraveling the role of glutamine in tumorigenesis. *Seminars in Cell & Developmental Biology, 23*, 362–369.

DeBerardinis, R. J., Mancuso, A., Daikhin, E., Nissim, I., Yudkoff, M., Wehrli, S., et al. (2007). Beyond aerobic glycolysis: Transformed cells can engage in glutamine metabolism that exceeds the requirement for protein and nucleotide synthesis. *Proceedings of the National Academy of Sciences of the United States of America, 104*, 19345–19350.

Dickinson, D. A., & Forman, H. J. (2002). Glutathione in defense and signaling: Lessons from a small thiol. *Annals of the New York Academy of Sciences, 973*, 488–504.

Diehn, M., Cho, R. W., Lobo, N. A., Kalisky, T., Dorie, M. J., Kulp, A. N., et al. (2009). Association of reactive oxygen species levels and radioresistance in cancer stem cells. *Nature, 458*, 780–783.

Dinescu, A., Brown, T. R., Barelier, S., Cundari, T. R., & Anderson, M. E. (2010). The role of the glycine triad in human glutathione synthetase. *Biochemical and Biophysical Research Communications*, *400*, 511–516.

Dinescu, A., Cundari, T. R., Bhansali, V. S., Luo, J. L., & Anderson, M. E. (2004). Function of conserved residues of human glutathione synthetase: Implications for the ATP-grasp enzymes. *The Journal of Biological Chemistry*, *279*, 22412–22421.

Doulias, P. T., Tenopoulou, M., Greene, J. L., Raju, K., & Ischiropoulos, H. (2013). Nitric oxide regulates mitochondrial fatty acid metabolism through reversible protein S-nitrosylation. *Science Signaling*, *6*, rs1.

Fan, X., Liu, X., Hao, S., Wang, B., Robinson, M. L., & Monnier, V. M. (2012). The LEGSKO mouse: A mouse model of age-related nuclear cataract based on genetic suppression of lens glutathione synthesis. *PLoS One*, *7*, e50832.

Fawaz, M. V., Topper, M. E., & Firestine, S. M. (2011). The ATP-grasp enzymes. *Bioorganic Chemistry*, *39*, 185–191.

Franklin, C. C., Backos, D. S., Mohar, I., White, C. C., Forman, H. J., & Kavanagh, T. J. (2009). Structure, function, and post-translational regulation of the catalytic and modifier subunits of glutamate cysteine ligase. *Molecular Aspects of Medicine*, *30*, 86–98.

Franklin, C. C., Krejsa, C. M., Pierce, R. H., White, C. C., Fausto, N., & Kavanagh, T. J. (2002). Caspase-3-dependent cleavage of the glutamate-L-cysteine ligase catalytic subunit during apoptotic cell death. *The American Journal of Pathology*, *160*, 1887–1894.

Fraser, J. A., Kansagra, P., Kotecki, C., Saunders, R. D., & McLellan, L. I. (2003). The modifier subunit of Drosophila glutamate-cysteine ligase regulates catalytic activity by covalent and noncovalent interactions and influences glutathione homeostasis in vivo. *The Journal of Biological Chemistry*, *278*, 46369–46377.

Frey, I. M., Rubio-Aliaga, I., Siewert, A., Sailer, D., Drobyshev, A., Beckers, J., et al. (2007). Profiling at mRNA, protein, and metabolite levels reveals alterations in renal amino acid handling and glutathione metabolism in kidney tissue of Pept2-/- mice. *Physiological Genomics*, *28*, 301–310.

Friesen, R. W., Novak, E. M., Hasman, D., & Innis, S. M. (2007). Relationship of dimethylglycine, choline, and betaine with oxoproline in plasma of pregnant women and their newborn infants. *The Journal of Nutrition*, *137*, 2641–2646.

Gamcsik, M. P., Kasibhatla, M. S., Teeter, S. D., & Colvin, O. M. (2012). Glutathione levels in human tumors. *Biomarkers: Biochemical Indicators of Exposure, Response, and Susceptibility to Chemicals*, *17*, 671–691.

Gao, P., Tchernyshyov, I., Chang, T. C., Lee, Y. S., Kita, K., Ochi, T., et al. (2009). c-Myc suppression of miR-23a/b enhances mitochondrial glutaminase expression and glutamine metabolism. *Nature*, *458*, 762–765.

Goebel, G., Berger, R., Strasak, A. M., Egle, D., Muller-Holzner, E., Schmidt, S., et al. (2012). Elevated mRNA expression of CHAC1 splicing variants is associated with poor outcome for breast and ovarian cancer patients. *British Journal of Cancer*, *106*, 189–198.

Gogos, A., & Shapiro, L. (2002). Large conformational changes in the catalytic cycle of glutathione synthase. *Structure*, *10*, 1669–1676.

Griffith, O. W., Bridges, R. J., & Meister, A. (1978). Evidence that the gamma-glutamyl cycle functions in vivo using intracellular glutathione: Effects of amino acids and selective inhibition of enzymes. *Proceedings of the National Academy of Sciences of the United States of America*, *75*, 5405–5408.

Griffith, O. W., & Mulcahy, R. T. (1999). The enzymes of glutathione synthesis: Gamma-glutamylcysteine synthetase. *Advances in Enzymology and Related Areas of Molecular Biology*, *73*, 209–267, xii.

Hama, S., Arata, M., Nakamura, I., Kasetani, T., Itakura, S., Tsuchiya, H., et al. (2012). Prevention of tumor growth by needle-free jet injection of anti-C7orf24 siRNA. *Cancer Gene Therapy*, *19*, 553–557.

Hanigan, M. H., Gallagher, B. C., Townsend, D. M., & Gabarra, V. (1999). Gamma-glutamyl transpeptidase accelerates tumor growth and increases the resistance of tumors to cisplatin in vivo. *Carcinogenesis, 20*, 553–559.

Hanigan, M. H., & Ricketts, W. A. (1993). Extracellular glutathione is a source of cysteine for cells that express gamma-glutamyl transpeptidase. *Biochemistry, 32*, 6302–6306.

Harding, J. J. (1970). Free and protein-bound glutathione in normal and cataractous human lenses. *The Biochemical Journal, 117*, 957–960.

Hayes, J. D., Flanagan, J. U., & Jowsey, I. R. (2005). Glutathione transferases. *Annual Review of Pharmacology and Toxicology, 45*, 51–88.

Heisterkamp, N., Groffen, J., Warburton, D., & Sneddon, T. P. (2008). The human gamma-glutamyltransferase gene family. *Human Genetics, 123*, 321–332.

Huang, C. S., Chang, L. S., Anderson, M. E., & Meister, A. (1993). Catalytic and regulatory properties of the heavy subunit of rat kidney gamma-glutamylcysteine synthetase. *The Journal of Biological Chemistry, 268*, 19675–19680.

Iida, T., Kijima, H., Urata, Y., Goto, S., Ihara, Y., Oka, M., et al. (2001). Hammerhead ribozyme against gamma-glutamylcysteine synthetase sensitizes human colonic cancer cells to cisplatin by down-regulating both the glutathione synthesis and the expression of multidrug resistance proteins. *Cancer Gene Therapy, 8*, 803–814.

Ikeda, Y., & Taniguchi, N. (2005). Gene expression of gamma-glutamyltranspeptidase. *Methods in Enzymology, 401*, 408–425.

Ishimoto, T., Nagano, O., Yae, T., Tamada, M., Motohara, T., Oshima, H., et al. (2011). CD44 variant regulates redox status in cancer cells by stabilizing the xCT subunit of system xc(-) and thereby promotes tumor growth. *Cancer cell, 19*, 387–400.

Jackson, A. A., Persaud, C., Meakins, T. S., & Bundy, R. (1996). Urinary excretion of 5-L-oxoproline (pyroglutamic acid) is increased in normal adults consuming vegetarian or low protein diets. *The Journal of Nutrition, 126*, 2813–2822.

Jain, M., Nilsson, R., Sharma, S., Madhusudhan, N., Kitami, T., Souza, A. L., et al. (2012). Metabolite profiling identifies a key role for glycine in rapid cancer cell proliferation. *Science, 336*, 1040–1044.

Jaramillo, M. C., & Zhang, D. D. (2013). The emerging role of the Nrf2-Keap1 signaling pathway in cancer. *Genes and Development, 27*, 2179–2191.

Joo, N. E., Ritchie, K., Kamarajan, P., Miao, D., & Kapila, Y. L. (2012). Nisin, an apoptogenic bacteriocin and food preservative, attenuates HNSCC tumorigenesis via CHAC1. *Cancer Medicine, 1*, 295–305.

Kageyama, S., Iwaki, H., Inoue, H., Isono, T., Yuasa, T., Nogawa, M., et al. (2007). A novel tumor-related protein, C7orf24, identified by proteome differential display of bladder urothelial carcinoma. *Proteomics Clinical Applications, 1*, 192–199.

Keillor, J. W., Castonguay, R., & Lherbet, C. (2005). Gamma-glutamyl transpeptidase substrate specificity and catalytic mechanism. *Methods in Enzymology, 401*, 449–467.

Kelley, L. A., & Sternberg, M. J. (2009). Protein structure prediction on the Web: A case study using the Phyre server. *Nature Protocols, 4*, 363–371.

Kim, J.-W., Lee, G., Moon, S.-M., Park, M.-J., Hong, S. K., Ahn, Y.-H., et al. (2010). Metabolomic screening and star pattern recognition by urinary amino acid profile analysis from bladder cancer patients. *Metabolomics, 6*, 202–206.

Kinlough, C. L., Poland, P. A., Bruns, J. B., & Hughey, R. P. (2005). Gamma-glutamyltranspeptidase: Disulfide bridges, propeptide cleavage, and activation in the endoplasmic reticulum. *Methods in Enzymology, 401*, 426–449.

Krejsa, C. M., Franklin, C. C., White, C. C., Ledbetter, J. A., Schieven, G. L., & Kavanagh, T. J. (2010). Rapid activation of glutamate cysteine ligase following oxidative stress. *The Journal of Biological Chemistry, 285*, 16116–16124.

Kumar, A., & Bachhawat, A. K. (2010). OXP1/YKL215c encodes an ATP-dependent 5-oxoprolinase in Saccharomyces cerevisiae: Functional characterization, domain structure and identification of actin-like ATP-binding motifs in eukaryotic 5-oxoprolinases. *FEMS Yeast Research, 10*, 394–401.

Kumar, A., Tikoo, S., Maity, S., Sengupta, S., Sengupta, S., Kaur, A., et al. (2012). Mammalian proapoptotic factor ChaC1 and its homologues function as gamma-glutamyl cyclotransferases acting specifically on glutathione. *EMBO Reports, 13*, 1095–1101.

Lam, B. K., Penrose, J. F., Freeman, G. J., & Austen, K. F. (1994). Expression cloning of a cDNA for human leukotriene C4 synthase, an integral membrane protein conjugating reduced glutathione to leukotriene A4. *Proceedings of the National Academy of Sciences of the United States of America, 91*, 7663–7667.

Le, A., Lane, A. N., Hamaker, M., Bose, S., Gouw, A., Barbi, J., et al. (2012). Glucose-independent glutamine metabolism via TCA cycling for proliferation and survival in B cells. *Cell Metabolism, 15*, 110–121.

Lieberman, M. W., Wiseman, A. L., Shi, Z. Z., Carter, B. Z., Barrios, R., Ou, C. N., et al. (1996). Growth retardation and cysteine deficiency in gamma-glutamyl transpeptidase-deficient mice. *Proceedings of the National Academy of Sciences of the United States of America, 93*, 7923–7926.

Lillig, C. H., Berndt, C., & Holmgren, A. (2008). Glutaredoxin systems. *Biochimica et Biophysica Acta, 1780*, 1304–1317.

Lim, J., Lawson, G. W., Nakamura, B. N., Ortiz, L., Hur, J. A., Kavanagh, T. J., et al. (2013). Glutathione-deficient mice have increased sensitivity to transplacental benzo[a]pyrene-induced premature ovarian failure and ovarian tumorigenesis. *Cancer Research, 73*, 908–917.

Locasale, J. W. (2013). Serine, glycine and one-carbon units: Cancer metabolism in full circle. *Nature Reviews Cancer, 13*, 572–583.

Locasale, J. W., Grassian, A. R., Melman, T., Lyssiotis, C. A., Mattaini, K. R., Bass, A. J., et al. (2011). Phosphoglycerate dehydrogenase diverts glycolytic flux and contributes to oncogenesis. *Nature Genetics, 43*, 869–874.

Lu, S. C. (2009). Regulation of glutathione synthesis. *Molecular Aspects of Medicine, 30*, 42–59.

Lu, S. C. (2013). Glutathione synthesis. *Biochimica et Biophysica Acta, 1830*, 3143–3153.

Lu, S. C., Kuhlenkamp, J., Garcia-Ruiz, C., & Kaplowitz, N. (1991). Hormone-mediated down-regulation of hepatic glutathione synthesis in the rat. *The Journal of Clinical Investigation, 88*, 260–269.

Lu, S. C., & Mato, J. M. (2012). S-adenosylmethionine in liver health, injury, and cancer. *Physiological Reviews, 92*, 1515–1542.

Luo, J. L., Huang, C. S., Babaoglu, K., & Anderson, M. E. (2000). Novel kinetics of mammalian glutathione synthetase: Characterization of gamma-glutamyl substrate cooperative binding. *Biochemical and Biophysical Research Communications, 275*, 577–581.

Mannervik, B. (2012). Five decades with glutathione and the GSTome. *The Journal of Biological Chemistry, 287*, 6072–6083.

Masuda, Y., Maeda, S., Watanabe, A., Sano, Y., Aiuchi, T., Nakajo, S., et al. (2006). A novel 21-kDa cytochrome c-releasing factor is generated upon treatment of human leukemia U937 cells with geranylgeraniol. *Biochemical and Biophysical Research Communications, 346*, 454–460.

Mayatepek, E. (1999). 5-Oxoprolinuria in patients with and without defects in the gamma-glutamyl cycle. *European Journal of Pediatrics, 158*, 221–225.

Meakins, T. S., Persaud, C., & Jackson, A. A. (1998). Dietary supplementation with L-methionine impairs the utilization of urea-nitrogen and increases 5-L-oxoprolinuria in normal women consuming a low protein diet. *The Journal of Nutrition, 128*, 720–727.

Meister, A. (1973). On the enzymology of amino acid transport. *Science, 180*, 33–39.

Meister, A. (1974). The gamma-glutamyl cycle. Diseases associated with specific enzyme deficiencies. *Annals of Internal Medicine, 81*, 247–253.

Meister, A., & Anderson, M. E. (1983). Glutathione. *Annual Review of Biochemistry, 52*, 711–760.

Metges, C. C., Yu, Y. M., Cai, W., Lu, X. M., Wong, S., Regan, M. M., et al. (2000). Oxoproline kinetics and oxoproline urinary excretion during glycine- or sulfur amino acid-free diets in humans. *American Journal of Physiology, Endocrinology and Metabolism*, *278*, E868–E876.

Mitsuishi, Y., Taguchi, K., Kawatani, Y., Shibata, T., Nukiwa, T., Aburatani, H., et al. (2012). Nrf2 redirects glucose and glutamine into anabolic pathways in metabolic reprogramming. *Cancer cell*, *22*, 66–79.

Mosharov, E., Cranford, M. R., & Banerjee, R. (2000). The quantitatively important relationship between homocysteine metabolism and glutathione synthesis by the transsulfuration pathway and its regulation by redox changes. *Biochemistry*, *39*, 13005–13011.

Mungrue, I. N., Pagnon, J., Kohannim, O., Gargalovic, P. S., & Lusis, A. J. (2009). CHAC1/MGC4504 is a novel proapoptotic component of the unfolded protein response, downstream of the ATF4-ATF3-CHOP cascade. *Journal of Immunology*, *182*, 466–476.

Nanjo, S., Asada, K., Yamashita, S., Nakajima, T., Nakazawa, K., Maekita, T., et al. (2012). Identification of gastric cancer risk markers that are informative in individuals with past H. pylori infection. *Gastric Cancer*, *15*, 382–388.

Naumov, V. A., Generozov, E. V., Zaharjevskaya, N. B., Matushkina, D. S., Larin, A. K., Chernyshov, S. V., et al. (2013). Genome-scale analysis of DNA methylation in colorectal cancer using Infinium HumanMethylation450 BeadChips. *Epigenetic*, *8*, 921–934.

Nicklin, P., Bergman, P., Zhang, B., Triantafellow, E., Wang, H., Nyfeler, B., et al. (2009). Bidirectional transport of amino acids regulates mTOR and autophagy. *Cell*, *136*, 521–534.

Oakley, A. J., Coggan, M., & Board, P. G. (2010). Identification and characterization of gamma-glutamylamine cyclotransferase, an enzyme responsible for gamma-glutamyl-epsilon-lysine catabolism. *The Journal of Biological Chemistry*, *285*, 9642–9648.

Oakley, A. J., Yamada, T., Liu, D., Coggan, M., Clark, A. G., & Board, P. G. (2008). The identification and structural characterization of C7orf24 as gamma-glutamyl cyclotransferase. An essential enzyme in the gamma-glutamyl cycle. *The Journal of Biological Chemistry*, *283*, 22031–22042.

Ochi, T. (1995). Hydrogen peroxide increases the activity of gamma-glutamylcysteine synthetase in cultured Chinese hamster V79 cells. *Archives of Toxicology*, *70*, 96–103.

Ochi, T. (1996). Menadione causes increases in the level of glutathione and in the activity of gamma-glutamylcysteine synthetase in cultured Chinese hamster V79 cells. *Toxicology*, *112*, 45–55.

Ohkama-Ohtsu, N., Oikawa, A., Zhao, P., Xiang, C., Saito, K., & Oliver, D. J. (2008). A gamma-glutamyl transpeptidase-independent pathway of glutathione catabolism to glutamate via 5-oxoproline in Arabidopsis. *Plant Physiology*, *148*, 1603–1613.

Ohno, Y., Hattori, A., Ueda, M., Kageyama, S., Yoshiki, T., & Kakeya, H. (2011). Multiple NF-Y-binding CCAAT boxes are essential for transcriptional regulation of the human C7orf24 gene, a novel tumor-associated gene. *The FEBS Journal*, *278*, 4088–4099.

Oinonen, C., & Rouvinen, J. (2000). Structural comparison of Ntn-hydrolases. *Protein Science*, *9*, 2329–2337.

Okada, T., Suzuki, H., Wada, K., Kumagai, H., & Fukuyama, K. (2006). Crystal structures of gamma-glutamyltranspeptidase from Escherichia coli, a key enzyme in glutathione metabolism, and its reaction intermediate. *Proceedings of the National Academy of Sciences of the United States of America*, *103*, 6471–6476.

Okada, T., Suzuki, H., Wada, K., Kumagai, H., & Fukuyama, K. (2007). Crystal structure of the gamma-glutamyltranspeptidase precursor protein from Escherichia coli—Structural changes autocatalytic processing and implications for the maturation mechanism. *The Journal of Biological Chemistry*, *282*, 2433–2439.

Orlowski, M., & Meister, A. (1970). The gamma-glutamyl cycle: A possible transport system for amino acids. *Proceedings of the National Academy of Sciences of the United States of America*, *67*, 1248–1255.

Orlowski, M., & Meister, A. (1971). Partial reactions catalyzed by -glutamylcysteine synthetase and evidence for an activated glutamate intermediate. *The Journal of Biological Chemistry, 246*, 7095–7105.

Orlowski, M., Richman, P. G., & Meister, A. (1969). Isolation and properties of gamma-L-glutamylcyclotransferase from human brain. *Biochemistry, 8*, 1048–1055.

Palekar, A. G., Tate, S. S., & Meister, A. (1974). Formation of 5-oxoproline from glutathione in erythrocytes by the gamma-glutamyltranspeptidase-cyclotransferase pathway. *Proceedings of the National Academy of Sciences of the United States of America, 71*, 293–297.

Persaud, C., Forrester, T., & Jackson, A. A. (1996). Urinary excretion of 5-L-oxoproline (pyroglutamic acid) is increased during recovery from severe childhood malnutrition and responds to supplemental glycine. *The Journal of Nutrition, 126*, 2823–2830.

Pompella, A., Corti, A., Paolicchi, A., Giommarelli, C., & Zunino, F. (2007). Gamma-glutamyltransferase, redox regulation and cancer drug resistance. *Current Opinion in Pharmacology, 7*, 360–366.

Quintana-Cabrera, R., & Bolanos, J. P. (2013). Glutathione and gamma-glutamylcysteine in the antioxidant and survival functions of mitochondria. *Biochemical Society Transactions, 41*, 106–110.

Richman, P. G., & Meister, A. (1975). Regulation of gamma-glutamyl-cysteine synthetase by nonallosteric feedback inhibition by glutathione. *The Journal of Biological Chemistry, 250*, 1422–1426.

Ristoff, E., & Larsson, A. (2007). Inborn errors in the metabolism of glutathione. *Orphanet Journal of Rare Diseases, 2*, 16.

Sasaki, H., Sato, H., Kuriyama-Matsumura, K., Sato, K., Maebara, K., Wang, H., et al. (2002). Electrophile response element-mediated induction of the cystine/glutamate exchange transporter gene expression. *The Journal of Biological Chemistry, 277*, 44765–44771.

Sato, H., Shiiya, A., Kimata, M., Maebara, K., Tamba, M., Sakakura, Y., et al. (2005). Redox imbalance in cystine/glutamate transporter-deficient mice. *The Journal of Biological Chemistry, 280*, 37423–37429.

Seelig, G. F., Simondsen, R. P., & Meister, A. (1984). Reversible dissociation of gamma-glutamylcysteine synthetase into two subunits. *The Journal of Biological Chemistry, 259*, 9345–9347.

Sekura, R., & Meister, A. (1974). Glutathione turnover in the kidney; considerations relating to the gamma-glutamyl cycle and the transport of amino acids. *Proceedings of the National Academy of Sciences of the United States of America, 71*, 2969–2972.

Shi, Z. Z., Osei-Frimpong, J., Kala, G., Kala, S. V., Barrios, R. J., Habib, G. M., et al. (2000). Glutathione synthesis is essential for mouse development but not for cell growth in culture. *Proceedings of the National Academy of Sciences of the United States of America, 97*, 5101–5106.

Shibayama, K., Wachino, J., Arakawa, Y., Saidijam, M., Rutherford, N. G., & Henderson, P. J. F. (2007). Metabolism of glutamine and glutathione via gamma-glutamyltranspeptidase and glutamate transport in Helicobacter pylori: Possible significance in the pathophysiology of the organism. *Molecular Microbiology, 64*, 396–406.

Siitonen, T., Alaruikka, P., Mantymaa, P., Savolainen, E. R., Kavanagh, T. J., Krejsa, C. M., et al. (1999). Protection of acute myeloblastic leukemia cells against apoptotic cell death by high glutathione and gamma-glutamylcysteine synthetase levels during etoposide-induced oxidative stress. *Annals of Oncology, 10*, 1361–1367.

Slavens, K. D., Brown, T. R., Barakat, K. A., Cundari, T. R., & Anderson, M. E. (2011). Valine 44 and valine 45 of human glutathione synthetase are key for subunit stability and negative cooperativity. *Biochemical and Biophysical Research Communications, 410*, 597–601.

Snoke, J. E., & Bloch, K. (1955). Studies on the mechanism of action of glutathione synthetase. *The Journal of Biological Chemistry, 213*, 825–835.

Sousa Silva, M., Gomes, R. A., Ferreira, A. E., Ponces Freire, A., & Cordeiro, C. (2013). The glyoxalase pathway: The first hundred years... and beyond. *The Biochemical Journal, 453*, 1–15.

Sporn, M. B., & Liby, K. T. (2012). NRF2 and cancer: The good, the bad and the importance of context. *Nature Reviews Cancer, 12*, 564–571.

Stokke, O., Marstein, S., Jellum, E., & Lie, S. O. (1982). Accumulation of pyroglutamic acid (5-oxoproline) in homocystinuria. *Scandinavian Journal of Clinical and Laboratory Investigation, 42*, 361–369.

Strasak, A. M., Pfeiffer, R. M., Klenk, J., Hilbe, W., Oberaigner, W., Gregory, M., et al. (2008). Prospective study of the association of gamma-glutamyltransferase with cancer incidence in women. *International Journal of Cancer, 123*, 1902–1906.

Sullivan, L. B., Martinez-Garcia, E., Nguyen, H., Mullen, A. R., Dufour, E., Sudarshan, S., et al. (2013). The proto-oncometabolite fumarate binds glutathione to amplify ROS-dependent signaling. *Molecular Cell, 51*, 236–248.

Sun, W. M., Huang, Z. Z., & Lu, S. C. (1996). Regulation of gamma-glutamylcysteine synthetase by protein phosphorylation. *The Biochemical Journal, 320*(Pt. 1), 321–328.

Suzuki, H., & Kumagai, H. (2002). Autocatalytic processing of gamma-glutamyltranspeptidase. *The Journal of Biological Chemistry, 277*, 43536–43543.

Suzuki, S., Tanaka, T., Poyurovsky, M. V., Nagano, H., Mayama, T., Ohkubo, S., et al. (2010). Phosphate-activated glutaminase (GLS2), a p53-inducible regulator of glutamine metabolism and reactive oxygen species. *Proceedings of the National Academy of Sciences of the United States of America, 107*, 7461–7466.

Takemura, K., Kawachi, H., Eishi, Y., Kitagaki, K., Negi, M., Kobayashi, M., et al. (2014). Gamma-glutamylcyclotransferase as a novel immunohistochemical biomarker for the malignancy of esophageal squamous tumors. *Human Pathology, 45*, 331–341.

Tang, F., Xie, C., Huang, D., Wu, Y., Zeng, M., Yi, L., et al. (2011). Novel potential markers of nasopharyngeal carcinoma for diagnosis and therapy. *Clinical Biochemistry, 44*, 711–718.

Tattoli, I., Sorbara, M. T., Vuckovic, D., Ling, A., Soares, F., Carneiro, L. A., et al. (2012). Amino acid starvation induced by invasive bacterial pathogens triggers an innate host defense program. *Cell Host & Microbe, 11*, 563–575.

Thompson, G. A., & Meister, A. (1975). Utilization of L-cystine by the gamma-glutamyl transpeptidase-gamma-glutamyl cyclotransferase pathway. *Proceedings of the National Academy of Sciences of the United States of America, 72*, 1985–1988.

Townsend, D. M., & Tew, K. D. (2003). The role of glutathione-S-transferase in anti-cancer drug resistance. *Oncogene, 22*, 7369–7375.

Traverso, N., Ricciarelli, R., Nitti, M., Marengo, B., Furfaro, A. L., Pronzato, M. A., et al. (2013). Role of glutathione in cancer progression and chemoresistance. *Oxidative Medicine and Cellular Longevity, 2013*, 972913.

Tu, Z., & Anders, M. W. (1998). Identification of an important cysteine residue in human glutamate-cysteine ligase catalytic subunit by site-directed mutagenesis. *The Biochemical Journal, 336*(Pt. 3), 675–680.

Uejima, D., Nishijo, K., Kajita, Y., Ishibe, T., Aoyama, T., Kageyama, S., et al. (2011). Involvement of cancer biomarker C7orf24 in the growth of human osteosarcoma. *Anticancer Research, 31*, 1297–1305.

Van Der Werf, P., Griffith, O. W., & Meister, A. (1975). 5-Oxo-L-prolinase (L-pyroglutamate hydrolase). Purification and catalytic properties. *The Journal of Biological Chemistry, 250*, 6686–6692.

Van der Werf, P., Orlowski, M., & Meister, A. (1971). Enzymatic conversion of 5-oxo-L-proline (L-pyrrolidone carboxylate) to L-glutamate coupled with cleavage of adenosine triphosphate to adenosine diphosphate, a reaction in the -glutamyl cycle. *Proceedings of the National Academy of Sciences of the United States of America, 68*, 2982–2985.

Van Der Werf, P., Stephani, R. A., & Meister, A. (1974). Accumulation of 5-oxoproline in mouse tissues after inhibition of 5-oxoprolinase and administration of amino acids: Evidence for function of the gamma-glutamyl cycle. *Proceedings of the National Academy of Sciences of the United States of America*, *71*, 1026–1029.

Wang, W., & Ballatori, N. (1998). Endogenous glutathione conjugates: Occurrence and biological functions. *Pharmacological Reviews*, *50*, 335–356.

Watanabe, T., Abe, K., Ishikawa, H., & Iijima, Y. (2004). Bovine 5-oxo-L-prolinase: Simple assay method, purification, cDNA cloning, and detection of mRNA in the coronary artery. *Biological & Pharmaceutical Bulletin*, *27*, 288–294.

Wellner, V. P., Sekura, R., Meister, A., & Larsson, A. (1974). Glutathione synthetase deficiency, an inborn error of metabolism involving the gamma-glutamyl cycle in patients with 5-oxoprolinuria (pyroglutamic aciduria). *Proceedings of the National Academy of Sciences of the United States of America*, *71*, 2505–2509.

Welsch, D. J., Creely, D. P., Hauser, S. D., Mathis, K. J., Krivi, G. G., & Isakson, P. C. (1994). Molecular cloning and expression of human leukotriene-C4 synthase. *Proceedings of the National Academy of Sciences of the United States of America*, *91*, 9745–9749.

West, M. B., Chen, Y., Wickham, S., Heroux, A., Cahill, K., Hanigan, M. H., et al. (2013). Novel insights into eukaryotic gamma-glutamyltranspeptidase 1 from the crystal structure of the glutamate-bound human enzyme. *The Journal of Biological Chemistry*, *288*, 31902–31913.

Williamson, J. M., & Meister, A. (1982). New substrates of 5-oxo-L-prolinase. *The Journal of Biological Chemistry*, *257*, 12039–12042.

Willis, M. N., Liu, Y., Biterova, E. I., Simpson, M. A., Kim, H., Lee, J., et al. (2011). Enzymatic defects underlying hereditary glutamate cysteine ligase deficiency are mitigated by association of the catalytic and regulatory subunits. *Biochemistry*, *50*, 6508–6517.

Winkler, A., Njalsson, R., Carlsson, K., Elgadi, A., Rozell, B., Abraham, L., et al. (2011). Glutathione is essential for early embryogenesis—Analysis of a glutathione synthetase knockout mouse. *Biochemical and Biophysical Research Communications*, *412*, 121–126.

Yang, Y., Dieter, M. Z., Chen, Y., Shertzer, H. G., Nebert, D. W., & Dalton, T. P. (2002). Initial characterization of the glutamate-cysteine ligase modifier subunit Gclm(-/-) knockout mouse. Novel model system for a severely compromised oxidative stress response. *The Journal of Biological Chemistry*, *277*, 49446–49452.

Yoshimoto, T., Soberman, R. J., Lewis, R. A., & Austen, K. F. (1985). Isolation and characterization of leukotriene C4 synthetase of rat basophilic leukemia cells. *Proceedings of the National Academy of Sciences of the United States of America*, *82*, 8399–8403.

Zhang, W., Braun, A., Bauman, Z., Olteanu, H., Madzelan, P., & Banerjee, R. (2005). Expression profiling of homocysteine junction enzymes in the NCI60 panel of human cancer cell lines. *Cancer Research*, *65*, 1554–1560.

Zhang, W., Trachootham, D., Liu, J., Chen, G., Pelicano, H., Garcia-Prieto, C., et al. (2012). Stromal control of cystine metabolism promotes cancer cell survival in chronic lymphocytic leukaemia. *Nature Cell Biology*, *14*, 276–286.

CHAPTER THREE

Gamma-Glutamyl Transpeptidase: Redox Regulation and Drug Resistance

Marie H. Hanigan[1]
Department of Cell Biology, University of Oklahoma Health Sciences Center, Oklahoma City, Oklahoma, USA
[1]Corresponding author: e-mail address: marie-hanigan@ouhsc.edu

Contents

Abstract

The expression of gamma-glutamyl transpeptidase (GGT) is essential to maintaining cysteine levels in the body. GGT is a cell surface enzyme that hydrolyzes the gamma-glutamyl bond of extracellular reduced and oxidized glutathione, initiating their cleavage into glutamate, cysteine (cystine), and glycine. GGT is normally expressed

Advances in Cancer Research, Volume 122
ISSN 0065-230X
http://dx.doi.org/10.1016/B978-0-12-420117-0.00003-7

on the apical surface of ducts and glands, salvaging the amino acids from glutathione in the ductal fluids. GGT in tumors is expressed over the entire cell membrane and provides tumors with access to additional cysteine and cystine from reduced and oxidized glutathione in the blood and interstitial fluid. Cysteine is rate-limiting for glutathione synthesis in cells under oxidative stress. The induction of GGT is observed in tumors with elevated levels of intracellular glutathione. Studies in models of hepatocarcinogenesis show that GGT expression in foci of preneoplastic hepatocytes provides a selective advantage to the cells during tumor promotion with agents that deplete intracellular glutathione. Similarly, expression of GGT in tumors enables cells to maintain elevated levels of intracellular glutathione and to rapidly replenish glutathione during treatment with prooxidant anticancer therapy. In the clinic, the expression of GGT in tumors is correlated with drug resistance. The inhibitors of GGT block GGT-positive tumors from accessing the cysteine in extracellular glutathione. They also inhibit GGT activity in the kidney, which results in the excretion of GSH in the urine and a rapid decrease in blood cysteine levels, leading to depletion of intracellular GSH in both GGT-positive and GGT-negative tumors. GGT inhibitors are being developed for clinical use to sensitize tumors to chemotherapy.

1. INTRODUCTION

In 1985, gamma-glutamyl transpeptidase (GGT, aka gamma-glutamyl transferase) was first proposed to play a role in tumor formation (Hanigan & Pitot, 1985b). Preneoplastic liver foci in rats treated with chemical carcinogens were identified by their expression of GGT (Goldsworthy, Hanigan, & Pitot, 1986). We proposed that the expression of GGT provided the cells within the foci a selective growth advantage during the promotion phase of carcinogenesis (Hanigan & Pitot, 1985b). This hypothesis was based on the observation that the treatment regimens, which gave rise to GGT-positive liver foci, all included promoting compounds that depleted glutathione (GSH). The hypothesis was that GGT, which is localized to the cell surface, cleaved extracellular GSH, thereby providing the cell with the amino acids necessary for intracellular GSH synthesis. GGT activity enabled the cells to maintain their intracellular GSH levels, thus resisting the toxicity of the promoting compounds and enabling them to respond to the proliferative signals triggered by the carcinogenic regimen. Now, decades later, there is a great deal of new information about the enzyme that supports this hypothesis. These data will be reviewed in this chapter. Further, studies from many laboratories have demonstrated that this same mechanism, through which GGT was proposed to contribute to the preneoplastic cell's resistance to the toxicity of promoting agents, also confers GGT-positive tumors with resistance to prooxidant anticancer therapy. GGT enhances the tumor's access to cysteine, thereby increasing the intracellular GSH level. This enables the tumors to maintain their redox balance during the onslaught of reactive oxygen species (ROS) generated by the prooxidant anticancer therapies and to avoid death via the cell death pathways triggered by oxidative stress. Clinical studies have shown a correlation between GGT expression in human tumors and their resistance to therapy. Studies in cell culture and data from animal models have provided information on the signaling and regulatory pathways that underlie this correlation. To understand the relationship between GGT expression and drug resistance, it is necessary to first understand the role of GGT in normal physiology. This chapter will review the current information about GGT and its role in redox regulation, its expression in tumors, its induction by toxins including many of the most commonly used chemotherapy agents, and a strategy to use GGT inhibitors to overcome the resistance of both GGT-positive and GGT-negative tumors to prooxidant anticancer therapy.

2. EXPRESSION OF GGT AND DRUG RESISTANCE IN HUMAN TUMORS

Clinical studies show a strong correlation between the expression of GGT in tumors and poor survival. Our study of 451 human tumors prior to treatment showed that GGT is induced during the development of many tumors (Hanigan, Frierson, Swanson, & De Young, 1999). Tumors derived from ductal epithelial cells that normally express GGT were generally strongly GGT-positive (Hanigan & Frierson, 1996). These include liver, renal, prostatic, pancreatic, and breast carcinomas (Table 3.1). Due to the ubiquitous expression of GGT in liver, renal, prostatic and pancreatic cancers, it is not possible to study the effect of GGT on the clinical outcome of these tumors. However, studies in breast, ovarian, and other types of tumors demonstrated a positive correlation between GGT expression, tumor progression, and poor overall survival of the patients. GGT is normally expressed in the breast ductal epithelium and we found that all benign breast lesions were GGT-positive (Durham, Frierson, & Hanigan, 1997; Hanigan & Frierson, 1996). However, we observed that GGT expression was lost in some breast tumors (Durham et al., 1997). Bard et al. reported that GGT-positive breast tumors were more likely to be estrogen receptor-negative tumors, which have a poor prognosis, and were more likely to have metastasized at the time of diagnosis, suggesting that

Table 3.1 Expression of GGT1 in human tumors

Tumor type	GGT+	GGT−	GGT-positive (%)
Hepatocellular carcinoma[a]	9	0	100
Renal cell carcinoma[a]	6	0	100
Prostatic adenocarcinoma[a]	70	1	99
Pancreatic adenocarcinoma[a]	10	1	91
Invasive breast carcinoma[a]	60	21	74
Ovarian adenocarcinoma[a]	16	6	73
Ovarian adenocarcinoma [b]	33	12	73

[a]Data from Hanigan, Frierson, et al. (1999).
[b]Data from Hanigan, Frierson, Brown, Lovell, and Taylor (1994).

expression of GGT provided a selective advantage to the tumor (Bard, Noel, Chauvin, & Quash, 1986). GGT is also induced prior to therapy in some human tumors derived from tissues that do not normally express GGT, such as ovarian adenocarcinomas and soft tissue sarcomas. Among these tumors, it has been observed that expression of GGT correlates with higher-grade and drug-resistant tumors. In a study of 634 ovarian cancer patients, Grimm et al. found that high pretherapeutic expression of GGT served as a prognostic indicator of worse overall survival (Grimm et al., 2013). A smaller study of patients with stage III and IV ovarian cancer found that GGT-positive and GGT-negative tumors had a similar response to initial therapy, but patients with GGT-positive tumors had a poorer 2-year survival rate than those with GGT-negative tumors (Hanigan, Frierson, & Taylor, 1998). In soft tissue sarcomas, GGT activity is higher in high-grade tumors and in metastases than in low-grade tumors (Hochwald, Rose, Brennan, & Burt, 1997). The induction of GGT has been observed in the drug-resistant tumors that arise during treatment. In a set of ovarian cancer cell lines derived from a patient before and after the development of resistance to cisplatin, chlorambucil, and 5-fluorouracil, drug-resistant cells exhibited a 6.5-fold induction of GGT activity (Lewis, Hayes, & Wolf, 1988).

As described in detail throughout this chapter, the mechanism underlying the correlation between expression of GGT and drug resistance is the ability of GGT to cleave extracellular GSH and thereby provide cells with an additional source of cysteine with which to increase intracellular GSH levels. The cell lines derived before and after the development of drug resistance had elevated the levels of both GGT and GSH (Lewis et al., 1988).

The induction of both intracellular GSH levels and GGT was also observed in human tumor cell lines selected in culture for resistance to chemotherapy (Godwin et al., 1992).

3. STRUCTURE OF GGT

In eukaryotes, GGT is a cell surface glycoprotein. It is anchored in the cell membrane via a single N-terminus transmembrane domain. All of the catalytic activity is within the extracellular domain of the protein (Ikeda, Fujii, Taniguchi, & Meister, 1995). Human GGT is synthesized as a 569 amino acid propeptide (Rajpert-De Meyts, Heisterkamp, & Groffen, 1988). The propeptide is enzymatically inactive, but is activated by auto-cleavage into two subunits (West et al., 2010; West et al., 2011). The auto-cleavage is dependent on an internal nucleophilic threonine (Thr-381), indicating that GGT is a member of the N-terminal nucleophilic (Ntn) hydrolase family, a group of enzymes with similar autocleavage and kinetic mechanisms but no sequence homology (Brannigan et al., 1995; Galperin & Koonin, 2012). N-linked glycosylation is essential for proper folding, auto-cleavage, and activation of human GGT, although the enzyme retains its activity when the glycans are removed from the mature, enzymatically active heterodimer (West, Chen, et al., 2013; West et al., 2011).

We recently solved the crystal structure of human GGT, the first crystal structure for any eukaryotic GGT (West, Chen, et al., 2013). The structure revealed a heterodimer with a stacked α-β-β-α core structure, a common element found in Ntn hydrolases (Galperin & Koonin, 2012). A surface representation of the enzyme reveals that the large subunit is wrapped around the small subunit (Fig. 3.1). Human GGT has seven N-linked glycosylation sites. The crystal structure showed that all seven sites are located on the external surface of the protein (Fig. 3.1). Additional studies have shown site-specific and tissue-specific glycan compositional patterns on human GGT (West et al., 2010).

The crystal structure of human GGT identified the substrate channel with the catalytic nucleophile (Thr-381) in the deepest part of the cleft. The relatively open active site in human GGT differs from the more restricted access found in most prokaryotic GGTs (Morrow, Williams, Sand, Boanca, & Barycki, 2007; Okada, Suzuki, Wada, Kumagai, & Fukuyama, 2006). In addition, we observed two different conformations for the side chain of the catalytic residue, Thr-381, suggesting that the side chain moves during substrate binding.

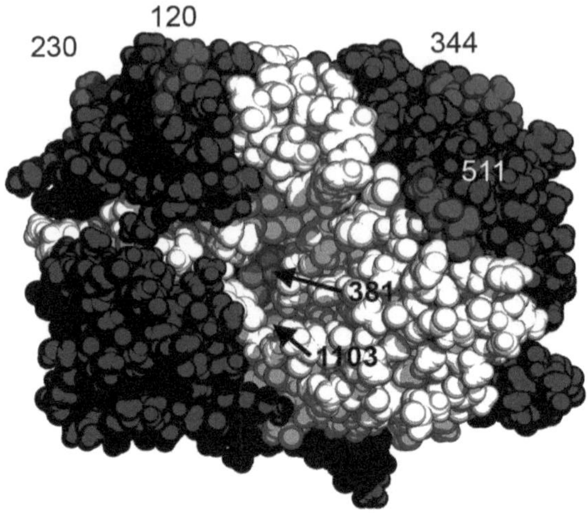

Figure 3.1 The van der Waals surface of human GGT with the active site cleft facing the viewer. The large subunit (dark gray) and the small subunit (white) are shown with the catalytic Thr-381 (red) in the deepest part of the small subunit cleft. Four of the seven potential glycosylation sites are seen in this orientation, asparagines 511, 120, 230, and 344, and the basal *N*-acetyl glucosamine residue, which was identified in the crystal structure at each of these sites, is represented as dark orange van der Waals spheres. An anion-binding site (green) within the small subunit cleft is labeled 1103. *This figure was originally published in West, Chen, et al. (2013). ©The American Society for Biochemistry and Molecular Biology.* (See the color plate.)

4. BIOCHEMISTRY OF GGT-CATALYZED REACTIONS

GGT is localized to the cell surface and only cleaves extracellular substrates. GSH and oxidized GSH (GSSG) are the most abundant physiological substrates, although GGT cleaves any gamma-glutamyl substrate including GSH *S*-conjugates (Wickham, West, Cook, & Hanigan, 2011). The substrate glutamate moiety must be unrestricted except for the gamma-glutamyl bond (Fig. 3.2). GGT will not cleave gamma-glutamyl bonds formed with glutamate that is within a peptide or with glutamate bound to any other compound.

The physiological reaction catalyzed by GGT is the hydrolysis of gamma-glutamyl bonds (PetitClerc, Shiele, Bagrel, Mahassen, & Siest, 1980) (Fig. 3.2). This is a multistep reaction that proceeds by a mechanism common to Ntn hydrolases (Galperin & Koonin, 2012; Oinonen & Rouvinen, 2000). Thr-381 is the N-terminus residue of the small subunit and therefore has a free amino group. As the substrate binds in the active

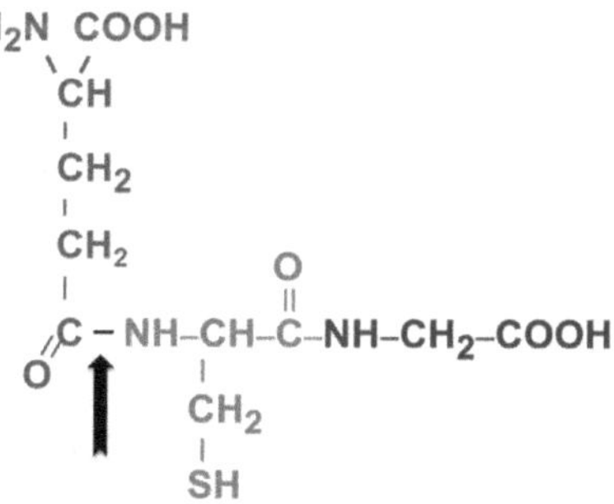

Figure 3.2 Gamma-glutamyl bond. The structure of glutathione is shown glutamate (green) bound via a γ-glutamyl bond (arrow) to cysteine (red) and glycine (blue).

site, the hydroxyl group on the side chain of Thr-381 is deprotonated by its free amino group (Castonguay et al., 2007). The side chain oxygen of Thr-381 initiates a nucleophilic attack on the carbonyl carbon of the gamma-glutamyl bond of the substrate. A transient, covalent acyl bond is formed between the gamma carbon of the gamma-glutamyl substrate and the hydroxyl (beta-oxygen) on the side chain of Thr-381 of GGT (Fig. 3.3) (Keillor, Castonguay, & Lherbet, 2005). This results in the release of all but the γ-glutamyl group of the substrate. Hydrolysis of the acyl bond is the rate-limiting step in the reaction and results in the release of glutamate. We have developed an assay that quantifies this activity at physiological pH by measuring the release of glutamate (Wickham, West, et al., 2011). This assay can be used to evaluate the K_m for any gamma-glutamyl substrate and the velocity that GGT hydrolyzes the substrate. With GSH as the substrate, cysteinylglycine is the first product released and glutamate is the second product. The K_m for GSH is 11 μM, consistent with the concentration of GSH in human serum and interstitial fluid (Wickham, West et al., 2011). The K_m for GSSG is only slightly lower at 9 μM; the rate of the reaction is the same for GSH and GSSG (Wickham, West et al., 2011).

There was confusion for many years regarding the physiological role of GGT. GGT was discovered as a glutathionase and its hydrolysis of the gamma-glutamyl bond was first reported in 1948 (Binkley & Nakamura, 1948). But several years later, it was reported that in the presence of high concentrations of a dipeptide acceptor, GGT could catalyze a transpeptidation reaction, transferring the gamma-glutamyl group from the substrate to an acceptor, yielding a new gamma-glutamyl compound (Hanes, Hird, & Isherwood, 1952). The enzyme, originally called glutathionase,

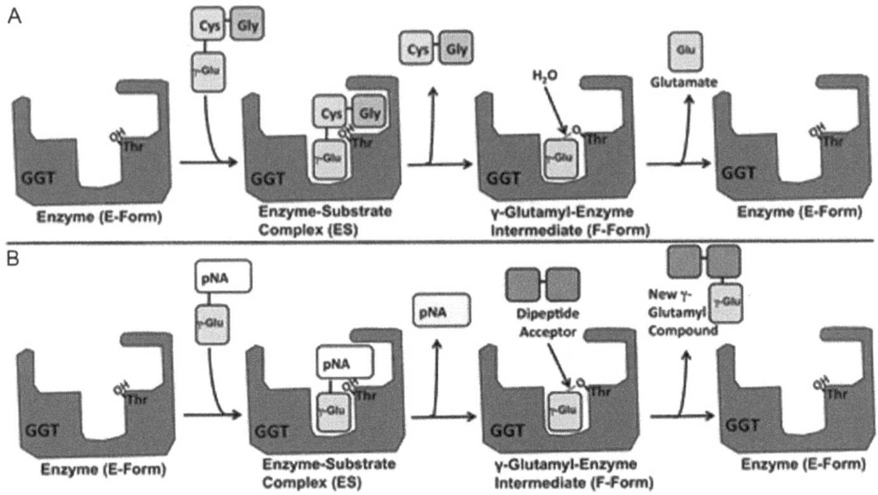

Figure 3.3 The hydrolysis and transpeptidation reactions catalyzed by GGT. (A) The physiological reaction catalyzed by GGT is the hydrolysis of gamma-glutamyl bonds. The hydrolysis reaction is shown with glutathione as the substrate and shows the release of glutamate, which is the product that is measured in the glutamate release assay for GGT activity. (B) The transpeptidation reaction requires high concentration of dipeptide acceptor and is favored at pH 8 and higher. The transpeptidation reaction is shown with L-gamma-glutamyl-p-nitroanilide (L-GpNA), the substrate used in the standard biochemical GGT assay, which monitors the release of pNA, which is yellow. *Adapted from a figure published in Wickham et al. (2013)@ The Biochemical Society.* (See the color plate.)

was renamed GGT (Hanes et al., 1952). Based on the transpeptidation reaction, GGT was proposed to be the central component of an amino acid transport system called the gamma-glutamyl cycle (Meister, 1973; Orlowski & Meister, 1970). However, kinetic and functional studies have shown that the physiological function of GGT is that of a hydrolase rather than a transpeptidase (Curthoys & Hughey, 1979; Woodlock et al., 1990). Yet confusion persists in the literature. Adding to the confusion is the fact that the standard GGT biochemical assay measures the transpeptidation reaction (Fig. 3.3). The assay includes millimolar concentrations of γ-glutamyl-p-nitroanilide (γ-GpNA) as a substrate and 10–40 mM glycylglycine as an acceptor (Orlowski & Meister, 1963; Tateishi, Higashi, Nomura, Naruse, & Nakashima, 1976). Although this is not the physiological reaction, this transpeptidation appears to be uniquely catalyzed by GGT, and this colorimetric assay can be used to easily quantify GGT.

5. FUNCTION OF GGT

5.1. In normal tissues and in tumors

GGT is expressed on the luminal surface of excretive and absorptive cells that line glands and ducts throughout the body, with the highest level of GGT activity in the kidney (Hanigan & Frierson, 1996). The development of strains of GGT knockout mice revealed the role of GGT in the distribution of cysteine throughout the body (Harding et al., 1997; Lieberman et al., 1996; Yamada, Tsuji, & Kunieda, 2013). GGT knockout mice excrete 2500-fold more GSH in their urine than wild-type mice (Lieberman et al., 1996). The glutathionuria is the result of the absence of GGT activity on the apical surface of the proximal tubules of the kidney. The cells lining the renal tubules are unable to take up GSH. In the GGT knockout mice, GSH remains intact in the glomerular filtrate as it transits the kidney and GSH is, thus, excreted in the urine. The continuous excretion of GSH results in a cysteine deficiency. The cysteine concentration in the plasma of GGT knockout mice is only 20% of that seen in wild-type mice (Lieberman et al., 1996). The mice fail to grow normally and die at approximately 10 weeks due to the cysteine deficiency. The GGT knockout mice can be rescued by supplementing their drinking water with *N*-acetylcysteine (Lieberman et al., 1996). Prior to death, the mice develop cataracts, have increased levels of oxidative DNA damage, and are susceptible to oxygen-induced lung injury, all symptoms of reduced intracellular GSH and redox stress (Barrios et al., 2001; Chevez-Barrios, Wiseman, Rojas, Ou, & Lieberman, 2000; Rojas, Valverde, Kala, Kala, & Lieberman, 2000). Glutathionuria has also been reported in the few patients that have been identified with a GGT deficiency (Wright, Stern, Ersser, & Patrick, 1980) (Schulman et al., 1975). Therefore, the expression of GGT is essential to maintain cysteine homeostasis and protect tissues against oxidative stress.

Many tumors express GGT, but they are not polarized and, therefore, express GGT on their entire cell surface (Hanigan, Frierson, et al., 1999). Unlike normal cells in which GGT only has access to substrates in ductal fluids, the GGT on tumor cells can cleave GSH in the interstitial fluid and blood. The expression of GGT provides tumor cells with an additional source of cysteine and cystine from the cleavage of extracellular GSH and GSSG. This has been demonstrated by several investigators using cells transfected with GGT. We showed that Hep 1–6 cells, a GGT-negative mouse liver tumor cell line, were unable to metabolize GSH in media nor were they able to take up GSH or GSH-derived cysteine (Hanigan, 1995). However, Hep 1–6 cells transfected with GGT rapidly metabolized ^{35}S-labeled GSH in the tissue culture media. Cysteinylglycine, a product of the GGT cleavage of GSH, was hydrolyzed by cell surface dipeptidases into cysteine, and glycine and the GSH-derived ^{35}S-labeled cysteine was taken up into the cells (Hanigan, 1995). In other studies, GGT-positive NIH3T3 cells and GGT-negative controls grew at a similar rate in standard tissue culture media, which contains more than three times the concentration of cysteine and cystine present in interstitial fluid. However, when the cells were incubated in cysteine-free media containing GSH, the GGT-positive cells were able to cleave the extracellular GSH and use the cysteine for continued growth, while the GGT-negative controls died (Hanigan & Ricketts, 1993).Similar studies in a human B-cell lymphoblastoid cell line transfected with GGT showed that the survival advantage provided by GGT for cells in cysteine-free medium was blocked by the inhibition of cystine and cysteine uptake, further demonstrating that GGT released the cysteine from GSH providing cysteine to the cell (Karp, Shimooku, & Lipsky, 2001). The inhibition of GGT with acivicin was used to demonstrate the ability of GGT to provide rat colon cancer cells with GSH-derived cysteine and maintain their intracellular GSH levels when cultured in cysteine-free media (Huseby et al., 2003). Studies in which intracellular GSH was depleted with diethyl maleate showed that GGT-transfected NIH3T3 fibroblasts cleaved extracellular GSH and replenished intracellular GSH levels more rapidly than the GGT-negative controls (Rajpert-De Meyts et al., 1992). Access to additional cysteine becomes critical in resisting the toxicity of prooxidant anticancer therapy as described in the succeeding text.

5.2. Other GGT substrates

The GGT knockout mice have also been used to investigate the role of GGT in metabolizing other gamma-glutamyl substrates. GGT is an essential enzyme in the renal mercapturic acid pathway, a detoxification pathway that metabolizes GSH *S*-conjugates to cysteine *S*-conjugates for excretion (Cooper & Hanigan, 2010). There are, however, some drugs that are conjugated to GSH and are activated to nephrotoxins via the renal mercapturic acid pathway (Anders & Dekant, 1998). Studies in GGT knockout mice and in culture have shown that the chemotherapy drug, cisplatin, is metabolized to a nephrotoxin through this pathway and the inhibition of GGT blocks the nephrotoxicity of this drug (Hanigan, Gallagher, Taylor, & Large, 1994; Hanigan et al., 2001; Townsend & Hanigan, 2002).

Stark et al. proposed that the GGT cleavage of GSH in the presence of transitional metals leads to oxidative damage of cell surface proteins and membrane lipids (Glass & Stark, 1997; Stark, Zeiger, & Pagano, 1993). The mechanism underlying their hypothesis is that sulfur on the cysteine becomes increasingly reactive as GSH is cleaved to cysteinylglycine and then to cysteine by GGT and dipeptidase. Transition metals catalyze autoxidation of sulfur via a Fenton reaction that results in the production of oxygen radicals. GGT has been shown to participate in generating oxidative damage in cell culture (Corti et al., 2009; Dominici et al., 1999; Stark et al., 1994). However, the reactions are unlikely to occur under physiological conditions due to the limited availability of the reactants in extracellular fluids. The exception may be reperfusion injury, where inhibition of GGT in rats has been shown to be protective (Yamamoto et al., 2011). In contrast, mouse studies show that lack of GGT activity leads to oxidative damage due to a reduced supply of cysteine to the tissues (Rojas et al., 2000).

5.3. Other GGT genes

In the human genome, there is a family of GGT genes (Heisterkamp, Groffen, Warburton, & Sneddon, 2008). GGT1, which is generally referred to simply as GGT, is located at chromosome 22q11 (Figlewicz et al., 1993; Morris et al., 1993). This region of chromosome 22 has undergone multiple duplications and rearrangements. Most of the human GGT genes contain coding regions for only fragments of the GGT1 protein. Courtay et al. found mRNA expressed from four human GGT genes, later identified as *GGTLC2*, *GGTLC3*, *GGTLC4P*, and *GGTLC5P*, all of which encode only the light chain of GGT and therefore would not have enzymatic activity since both subunits are needed for activity (Courtay, Heisterkamp, Siest, & Groffen, 1994; Heisterkamp et al., 2008). However, *GGT2*, a gene that is located in this region of chromosome 22 and is closely related to *GGT1*, encodes a full length protein (Heisterkamp et al., 2008). *GGT2* is likely a duplication of the *GGT1* gene that occurred late in evolution as *GGT2* is present in the human genome, but is not present in any other species including other primates (Heisterkamp et al., 2008). There is 97% nucleotide identity between *GGT1* and *GGT2* and 94% amino acid identity. *GGT2* mRNA has been detected and this has led to the assumption that *GGT2* encodes an enzymatically active protein (Auman et al., 2008; Moon et al., 2012). However, we have recently shown that GGT2 propeptides encoded by all three isoforms of GGT2 failed to autocleave, were enzymatically inactive, did not localize to the plasma membrane, and were rapidly degraded within the cell (West, Wickham, et al., 2013).

The only GGT gene in addition to *GGT1* that has been shown to encode a protein with enzymatic activity is *GGT5*, which has been identified in humans and rodents (Carter, Shi, Barrios, & Lieberman, 1998; Heisterkamp, Rajpert-De Meyts, Uribe, Forman, & Groffen, 1991). *GGT5* was originally identified in mice based on its ability to cleave the GSH *S*-conjugate leukotriene C_4 to leukotriene D_4 and is implicated in inflammation (Shi, Han, Habib, Matzuk, & Lieberman, 2001). Mouse GGT5 does not cleave GSH (Shi et al., 2001). The human *GGT5* gene has 40% amino acid identity to *GGT1*, but the second-order rate kinetics show that human GGT5 has less than 4% of the activity of GGT1 in hydrolyzing GSH, GSSG, and leukotriene C_4 (Heisterkamp et al., 1991; Wickham, West, et al., 2011). There is no published information on GGT5 expression in tumors. Therefore, this chapter will focus exclusively on GGT1, which will be referred to as GGT.

6. GSH AND CYSTEINE IN REDOX REGULATION

6.1. GSH and intracellular redox regulation

GSH is the most abundant nonprotein thiol in mammalian systems and the major redox buffer in the cell (Moriarty-Craige & Jones, 2004). GSH protects cellular components from oxidative damage of ROS, such as hydrogen peroxide and organic peroxides via GSH peroxidases, and also detoxifies the electrophilic metabolites of toxins, including chemotherapy drugs (Moriarty-Craige & Jones, 2004) (Fig. 3.4). The ratio of oxidized GSH (GSSG) to 2GSH can be used as a measure of the intracellular redox environment (Schafer & Buettner, 2001).

Prooxidant anticancer therapies are defined as those that stress the redox balance in the cell towards a more positive (oxidized) state. These include many of the most commonly used anticancer treatments including platinum-based compounds, alkylating agents, anthracyclines, and radiation (Mistry & Harrap, 1991). The GSH levels are often elevated in tumor cells prior to treatment because the high metabolic rate in tumors results in enhanced ROS production, which stresses their redox balance and induces GSH synthesis (Balendiran, Dabur, & Fraser, 2004; Cairns, Harris, & Mak, 2011; Chaiswing, Zhong, & Oberley, 2011; Kansanen, Kuosmanen, Leinonen, & Levonen, 2013; Pani, Galeotti, & Chiarugi, 2010). Many studies have shown a strong correlation between elevated levels of intracellular

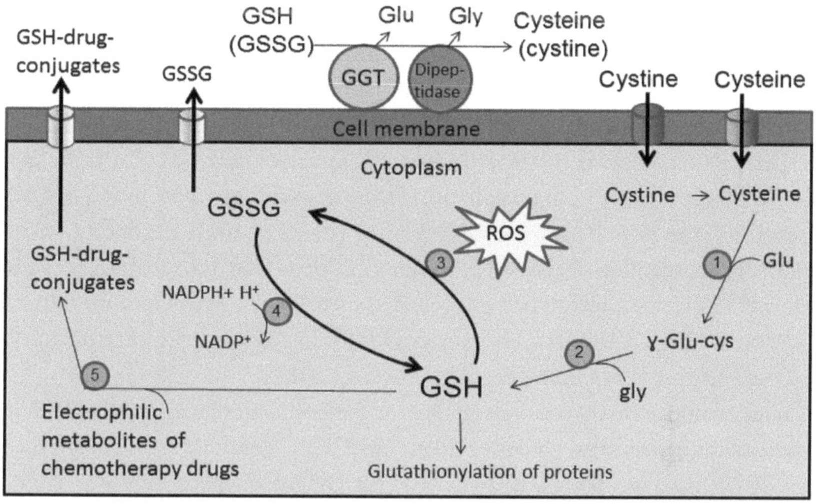

Figure 3.4 Gamma-glutamyl transpeptidase (GGT)-positive tumors cleave extracellular reduced and oxidized glutathione (GSH and GSSG), providing an additional source of cysteine for intracellular GSH synthesis. GGT cleaves glutamate from GSH and GSSG. The cysteinylglycine dipeptides can be cleaved by any of several dipeptidases that are present on the surface of the cell. The glutamate, glycine, cystine, and cysteine that are released from GSH and GSSG are transported into the cell by the standard amino acid transporters. Cystine is taken up by the x_c^- cystine/glutamate antiporter (red). Cysteine is taken up by the ACS transporter (brown). Once inside the cell, cystine is reduced to cysteine by the strongly reducing environment of the cytoplasm. The first step in GSH synthesis is catalyzed by gamma-glutamyl cysteine synthetase (1) and the second step is catalyzed by GSH synthetase (2). Tumors are under redox stress that can be further increased by prooxidant anticancer therapy. GSH levels are depleted in tumors by several pathways. GSH is oxidized to GSSG as part of an ROS detoxification system that is present in both the cytoplasm and mitochondria. GSH peroxidases (3) catalyze the oxidation of GSH to GSSG. GSSG can be reduced to GSH by GSH reductase (4) and NADPH. However, under extreme oxidative stress, GSSG is transported out of the cell by the multidrug resistance protein (MRP) transporters (blue). GSH is also depleted from the cell by binding to the electrophilic metabolites of chemotherapy drugs, which is catalyzed by GSH *S*-transferases (5). The GSH *S*-conjugates are transported out of the cell by the MRP transporters (blue). Intracellular GSH can also be depleted by binding to intercellular proteins, a process known as protein *S*-glutathionylation. (See the color plate.)

GSH and resistance to prooxidant chemotherapy (Balendiran et al., 2004; Estrela, Ortega, & Obrador, 2006) (Newkirk et al., 1997).

6.2. Increased requirement for GSH for tumors and cells under redox stress

During redox stress, there are multiple pathways by which toxins, including many chemotherapy drugs, deplete intracellular GSH (Fig. 3.4). GSH

S-transferases catalyze the conjugation of GSH to electrophilic drug metabolites, which prevents the drugs from binding to DNA and other nucleophilic cellular components (Tew & Townsend, 2012). These GSH *S*-conjugates are transported out of the cell by the multidrug resistance protein (MRP) transporters (Bachhawat, Thakur, Kaur, & Zulkifli, 2013; Cole & Deeley, 2006). In addition, GSH is oxidized to GSSG, which is excreted by the MRP transporters during times of high oxidative stress to prevent its accumulation (Cole & Deeley, 2006). During redox, stress intracellular GSH levels are further depleted as cysteine residues in some proteins undergo, *S*-glutathionylation, a process that can alter protein structure and function (Xiong, Uys, Tew, & Townsend, 2011).

Cells contain both redox-sensing and redox-signaling systems, which regulate differentiation, proliferation, and cell death (Chaudhari, Ye, & Jang, 2012; Go & Jones, 2013). As prooxidant drugs deplete reduced GSH from the cell, the redox environment becomes increasingly oxidized, which triggers proapoptotic signaling pathways and other death pathways (Circu & Aw, 2010; Jones, 2010; Ortega, Salvador, & Estrela, 2011). *S*-glutathionylation is also a consequence of oxidative stress, and *S*-glutathionylation of substrates such as protein disulfide isomerase also triggers cell death through the activation of the unfolded protein response and disruption of other cell signaling pathways (Xiong et al., 2011). The drug-resistant cells with increased GSH synthesis are able to maintain intracellular GSH and avoid activation of the cell death pathways.

6.3. Replenishment of GSH is dependent on cysteine and cystine uptake

Most normal cells and tumors cannot take up intact GSH. Intracellular levels of GSH are normally maintained by feedback regulation of the rate-limiting enzyme in GSH synthesis, gamma-glutamyl cysteine synthetase (Meister, 1991). The cysteine and cystine concentrations in the serum and interstitial fluid are sufficient to enable cells to maintain intracellular GSH levels. However, under redox stress, increased amounts of GSH are synthesized and cysteine becomes rate-limiting for GSH synthesis (Ortega et al., 2011). Tumors are, therefore, dependent on enhanced uptake of cysteine and cystine to maintain their intracellular GSH levels (Fig. 3.4) (Ishimoto et al., 2011). Cysteine is taken up into the cells via the alanine, serine, and cysteine transporter (ASCT) (Scopelliti, Ryan, & Vandenberg, 2013). Cystine, the oxidized form of cysteine, is present in human serum and interstitial fluid at 66 μM, five times the concentration of cysteine, and is a major source of cysteine for the cell (Blanco et al., 2007; Davis, Quinlivan, Stacpoole, & Gregory, 2006). Cystine is transported into the cell via the x_c^- cystine/glutamate antiporter (Davis et al., 2006; Lewerenz et al., 2013).

Tumors have a high requirement for cysteine in order to maintain their intracellular GSH levels. Many tumors have enhanced uptake of cystine. xCT is the light chain of the x_c^- cystine/glutamate antiporter that transports cystine into the cell (Lewerenz et al., 2013). Overexpression of xCT is commonly observed in drug-resistant tumors and is a marker for poor survival in patients (Kinoshita et al., 2013; Takeuchi et al., 2013). Blocking cystine uptake sensitizes these tumor cells to therapy (Doxsee et al., 2007; Pham et al., 2010; Vene et al., 2011). xCT overexpression has been identified in tumor stem cells. Epithelial tumor stem cells express the epithelial splicing regulatory protein 1 (ESRP-1), a physiological regulator that maintains pluripotency in normal mouse and human epithelial stem cells (Fagoonee et al., 2013). The coexpression of ESRP-1 and CD44, a cell surface marker used to identity cancer stem cells, results in production of alternatively spliced variants of CD44 (Biddle, Gammon, Fazil, & Mackenzie, 2013; Warzecha, Shen, Xing, & Carstens, 2009). One of the CD44 variants, CD44v8-10, produces a highly metastatic and drug-resistant population of tumor cells (Nagano, Okazaki, & Saya, 2013). The mechanism by which CD44v8-10 affects metastasis and drug resistance is by stabilizing xCT, resulting in increased cystine uptake (Ishimoto et al., 2011; Lewerenz et al., 2013). Blocking cystine uptake negates the metastatic potential and drug resistance of CD44v8-10 expressing cells (Biddle et al., 2013; Ishimoto et al., 2011; Yoshikawa et al., 2013). There are no studies on the coregulation of cystine transport and expression of GGT. Both can increase the supply of cysteine to the cell. They could work in tandem, with GGT initiating the release of cystine from extracellular GSSG, providing more cystine for uptake into the cell.

7. THE ROLE OF GGT IN ENHANCING CYSTEINE AVAILABILITY AND DRUG RESISTANCE

A dramatic shift in the availability of cysteine occurs when GGT-positive cells are depolarized or GGT is induced in GGT-negative cells. Under these conditions, GGT is no longer restricted to substrates present in the fluids within ducts and glands as it is in normal tissues, but now can cleave both oxidized and reduced GSH in interstitial fluid, providing the cell access to the cystine and cysteine therein. Cameron et al. reported that in rats with neoplastic liver nodules and carcinomas, GSH is excreted by the normal hepatocytes, but up to 80% of the GSH produced in the liver was cleaved by the GGT-positive nodules and tumors (Cameron, Armstrong, Gunsekara, Varghese, & Speisky, 1991). When GGT was inhibited, GSH in the venous fluid increased fivefold (Cameron et al., 1991). The use of extracellular GSH by GGT-positive tumors *in vivo* was also demonstrated by measuring GSH and cysteine concentrations in arterial and venous blood flowing through GGT-positive tumors implanted in the ovary (Hochwald, Harrison, Rose, Anderson, & Burt, 1996). A single pass of the blood through the tumor resulted in a 69% decrease in the serum GSH concentration, a significantly higher utilization rate than that observed in the systemic circulation. Administration of a GGT inhibitor blocked GSH degradation.

During the onslaught of toxins that deplete GSH including many chemotherapy drugs, expression of GGT provides cells with additional cysteine that is essential to overcoming the toxicity of the drug. In addition, the elevated levels of intracellular GSH maintains the redox status within the cells, enabling them to respond to the proliferative and differentiation signals present in the tissue following injury from the toxin. This was elegantly demonstrated in a study of coumarin-induced Clara cell toxicity in the lung (Vassallo et al., 2010). Clara cells normally express GGT. A single treatment with coumarin killed the Clara cells in the bronchiolar epithelium of both wild-type mice and GGT knockout mice within 24 h. However, during a 12-day period with repeated coumarin dosing, the Clara cells in the bronchial epithelium of the wild-type mice regenerated, whereas those in the GGT knockout mice did not. An important component of this experiment was the fact that the GGT knockout mice were fed with NAC in their drinking water throughout the study, they had normal levels of cysteine in their serum, and they were not cysteine-deficient. The GSH concentration in the lungs of the untreated controls was the same for the wild-type and GGT knockout mice. In mice, Clara cells metabolize coumarin to an epoxide that is detoxified by conjugation to GSH (Vassallo, Hicks, Born, & Daston, 2004). GSH is rapidly depleted from the lung during metabolism of epoxides to GSH *S*-conjugates (Warren, Brown, & Buckpitt, 1982). The Clara cells that grew out in the wild-type mice had 13-fold higher GGT activity and 3.3 times the intracellular GSH concentration compared to the Clara cells in untreated controls. In GGT knockout mice, the cells were unable to sustain adequate concentrations of GSH and could not regenerate.

Availability of cysteine can limit the growth of rapidly dividing tumors in mice, while expression of GGT can provide additional cysteine for an enhanced growth rate. Two GGT-negative skin cell lines that produce papillomas in nude mice were transfected with GGT. When injected into nude mice, the GGT-positive clones grew more rapidly and resulted in more malignant tumors than the vector-transfected controls (Slaga, Budunova, Gimenez-Conti, & Aldaz, 1996). Similarly, the human PC3 tumor cell line transfected with GGT grew more rapidly in mice than its GGT-negative vector-transfected control (Hanigan, Gallagher, Townsend, & Gabarra, 1999). Generally, rapidly growing tumors are more sensitive to alkylating agents; however, expression of GGT in PC3 tumors and access to additional cysteine not only increased the growth rate but also increased their resistance to chemotherapy (Hanigan, Frierson, et al., 1999).

GGT is induced in mouse cells selected in culture for resistance to alkylating agents. Ahmad et al. selected murine L1210 leukemia cells for resistance to L-phenylalanine mustard and found that the drug-resistant line had threefold more intracellular GSH and threefold higher GGT activity (Ahmad, Okine, Wood, Aljian, & Vistica, 1987). The drug-resistant cells metabolized more extracellular GSH and took up more GSH-derived ^{35}S-cysteine. The authors concluded that the increased level of GGT in the drug-resistant cells provided the cells with more cysteine with which to maintain their elevated intracellular GSH concentration and resist to toxicity of the alkylating agent.

GGT and other GSH-related enzymes are induced in human cells selected in culture for resistance to prooxidant chemotherapeutic drugs. Selection of the human ovarian tumor cell lines, A2780 and A1847, for resistance to cisplatin yielded highly resistant clones that had 13- to 50-fold increased levels of intracellular GSH, enhanced expression of gamma-glutamyl cysteine synthetase, and a 15- to 40-fold increase in GGT mRNA (Godwin et al., 1992). Other investigators showed that the GGT mRNA and GGT activity in human ovarian cell lines increased within 48 h of exposure to cisplatin (Oguchi et al., 1994).

Tissue culture studies on the role of GGT in cells under oxidative stress can be misleading if the cysteine levels in the media are not adjusted to physiological levels. The concentration of cysteine and cystine in standard tissue culture media ranges from 200 μM cystine (equivalent to 400 μM cysteine) in DMEM and RPMI-1640 to 990 μM cysteine in L-15 medium. In contrast, the concentration of cysteine and cystine in serum and interstitial fluids in humans is 12 and 66 μM, respectively (Blanco et al., 2007). Most tissue culture media does not contain GSH with the exception of RPMI-1640, which is formulated with 3.2 μM GSH. As described earlier in the text, GGT-positive PC3 cells grew faster in mice than their GGT-negative controls due to their additional supply of cysteine, but when these cells were cultured in RPMI-1640 media containing 200 μM cystine, both cell lines grew at the same rate (Hanigan, Frierson, et al., 1999). However, studies of cell growth in tissue culture have shown that media with lower levels of cysteine/cystine can be used to demonstrate the effect of GGT activity on increasing the cysteine supply to the cells. Incubation of Hep 1–6 cells in media with various concentrations of cysteine showed that in media containing less than 80 μM cysteine, the GGT-positive cells grew more rapidly than the GGT-negative cells (Hanigan, 1995). Analysis of media that was on the cells for 24 h showed that GSH was being secreted into the media by these liver-derived cells. GSH accumulated in the media on the GGT-negative cells, but could not be detected in the media on the GGT-positive cells because it was being degraded by the GGT. Salvaging the cysteine from the secreted GSH provided the GGT-positive cells with access to more cysteine and allowed for increased growth under conditions of limiting cysteine (Hanigan, 1995). A potential complication that can arise while investigating the effect of GGT or extracellular GSH in cell culture studies is the presence of GGT activity in some lots of commercial bovine serum (Hanigan, Brown, & Ricketts, 1993). It is essential that investigators planning these types of studies screen aliquots of the serum for GGT activity prior to purchasing it.

Studies done in tissue culture investigating the effect of transfection of GGT on drug sensitivity generally show no effect when the studies are done in complete media because there is no advantage to accessing additional cysteine when the cells are cultured in the very high level of cysteine present in tissue culture media (Bailey, Gipp, & Mulcahy, 1994). Some studies have reported a protective effect of GGT expression against the toxicity of chemotherapy in human tumor cells cultured in complete media, likely due to the ability of the GGT-positive cells to cleave GSH and GSSG they have excreted into the media. The toxicity of cisplatin and doxorubicin towards HepG2 cells was increased when GGT was inhibited by acivicin, but the effect was very modest compared to the potentiation of cytotoxicity observed in cells cultured in physiological concentrations of cysteine (Kwiecien, Rokita, Lorenc-Koci, Sokolowska, & Wlodek, 2007).

The protective effect of GGT against toxicity can be demonstrated in cells grown in standard media when the cells are under extreme oxidative stress. Shi et al. used GGT-transfected NIH3T3 cells to investigate the role of GGT in cells under oxidative stress (Shi, Gozal, Choy, & Forman, 1993). The GGT-positive and control GGT-negative cells were treated with

2,3-dimethoxy-1,4-naphthoquinone (DMNQ), which generates H_2O_2 intracellularly through redox cycling and caused a dose-dependent decrease in intracellular GSH in both cell lines. Addition of GSH to the media resulted in maintenance of intracellular GSH and resistance to DMNQ-induced toxicity in the GGT-positive cells, but not the control cells. Treatment of the GGT-positive cells with an inhibitor of GGT blocked the protective effect of the extracellular GSH. This study demonstrates that in cells under oxidative stress, expression of GGT provided cells with access to an additional source of cysteine that helped maintain intracellular GSH and increased resistance to the toxicity of the drug.

8. REDOX REGULATION OF GGT

Expression of GGT and its role in redox regulation have been studied extensively in rats, mice, and humans. However, there are differences in the expression and regulation of GGT among these species that are often not recognized, but provide insight into GGT regulation and function. Therefore, we will consider the data from each species separately.

8.1. Redox regulation of GGT expression in rats

In the rat, GGT is a single-copy gene and its expression is regulated by five tandemly arranged promoters that give rise to seven different transcripts, all of which encode the same protein, but have distinct 5′ untranslated regions (Chikhi, Holic, Guellaen, & Laperche, 1999). The promoters are tissue-specific and AP-1, AP-2, and nuclear factor-κB (NF-κB) binding sites have been identified in the promoter region (Chikhi et al., 1999). Few studies have been done in the rat *in vivo* to identify pathways that regulate GGT expression, although Yang et al. reported that electroporation of a dominant-negative ETV5 into the cells in an isolated inner segment of the epididymis resulted in a decreased expression of GGT mRNA IV (Yang, Fox, Kirby, Troan, & Hinton, 2006).

In vivo studies in rats demonstrated that oxidative stress in the lungs caused by hyperoxia, ozone, or nitrogen dioxide resulted in induction of GGT activity (Knickelbein et al., 1996; Takahashi, Oakes, et al., 1997; Takahashi, Takahashi, et al., 1997). Using primary isolated rat lung macrophages, Forman and Skelton showed that extracellular GSH and GGT were protective against hyperoxia-induced toxicity and inhibitors of GGT blocked the protective effect of extracellular GSH (Forman & Skelton, 1990).

Induction of GGT in rat liver has been studied extensively following treatment with toxins and hepatocarcinogens. In toxicology studies, GGT is induced in specific zones within the liver, dependent on the site of metabolism of the administered compound. For example, in rats fed ethoxyquin, GGT was inducted in the periportal hepatocytes (Mandel et al., 1987). Experimental protocols for inducing hepatocellular carcinomas in rats consist of a single treatment with a mutagen followed by repeated doses of a promoting agent (Goldsworthy et al., 1986). During the promotion phase, distinct foci of hepatocytes with altered levels of several enzymes can be identified. The most commonly used marker for identification of these enzyme-altered foci is the induction of GGT. We showed that the GGT-positive hepatocytes from carcinogen-treated rats have reduced levels of cytochrome P450 enzymes associated with metabolic activation of liver carcinogens (Hanigan & Pitot, 1985a). Stenius et al. showed that these GGT-positive hepatocytes were resistant to GSH depletion and hydroquinone- or menadione-induced oxidative stress (Stenius, Rubin, Gullberg, & Hogberg, 1990). The GGT-positive hepatocytes are not polarized and GGT is expressed on their entire cell surface, enabling them to use GSH in the serum and interstitial fluid as an additional source of cysteine to increase their intracellular GSH levels. The GGT-positive hepatocytes have a selective advantage during the promotion phase of carcinogenesis as the liver is exposed repeatedly to promoting agents such as phenobarbital that induces oxidative stress and depletes intracellular GSH (Sies, Bartoli, Burk, & Waydhas, 1978). It is from within these enzyme-altered foci that the tumors arise. In protocols that use peroxisome proliferators as promoting agents, the GSH levels in the liver are not depleted and induction of GGT would not provide any selective advantage to the cells. The enzyme-altered foci and liver tumors in peroxisome-treated rats are GGT-negative (Rao, Lalwani, Scarpelli, & Reddy, 1982). These studies demonstrate a strong association between oxidative stress, depletion of GSH, induction of GGT, and the selective outgrowth of GGT-positive hepatocytes in the rat (Hanigan & Pitot, 1985b).

Despite the induction of GGT in preneoplastic liver lesions by many different carcinogen treatments, it has not been possible to study this induction in primary culture. When rat hepatocytes are isolated and put in cell culture, their GGT activity increases dramatically even in the absence of any carcinogen treatment. We found that GGT activity in rat hepatocytes increased 20-fold during the first 7 days in culture (Gallagher, Rudolph, Hinton, & Hanigan, 1998). Only GGT mRNA$_{III}$ is induced indicating that promoter

III is responsible for the induction (Gallagher et al., 1998). GGT $mRNA_{III}$ is the same GGT mRNA induced in rat liver tumors, and it contains the antioxidant response element consensus sequence and binding sites for NF-1 and NFY transcription factors (Brouillet et al., 1994). The oxygen levels in culture are much higher than the oxygen levels that hepatocytes are exposed to *in vivo*. These data suggest that the hyperoxia induces oxidative stress that triggers the induction of GGT. Interestingly, as the GGT levels in the hepatocytes increase, the cystathionase activity decreases (Meredith, 1987). Cystathionase enables the cells to synthesize cysteine from methionine and a lack of cystathionase would increase their requirement for extracellular sources of cysteine in order to maintain intracellular GSH levels.

Redox regulation of GGT in rat tissues other than hepatocytes has been studied in culture. A time- and dose-dependent increase in GGT mRNA and GGT activity was reported in CC531 rat colon carcinoma cells exposed to ionizing radiation, which caused the formation of reactive oxygen and nitrogen species (Pankiv, Moller, Bjorkoy, Moens, & Huseby, 2006). In the same cell line, menadione treatment induced both ROS and GGT activity (Mikkelsen, Mortensen, Laperche, & Huseby, 2002). In a subsequent study, the authors reported that ROS-induced GGT involved activation of Ras and was mediated through an AKT-, p38 MAPK-, and MEK1-dependent pathway (Pandur, Pankiv, Johannessen, Moens, & Huseby, 2007).

8.2. Redox regulation of GGT expression in mice

In the mouse, GGT is a single-copy gene that can be transcribed from seven different promoters that give rise to distinct mRNAs that encode the same protein but differ in their 5′ untranslated region (Lieberman et al., 1995). Induction of GGT is tissue-specific and murine GGT mRNA has been shown to undergo tissue-specific and developmentally regulated alternative splicing that gives rise to both active and enzymatically inactive GGT proteins (Chikhi et al., 1999) (Joyce-Brady, Jean, & Hughey, 2001). Binding sites for AP-1 and the redox-sensitive transcription factors AP-2 and NF-κB have been identified in the murine promoters (Chikhi et al., 1999).

Expression of GGT in response to oxidative stress has been observed in several tissues in the mouse. Using mouse embryo fibroblasts (MEFs) from Nrf2-deficient mice and wild-type controls, DeNicola et al. showed that expression of the oncogenic K-ras in wild-type MEFs resulted in an induction of GGT mRNA (DeNicola et al., 2011). In this study, expression of K-ras and Myc oncogenes increased the transcription of Nrf2 and induced a series of genes including GGT, all of which are important in the cellular response to ROS (DeNicola et al., 2011). In Nrf2-deficient cells, GGT mRNA was reduced and was not altered by expression of the oncogenic K-ras, indicating that K-ras was signaling through Nrf2 to induce GGT mRNA. This was confirmed in Nrf2 knockdown studies. GGT-deficient mice develop oxidant stress in the lung even under conditions of normoxia suggesting that GGT protects the lung from oxidative stress (Jean et al., 2002).

Induction of GGT is a commonly used marker for preneoplastic liver lesions in rats, but is rarely used in mice. Many strains of mice develop preneoplastic lesions and liver tumors spontaneously with age. The endogenous promoting agent for these tumors is unknown and GGT is not expressed in the preneoplastic lesions or tumors (Goldsworthy et al., 1986). However, GGT is induced in these lesions when they are subjected to oxidative stress. Administration of phenobarbital to aged C3HfB/HeN mice for 5–7 weeks resulted in an induction of GGT in the tumors and in centrilobular hepatocytes, which was reversible when the phenobarbital was withdrawn (Williams, Ohmori, Katayama, & Rice, 1980). Phenobarbital induces oxidative stress, depletes intracellular GSH, and is a commonly used promoting agent in rat hepatocarcinogenesis protocols that induce GGT-positive preneoplastic lesions and tumors (Goldsworthy et al., 1986; Sies et al., 1978). We found that different promoting agents used in initiation/promotion protocols in mice gave rise to preneoplastic foci that differed in their enzyme profile (Hanigan, Brown, et al., 1993, Hanigan, Winkler, et al., 1993). GGT-positive foci were present in mice treated with *ortho*-azoaminotoluene or safrole as promoting agents. Both *ortho*-azoaminotoluene and safrole require metabolic activation in the liver to mutagenic compounds that cause oxidative stress (Nakagawa, Suzuki, Nakajima, Ishii, & Ogata, 2009). The GGT-positive foci were independent from the glucose-6-phosphatase-deficient foci present in the same livers that were presumably being stimulated by endogenous promoters in the mice. Roomi et al. fed C3H mice with griseofulvin, a hepatocarcinogen with both initiating and promoting activities (Roomi et al., 2006). The preneoplastic foci were GGT-positive and expressed glutathione *S*-transferase μ When the mice were injected with a hepatotoxic dose of thioacetamide, the GGT-positive hepatocytes began to proliferate while necrosis was observed in the surrounding liver. Thioacetamide is metabolized in mouse liver by CYP2E and increases oxidative stress (Kang et al., 2008). The authors concluded that the GGT-positive, glutathione *S*-transferase μ-positive preneoplastic foci had a survival advantage over the normal hepatocytes and were able to respond to the proliferation stimulus of the compensatory hyperplasia following the hepatotoxicity (Roomi et al., 2006).

There is species specificity to the induction of GGT in primary cultures of hepatocytes. The concentration of GSH in the livers of rats and mice is similar (Vina, Perez, Furukawa, Palacin, & Vina, 1989). However, GGT is rapidly induced in rat hepatocytes when they are placed in culture, but there is no induction of GGT in mouse hepatocytes (Gallagher et al., 1998). This difference between rat and mouse hepatocytes is only one of several differences between the hepatocytes of these two species with regard to GGT and other enzymes involved in GSH metabolism. Rats are very sensitive to the hepatotoxicity of aflatoxin B_1 while mice are resistant. Primary cultures of hepatocytes from these two species show this same difference in susceptibility to aflatoxin B_1 toxicity (Hanigan & Laishes, 1984). The resistance of the mouse hepatocytes has been shown to be the result of the constitutively high level of expression of an alpha class GSH *S*-transferase that inactivates the electrophilic metabolite of aflatoxin B_1 that is formed in the liver of both rats and mice (Hayes, Judah, Neal, & Nguyen, 1992). Hepatocytes can lose their ability to synthesize cysteine during transformation as shown with the mouse-derived liver tumor cell line, Hep 1–6, which is unable to survive in cysteine-free media as described in the GGT-transfection studies earlier in the text (Hanigan, 1995).

8.3. Redox regulation of GGT expression in humans

The transcriptional regulation of GGT in humans is tissue-specific with multiple mRNAs encoding the same open reading frame but differing in their 5′ untranslated region (Daubeuf et al., 2001; Visvikis et al., 2001). The human GGT promoter contains binding sites for AP1, AP2, CREB, GRE, NF-κB, and two Sp1 binding sites (Visvikis et al., 2001; Reuter et al., 2009). The 5′ untranslated region of human GGT also contains multiple steroid modulatory elements (Diederich, Wellman, Visvikis, Puga, & Siest, 1993). Modulation of GGT expression by Ras was demonstrated with HT29 cells and DLD-1 cells, two human colon carcinoma cells lines (Pankiv et al., 2006). When HT29 cells, which express wild-type c-Ha-ras, were stably transfected with activated Ras, the level of GGT protein increased and GGT activity increased approximately twofold compared to vector-transfected controls. Conversely, DLD-1 cells, which express activated Ras (Ki-ras mutation), were transfected with the dominant-negative

RasN17 mutant, their level of GGT expression decreased, and their GGT activity was reduced to 50%.

Induction of GGT through redox-signaling pathways has been demonstrated in human cell lines. Ravuri et al. reported that treatment of HT29 cells, a human colon carcinoma cell line, with phorbol 12-myristate-13-acetate activated the NADPH oxidase (NOX) system, increased intracellular ROS levels, induced GGT mRNA, increased GGT protein expression, and elevated GGT activity (Ravuri, Svineng, Pankiv, & Huseby, 2011). In the same study, inhibition of NOX activity with apocynin downregulated GGT mRNA, protein levels, and GGT activity. Tumor necrosis factor alpha was shown to induce GGT expression through the NF-κB signaling pathway in the human chronic myelogenous leukemia cell lines K562 and MEG-01(Reuter et al., 2009). NF-κB and two Sp1 binding sites were involved in basal transcription of GGT in these cells (Reuter et al., 2009). Ripple et al. treated LNCaP cells, an androgen-responsive human prostate carcinoma cell line, with the synthetic androgen R1881 and found that GGT was induced at doses of R1881 that increased ROS in the cells (Ripple, Henry, Rago, & Wilding, 1997). Blocking the induction of ROS by ascorbic acid blocked the induction of GGT. In this study, treatment of LNCaP cells with H_2O_2 or menadione also induced GGT activity.

There is a large amount of publically available microarray data regarding expression of human GGT mRNA. However, studies of GGT mRNA levels in human tissues have been confounded by the expression of a closely related gene GGT2 described in the preceding text (see Section 5.3) (Heisterkamp et al., 2008). There is 97% nucleotide identity between GGT (aka GGT1) and GGT2 and 94% amino acid identity. The GGT2 gene is transcribed and GGT2 protein has been detected, but the protein has no enzymatic activity and is rapidly degraded within the cell (West, Wickham, et al., 2013). Some microarray platforms have probes that distinguish between GGT1 and GGT2, while others do not, creating confusion as to which mRNA is being measured. As a result, microarray data for GGT mRNA expression in humans must be carefully evaluated based on the platform and probes used. In addition, some siRNAs that have been used to block translation of GGT2 have a high degree of homology to GGT1, such that functions of GGT1 have been ascribed to GGT2 (Moon et al., 2012). Issues with GGT2 nomenclature, misidentified primers, and siRNA have been summarized in detail (West, Wickham, et al., 2013). Due to these issues, only those studies that show an increase in both GGT mRNA and activity have been cited in this chapter with regard to induction of GGT in human cells.

8.4. GGT activity in serum

In humans, a low level of GGT activity can be detected in the serum of healthy individuals. GGT is a cell surface enzyme that is anchored in the membrane by a single transmembrane domain. People who have what is considered to be normal levels of GGT in serum (less than 60 U/L) have a hydrophilic form of GGT in their serum. The source of this GGT is unclear. The level of GGT activity in the serum increases dramatically in patients with liver disease (Rosalki, 1975). Analysis of the GGT on non-denaturing gels shows that there are multiple forms of GGT in the serum of these patients including GGT bound to low-density lipoproteins (Kojima et al., 1980; Paolicchi et al., 2006). When analyzed on non-denaturing gels, there is at least one unique form of GGT that is present in the serum of patients with hepatocellular carcinomas that is not present in patients with benign liver disease (Sawabu et al., 1983; Xu et al., 1992; Yao et al., 1998). The distinguishing characteristic is likely a glycan. We have reported tissue-specific glycosylation of GGT (West et al., 2010). The unique tumor form of GGT may be useful as the basis of a serum assay to aid in the diagnosis of hepatocellular carcinoma (Kobata & Amano, 2005). In most cancer patients, except those with liver tumors, the level of GGT in serum does not increase even in patients with GGT-positive tumors unless the tumor (GGT-positive or GGT-negative) has metastasized to the liver (Whitfield, 2001). The low level of GGT in the serum of healthy individuals and most cancer patients may cleave some of GSH in the serum. Nonetheless, the steady-state level of GSH in human serum ranges from 5 to 20 μM, indicating that the GGT activity in the serum is too low to deplete the serum of GSH (Davis et al., 2006).

9. OVERCOMING RESISTANCE TO PROOXIDANT ANTICANCER THERAPY BY INHIBITING GGT

Reducing the intracellular GSH concentration sensitizes tumors to many diverse chemotherapy drugs (Butturini et al., 2013; Calvert, Yao, Hamilton, & O'Dwyer, 1998; Estrela et al., 2006; Hamaguchi et al., 1993; Maciag et al., 2013; Ortega et al., 2011; Ruoso & Hedley, 2004; Traverso et al., 2013) (Crook, Souhami, Whyman, & McLean, 1986). To date, two different approaches have been evaluated clinically to try to reduce the synthesis of intracellular GSH. Sulfasalazine, a drug that blocks the cystine transporter (but not the cysteine transporter), increased the sensitivity of human glioma cells to chemotherapy and inhibited growth of human prostate tumors in nude mice (Doxsee et al., 2007; Pham et al., 2010). Sulfasalazine selectively killed CD44v-positive tumor cells in mice bearing head and neck tumors (Yoshikawa et al., 2013). Sulfasalazine is FDA-approved for use in the treatment of several conditions including inflammatory bowel disease. However, a phase I clinical trial using continuous daily dosing of sulfasalazine alone for the treatment of advanced gliomas proved neurotoxic and was terminated early (Robe et al., 2009). Buthionine sulfoximine (BSO) that inhibits the initial enzyme in GSH synthesis was evaluated in a phase I trial with melphalan (Bailey et al., 1997). Continuous dosing of BSO for 72 h in combination with melphalan resulted in severe myelosuppression. Neither of these treatments has advanced to clinical practice.

Inhibition of GGT is a novel approach to reducing intracellular GSH and sensitizing tumors to chemotherapy without the toxicities observed with sulfasalazine and BSO. Studies in mice have shown that administration of an inhibitor of GGT rapidly induces glutathionuria, decreases the cysteine levels in the tumor, and reduces the intracellular concentration of GSH in tumors. One hour after administering an inhibitor of GGT to mice, the concentration of GSH in the urine increased 1800-fold (Griffith & Meister, 1979). Administration of a GGT inhibitor to a mouse bearing a cervical tumor resulted in a 50% drop in the cysteine concentration in the tumor within 4 h (Ruoso & Hedley, 2004). Benlloch et al. showed that inhibiting GGT with acivicin in a model system of melanoma metastasis to the liver in the presence of normal extracellular levels of cysteine resulted in a 50% reduction in the GSH concentration in the tumors (Benlloch et al., 2005).

A significant difference between the effects of a GGT inhibitor and BSO is that when GGT is inhibited, the GSH concentration in the serum increases (Griffith & Meister, 1979). The GSH concentration in the serum of wild-type mice is 28 μM, while the serum of the GGT knockout mice contains 175 μM GSH (Lieberman et al., 1996). It is not clear whether the increased concentration of GSH in the blood of GGT knockout mice is due to increased synthesis and secretion of GSH by the liver or due to decreased catabolism of GSH as it circulates through the body (Anderson, Bridges, & Meister, 1980; Lauterburg, Adams, & Mitchell, 1984). However, cysteine depletion from serum that contains high levels of GSH does not appear to be toxic to blood cells as myelosuppression has not been observed in GGT-deficient mice (Lieberman et al., 1996). In fact, inhibiting GGT has been shown to block the toxic side effects of several chemotherapy drugs. We discovered that inhibiting GGT protects against the dose-limiting nephrotoxicity of the widely used chemotherapy drug, cisplatin (Hanigan, Frierson, et al., 1994; Hanigan et al., 2001). We found that cisplatin is metabolized to a nephrotoxin via the GGT-dependent mercapturic acid pathway (Townsend, Deng, Zhang, Lapus, & Hanigan, 2003). Other investigators have shown that pulmonary fibrosis, a toxic side effect of bleomycin, is attenuated in GGT knockout mice and that surprisingly, it is the cysteine deficiency that is protective (Pardo et al., 2003).

9.1. GGT inhibitors

Tumors are under oxidative stress due to their increased metabolic rate (Pani et al., 2010). They have higher intracellular GSH concentrations, utilize more cysteine than normal cells, and therefore are more sensitive to cysteine depletion than normal cells. Administration of an inhibitor of GGT to a patient for several hours prior to therapy would reduce the supply of cysteine and cystine available to both GGT-positive and GGT-negative tumors. This would sensitize the tumors to cytotoxic chemotherapy, radiation, and other prooxidant anticancer therapies. Systemic GGT inhibition would sensitize GGT-positive tumors not only by reducing cysteine concentrations in the blood but also by blocking their unique ability to use extracellular GSH as an additional source of cysteine.

The most potent GGT inhibitors are glutamate analogs. They include acivicin, diazonorleucine, and L-azaserine (Ahluwalia, Grem, Hao, & Cooney, 1990; Tate & Meister, 1978). However, they also inhibit glutamate metabolizing enzymes and, at doses needed to inhibit GGT, they are too toxic for clinical use (Ahluwalia et al., 1990; Hidalgo et al., 1998). Two new classes of GGT inhibitors have been reported. One is gamma-phosphono glutamate analogs, which are irreversible inhibitors of human GGT (Han, Hiratake, Kamiyama, & Sakata, 2007). The lead compound, GGsTop, has been used to inhibit renal GGT in rats (Yamamoto et al., 2011). These compounds are glutamate analogs and toxicity data have not been reported. We have identified a group of GGT inhibitors that are derivatives of our initial lead compound *N*-[5-(4-methoxybenzyl)-1,3,4-thiadiazol-2-yl]benzenesulfonamide, which we named OU749 (King, West, Cook, & Hanigan, 2009; Wickham, Regan, et al., 2011; Wickham et al., 2013). These compounds are uncompetitive inhibitors of human GGT with low toxicity and a large therapeutic window when evaluated in dividing human cells in culture (Wickham et al., 2013). These inhibitors are under further development.

10. Summary

Expression of GGT is essential in maintaining the cysteine levels in the body. Induction of GGT expression in response to redox stress provides the cell with access to additional cysteine, which becomes rate-limiting for intracellular GSH synthesis. Our understanding of the role of GGT in redox regulation and its induction in cells under redox stress will aid in the development of new anticancer therapies. GGT first came to the attention of cancer researchers as a biomarker for preneoplastic lesions in rodent models of hepatocarcinogenesis. Subsequent studies showed that the expression of GGT in the preneoplastic lesions provided a selective survival and proliferative advantage during treatment with promoting agents that deplete intracellular GSH. Clinical studies of GGT expression in human tumors revealed a correlation between GGT expression and poor patient survival. Expression of GGT on the entire surface of tumor cells provides the tumor with access to additional cysteine and cystine from GSH and GSSG in the interstitial fluid. This additional pool of cysteine allows the GGT-positive tumors to maintain higher levels of intracellular GSH, enhancing their resistance to prooxidant anticancer therapy. Inhibition of GGT is a new approach to overcoming drug resistance in tumors. Inhibition of GGT several hours prior to the administration of chemotherapy would not only block the access of GGT-positive tumors to cysteine from extracellular GSH but also inhibit GGT in the kidney, resulting in glutathionuria and a rapid reduction in the levels of cysteine in the blood. As a result, GSH concentrations in both GGT-positive and GGT-negative tumors would decrease and the tumors would be sensitized to the therapy. Short-term reduction of total body cysteine would not be detrimental to health, but would enhance the effectiveness of prooxidant chemotherapy. We have recently solved the crystal structure of human GGT that will accelerate the development of GGT inhibitors for clinical use.

ACKNOWLEDGMENTS

We gratefully acknowledge Dr. Stephanie Wickham's assistance in preparing Fig. 3.3. This work was supported in part by the National Institutes of Health Grants P20GM103640 (an Institutional Development Award (IDeA)).

REFERENCES

Ahmad, S., Okine, L., Wood, R., Aljian, J., & Vistica, D. T. (1987). gamma-Glutamyl transpeptidase (gamma-GT) and maintenance of thiol pools in tumor cells resistant to alkylating agents. *Journal of Cellular Physiology*, *131*(2), 240–246.

Anders, M. W., & Dekant, W. (1998). Glutathione-dependent bioactivation of haloalkenes. *Annual Review of Pharmacology and Toxicology*, *38*, 501–537.

Anderson, M. E., Bridges, R. J., & Meister, A. (1980). Direct evidence for inter-organ transport of glutathione and that the non-filtration renal mechanism for glutathione utilization involves gamma-glutamyl transpeptidase. *Biochemical and Biophysical Research Communications*, *96*(2), 848–853.

Auman, J. T., Church, R., Lee, S. Y., Watson, M. A., Fleshman, J. W., & McLeod, H. L. (2008). Celecoxib pre-treatment in human colorectal adenocarcinoma patients is associated with gene expression alterations suggestive of diminished cellular proliferation. *European Journal of Cancer*, *44*(12), 1754–1760.

Bachhawat, A. K., Thakur, A., Kaur, J., & Zulkifli, M. (2013). Glutathione transporters. *Biochimica et Biophysica Acta*, *1830*(5), 3154–3164.

Bailey, H. H., Gipp, J. J., & Mulcahy, R. T. (1994). Increased expression of gamma-glutamyl transpeptidase in transfected tumor cells and its relationship to drug sensitivity. *Cancer Letters*, *87*(2), 163–170.

Bailey, H. H., Ripple, G., Tutsch, K. D., Arzoomanian, R. Z., Alberti, D., Feierabend, C., et al. (1997). Phase I study of continuous-infusion L-S, R-buthionine sulfoximine with intravenous melphalan. *Journal of the National Cancer Institute*, *89*(23), 1789–1796.

Balendiran, G. K., Dabur, R., & Fraser, D. (2004). The role of glutathione in cancer. *Cell Biochemistry and Function*, *22*(6), 343–352.

Bard, S., Noel, P., Chauvin, F., & Quash, G. (1986). gamma-Glutamyltranspeptidase activity in human breast lesions: An unfavourable prognostic sign. *British Journal of Cancer*, *53*(5), 637–642.

Barrios, R., Shi, Z. Z., Kala, S. V., Wiseman, A. L., Welty, S. E., Kala, G., et al. (2001). Oxygen-induced pulmonary injury in gamma-glutamyl transpeptidase-deficient mice. *Lung*, *179*(5), 319–330.

Benlloch, M., Ortega, A., Ferrer, P., Segarra, R., Obrador, E., Asensi, M., et al. (2005). Acceleration of glutathione efflux and inhibition of gamma-glutamyltranspeptidase sensitize metastatic B16 melanoma cells to endothelium-induced cytotoxicity. *The Journal of Biological Chemistry*, *280*(8), 6950–6959.

Biddle, A., Gammon, L., Fazil, B., & Mackenzie, I. C. (2013). CD44 staining of cancer stem-like cells is influenced by down-regulation of CD44 variant isoforms and up-regulation of the standard CD44 isoform in the population of cells that have undergone epithelial-to-mesenchymal transition. *PLoS One*, *8*(2), e57314.

Binkley, F., & Nakamura, K. (1948). Metabolism of glutathione; hydrolysis by tissues of the rat. *The Journal of Biological Chemistry*, *173*(1), 411–421.

Blanco, R. A., Ziegler, T. R., Carlson, B. A., Cheng, P. Y., Park, Y., Cotsonis, G. A., et al. (2007). Diurnal variation in glutathione and cysteine redox states in human plasma. *The American Journal of Clinical Nutrition*, *86*(4), 1016–1023.

Brannigan, J. A., Dodson, G., Duggleby, H. J., Moody, P. C., Smith, J. L., Tomchick, D. R., et al. (1995). A protein catalytic framework with an N-terminal nucleophile is capable of self-activation. *Nature*, *378*(6555), 416–419.

Brouillet, A., Darbouy, M., Okamoto, T., Chobert, M. N., Lahuna, O., Garlatti, M., et al. (1994). Functional characterization of the rat gamma-glutamyl transpeptidase promoter that is expressed and regulated in the liver and hepatoma cells. *The Journal of Biological Chemistry*, *269*(21), 14878–14884.

Butturini, E., de Prati, A. C., Chiavegato, G., Rigo, A., Cavalieri, E., Darra, E., et al. (2013). Mild oxidative stress induces S-glutathionylation of STAT3 and enhances

chemosensitivity of tumoral cells to chemotherapeutic drugs. *Free Radical Biology and Medicine, 65*, 1322–1330.
Cairns, R. A., Harris, I. S., & Mak, T. W. (2011). Regulation of cancer cell metabolism. *Nature Reviews. Cancer, 11*(2), 85–95.
Calvert, P., Yao, K. S., Hamilton, T. C., & O'Dwyer, P. J. (1998). Clinical studies of reversal of drug resistance based on glutathione. *Chemico-Biological Interactions, 111–112*, 213–224.
Cameron, R. G., Armstrong, D., Gunsekara, A., Varghese, G., & Speisky, H. (1991). Utilization of circulating glutathione by nodular and cancerous intact rat liver. *Carcinogenesis, 12*(12), 2369–2372.
Carter, B. Z., Shi, Z. Z., Barrios, R., & Lieberman, M. W. (1998). gamma-Glutamyl leukotrienase, a gamma-glutamyl transpeptidase gene family member, is expressed primarily in spleen. *The Journal of Biological Chemistry, 273*(43), 28277–28285.
Castonguay, R., Halim, D., Morin, M., Furtos, A., Lherbet, C., Bonneil, E., et al. (2007). Kinetic characterization and identification of the acylation and glycosylation sites of recombinant human gamma-glutamyltranspeptidase. *Biochemistry, 46*(43), 12253–12262.
Chaiswing, L., Zhong, W., & Oberley, T. D. (2011). Distinct redox profiles of selected human prostate carcinoma cell lines: Implications for rational design of redox therapy. *Cancers, 3*(3), 3557–3584.
Chaudhari, P., Ye, Z., & Jang, Y. Y. (2012). Roles of reactive oxygen species in the fate of stem cells. *Antioxidants & Redox Signaling, 20*, 1881–1890.
Chevez-Barrios, P., Wiseman, A. L., Rojas, E., Ou, C. N., & Lieberman, M. W. (2000). Cataract development in gamma-glutamyl transpeptidase-deficient mice. *Experimental Eye Research, 71*(6), 575–582.
Chikhi, N., Holic, N., Guellaen, G., & Laperche, Y. (1999). Gamma-glutamyl transpeptidase gene organization and expression: A comparative analysis in rat, mouse, pig and human species. *Comparative Biochemistry and Physiology. Part B, Biochemistry and Molecular Biology, 122*(4), 367–380.
Circu, M. L., & Aw, T. Y. (2010). Reactive oxygen species, cellular redox systems, and apoptosis. *Free Radical Biology & Medicine, 48*(6), 749–762.
Cole, S. P., & Deeley, R. G. (2006). Transport of glutathione and glutathione conjugates by MRP1. *Trends in Pharmacological Sciences, 27*(8), 438–446.
Cooper, A. J. L., & Hanigan, M. H. (2010). Enzymes involved in processing glutathione conjugates. In C. A. McQueen (Ed.), *Comprehensive toxicology*, Vol. 4. (pp. 323–366). Oxford: Elsevier, Chapter 17.
Corti, A., Duarte, T. L., Giommarelli, C., De Tata, V., Paolicchi, A., Jones, G. D., et al. (2009). Membrane gamma-glutamyl transferase activity promotes iron-dependent oxidative DNA damage in melanoma cells. *Mutation Research, 669*(1–2), 112–121.
Courtay, C., Heisterkamp, N., Siest, G., & Groffen, J. (1994). Expression of multiple gamma-glutamyltransferase genes in man. *The Biochemical Journal, 297*(Pt 3), 503–508.
Crook, T. R., Souhami, R. L., Whyman, G. D., & McLean, A. E. (1986). Glutathione depletion as a determinant of sensitivity of human leukemia cells to cyclophosphamide. *Cancer Research, 46*(10), 5035–5038.
Curthoys, N. P., & Hughey, R. P. (1979). Characterization and physiological function of rat renal gamma-glutamyltranspeptidase. *Enzyme, 24*(6), 383–403.
Daubeuf, S., Accaoui, M. J., Pettersen, I., Huseby, N. E., Visvikis, A., & Galteau, M. M. (2001). Differential regulation of gamma-glutamyltransferase mRNAs in four human tumour cell lines. *Biochimica et Biophysica Acta, 1568*(1), 67–73.
Davis, S. R., Quinlivan, E. P., Stacpoole, P. W., & Gregory, J. F., 3rd. (2006). Plasma glutathione and cystathionine concentrations are elevated but cysteine flux is unchanged by dietary vitamin B-6 restriction in young men and women. *The Journal of Nutrition, 136*(2), 373–378.

DeNicola, G. M., Karreth, F. A., Humpton, T. J., Gopinathan, A., Wei, C., Frese, K., et al. (2011). Oncogene-induced Nrf2 transcription promotes ROS detoxification and tumorigenesis. *Nature, 475*(7354), 106–109.
Diederich, M., Wellman, M., Visvikis, A., Puga, A., & Siest, G. (1993). The 5' untranslated region of the human gamma-glutamyl transferase mRNA contains a tissue-specific active translational enhancer. *FEBS Letters, 332*(1–2), 88–92.
Dominici, S., Valentini, M., Maellaro, E., Del Bello, B., Paolicchi, A., Lorenzini, E., et al. (1999). Redox modulation of cell surface protein thiols in U937 lymphoma cells: The role of gamma-glutamyl transpeptidase-dependent H2O2 production and S-thiolation. *Free Radical Biology & Medicine, 27*(5–6), 623–635.
Doxsee, D. W., Gout, P. W., Kurita, T., Lo, M., Buckley, A. R., Wang, Y., et al. (2007). Sulfasalazine-induced cystine starvation: Potential use for prostate cancer therapy. *Prostate, 67*(2), 162–171.
Durham, J. R., Frierson, H. F., Jr., & Hanigan, M. H. (1997). Gamma-glutamyl transpeptidase immunoreactivity in benign and malignant breast tissue. *Breast Cancer Research and Treatment, 45*(1), 55–62.
Estrela, J. M., Ortega, A., & Obrador, E. (2006). Glutathione in cancer biology and therapy. *Critical Reviews in Clinical Laboratory Sciences, 43*(2), 143–181.
Fagoonee, S., Bearzi, C., Di Cunto, F., Clohessy, J. G., Rizzi, R., Reschke, M., et al. (2013). The RNA binding protein ESRP1 fine-tunes the expression of pluripotency-related factors in mouse embryonic stem cells. *PLoS One, 8*(8), e72300.
Figlewicz, D. A., Delattre, O., Guellaen, G., Krizus, A., Thomas, G., Zucman, J., et al. (1993). Mapping of human gamma-glutamyl transpeptidase genes on chromosome 22 and other human autosomes. *Genomics, 17*(2), 299–305.
Forman, H. J., & Skelton, D. C. (1990). Protection of alveolar macrophages from hyperoxia by gamma-glutamyl transpeptidase. *The American Journal of Physiology, 259*(2 Pt 1), L102–L107.
Gallagher, B. C., Rudolph, D. B., Hinton, B. T., & Hanigan, M. H. (1998). Differential induction of gamma-glutamyl transpeptidase in primary cultures of rat and mouse hepatocytes parallels induction during hepatocarcinogenesis. *Carcinogenesis, 19*(7), 1251–1255.
Galperin, M. Y., & Koonin, E. V. (2012). Divergence and convergence in enzyme evolution. *The Journal of Biological Chemistry, 287*(1), 21–28.
Glass, G. A., & Stark, A. A. (1997). Promotion of glutathione-gamma-glutamyl transpeptidase-dependent lipid peroxidation by copper and ceruloplasmin: The requirement for iron and the effects of antioxidants and antioxidant enzymes. *Environmental and Molecular Mutagenesis, 29*(1), 73–80.
Go, Y. M., & Jones, D. P. (2013). Thiol/disulfide redox states in signaling and sensing. *Critical Reviews in Biochemistry and Molecular Biology, 48*(2), 173–181.
Godwin, A. K., Meister, A., O'Dwyer, P. J., Huang, C. S., Hamilton, T. C., & Anderson, M. E. (1992). High resistance to cisplatin in human ovarian cancer cell lines is associated with marked increase of glutathione synthesis. *Proceedings of the National Academy of Sciences of the United States of America, 89*(7), 3070–3074.
Goldsworthy, T. L., Hanigan, M. H., & Pitot, H. C. (1986). Models of hepatocarcinogenesis in the rat—Contrasts and comparisons. *Critical Reviews in Toxicology, 17*(1), 61–89.
Ahluwalia, G. S., Grem, J. L., Hao, Z., & Cooney, D. A. (1990). Metabolism and action of amino acid analog anti-cancer agents. *Pharmacology & Therapeutics, 46*(2), 243–271.
Griffith, O. W., & Meister, A. (1979). Translocation of intracellular glutathione to membrane-bound gamma-glutamyl transpeptidase as a discrete step in the gamma-glutamyl cycle: Glutathionuria after inhibition of transpeptidase. *Proceedings of the National Academy of Sciences of the United States of America, 76*(1), 268–272.
Grimm, C., Hofstetter, G., Aust, S., Mutz-Dehbalaie, I., Bruch, M., Heinze, G., et al. (2013). Association of gamma-glutamyltransferase with severity of disease at diagnosis and prognosis of ovarian cancer. *British Journal of Cancer, 109*(3), 610–614.

Hamaguchi, K., Godwin, A. K., Yakushiji, M., O'Dwyer, P. J., Ozols, R. F., & Hamilton, T. C. (1993). Cross-resistance to diverse drugs is associated with primary cisplatin resistance in ovarian cancer cell lines. *Cancer Research, 53*(21), 5225–5232.

Han, L., Hiratake, J., Kamiyama, A., & Sakata, K. (2007). Design, synthesis, and evaluation of gamma-phosphono diester analogues of glutamate as highly potent inhibitors and active site probes of gamma-glutamyl transpeptidase. *Biochemistry, 46*(5), 1432–1447.

Hanes, C. S., Hird, F. J., & Isherwood, F. A. (1952). Enzymic transpeptidation reactions involving gamma-glutamyl peptides and alpha-amino-acyl peptides. *The Biochemical Journal, 51*(1), 25–35.

Hanigan, M. H. (1995). Expression of gamma-glutamyl transpeptidase provides tumor cells with a selective growth advantage at physiologic concentrations of cyst(e)ine. *Carcinogenesis, 16*(2), 181–185.

Hanigan, M. H., Brown, J. E., & Ricketts, W. A. (1993). Gamma-glutamyl transpeptidase, a glutathionase, is present in some cell culture grade bovine sera. *In Vitro Cellular Developmental Biology Animal, 29A*(11), 831–833.

Hanigan, M. H., & Frierson, H. F., Jr. (1996). Immunohistochemical detection of gamma-glutamyl transpeptidase in normal human tissue. *The Journal of Histochemistry and Cytochemistry, 44*(10), 1101–1108.

Hanigan, M. H., Frierson, H. F., Jr., Brown, J. E., Lovell, M. A., & Taylor, P. T. (1994). Human ovarian tumors express gamma-glutamyl transpeptidase. *Cancer Research, 54*(1), 286–290.

Hanigan, M. H., Frierson, H. F., Jr., Swanson, P. E., & De Young, B. R. (1999). Altered expression of gamma-glutamyl transpeptidase in human tumors. *Human Pathology, 30*(3), 300–305.

Hanigan, M. H., Frierson, H. F., Jr., & Taylor, P. T., Jr. (1998). Expression of gamma-glutamyl transpeptidase in stage III and IV ovarian surface epithelial carcinomas does not alter response to primary cisplatin-based chemotherapy. *American Journal of Obstetrics and Gynecology, 179*(2), 363–367.

Hanigan, M. H., Gallagher, B. C., Taylor, P. T., Jr., & Large, M. K. (1994). Inhibition of gamma-glutamyl transpeptidase activity by acivicin in vivo protects the kidney from cisplatin-induced toxicity. *Cancer Research, 54*(22), 5925–5929.

Hanigan, M. H., Gallagher, B. C., Townsend, D. M., & Gabarra, V. (1999). Gamma-glutamyl transpeptidase accelerates tumor growth and increases the resistance of tumors to cisplatin in vivo. *Carcinogenesis, 20*(4), 553–559.

Hanigan, H. M., & Laishes, B. A. (1984). Toxicity of aflatoxin B1 in rat and mouse hepatocytes in vivo and in vitro. *Toxicology, 30*(3), 185–193.

Hanigan, M. H., Lykissa, E. D., Townsend, D. M., Ou, C. N., Barrios, R., & Lieberman, M. W. (2001). Gamma-glutamyl transpeptidase-deficient mice are resistant to the nephrotoxic effects of cisplatin. *The American Journal of Pathology, 159*(5), 1889–1894.

Hanigan, M. H., & Pitot, H. C. (1985a). Activities of benzphetamine N-demethylase and aryl hydrocarbon hydroxylase in cells isolated from gamma-glutamyl transpeptidase-positive foci and surrounding liver. *Journal of the National Cancer Institute, 75*(6), 1107–1112.

Hanigan, M. H., & Pitot, H. C. (1985b). Gamma-glutamyl transpeptidase—Its role in hepatocarcinogenesis. *Carcinogenesis, 6*(2), 165–172.

Hanigan, M. H., & Ricketts, W. A. (1993). Extracellular glutathione is a source of cysteine for cells that express gamma-glutamyl transpeptidase. *Biochemistry, 32*(24), 6302–6306.

Hanigan, M. H., Winkler, M. L., & Drinkwater, N. R. (1993). Induction of three histochemically distinct populations of hepatic foci in C57BL/6 J mice. *Carcinogenesis, 14*(5), 1035–1040.

Harding, C. O., Williams, P., Wagner, E., Chang, D. S., Wild, K., Colwell, R. E., et al. (1997). Mice with genetic gamma-glutamyl transpeptidase deficiency exhibit glutathionuria, severe growth failure, reduced life spans, and infertility. *The Journal of Biological Chemistry*, *272*(19), 12560–12567.

Hayes, J. D., Judah, D. J., Neal, G. E., & Nguyen, T. (1992). Molecular cloning and heterologous expression of a cDNA encoding a mouse glutathione S-transferase Yc subunit possessing high catalytic activity for aflatoxin B1-8,9-epoxide. *The Biochemical Journal*, *285*(Pt 1), 173–180.

Heisterkamp, N., Groffen, J., Warburton, D., & Sneddon, T. P. (2008). The human gamma-glutamyltransferase gene family. *Human Genetics*, *123*(4), 321–332.

Heisterkamp, N., Rajpert-De Meyts, E., Uribe, L., Forman, H. J., & Groffen, J. (1991). Identification of a human gamma-glutamyl cleaving enzyme related to, but distinct from, gamma-glutamyl transpeptidase. *Proceedings of the National Academy of Sciences of the United States of America*, *88*(14), 6303–6307.

Hidalgo, M., Rodriguez, G., Kuhn, J. G., Brown, T., Weiss, G., MacGovren, J. P., et al. (1998). A Phase I and pharmacological study of the glutamine antagonist acivicin with the amino acid solution aminosyn in patients with advanced solid malignancies. *Clinical Cancer Research*, *4*(11), 2763–2770.

Hochwald, S. N., Harrison, L. E., Rose, D. M., Anderson, M., & Burt, M. E. (1996). gamma-Glutamyl transpeptidase mediation of tumor glutathione utilization in vivo. *Journal of the National Cancer Institute*, *88*(3–4), 193–197.

Hochwald, S. N., Rose, D. M., Brennan, M. F., & Burt, M. E. (1997). Elevation of glutathione and related enzyme activities in high-grade and metastatic extremity soft tissue sarcoma. *Annals of Surgical Oncology*, *4*(4), 303–309.

Huseby, N. E., Asare, N., Wetting, S., Mikkelsen, I. M., Mortensen, B., & Wellman, M. (2003). Role of gamma-glutamyltransferase in the homeostasis of glutathione during oxidative and nitrosative stress. *Biofactors*, *17*(1–4), 151–160.

Ikeda, Y., Fujii, J., Taniguchi, N., & Meister, A. (1995). Expression of an active glycosylated human gamma-glutamyl transpeptidase mutant that lacks a membrane anchor domain. *Proceedings of the National Academy of Sciences of the United States of America*, *92*(1), 126–130.

Ishimoto, T., Nagano, O., Yae, T., Tamada, M., Motohara, T., Oshima, H., et al. (2011). CD44 variant regulates redox status in cancer cells by stabilizing the xCT subunit of system xc(−) and thereby promotes tumor growth. *Cancer Cell*, *19*(3), 387–400.

Jean, J. C., Liu, Y., Brown, L. A., Marc, R. E., Klings, E., & Joyce-Brady, M. (2002). Gamma-glutamyl transferase deficiency results in lung oxidant stress in normoxia. *American Journal of Physiology Lung Cellular and Molecular Physiology*, *283*(4), L766–L776.

Jones, D. P. (2010). Redox sensing: Orthogonal control in cell cycle and apoptosis signalling. *Journal of Internal Medicine*, *268*(5), 432–448.

Joyce-Brady, M., Jean, J. C., & Hughey, R. P. (2001). gamma -glutamyltransferase and its isoform mediate an endoplasmic reticulum stress response. *The Journal of Biological Chemistry*, *276*(12), 9468–9477.

Kang, J. S., Wanibuchi, H., Morimura, K., Wongpoomchai, R., Chusiri, Y., Gonzalez, F. J., et al. (2008). Role of CYP2E1 in thioacetamide-induced mouse hepatotoxicity. *Toxicology and Applied Pharmacology*, *228*(3), 295–300.

Kansanen, E., Kuosmanen, S. M., Leinonen, H., & Levonen, A. L. (2013). The Keap1-Nrf2 pathway: Mechanisms of activation and dysregulation in cancer. *Redox Biology*, *1*(1), 45–49.

Karp, D. R., Shimooku, K., & Lipsky, P. E. (2001). Expression of gamma-glutamyl transpeptidase protects ramos B cells from oxidation-induced cell death. *The Journal of Biological Chemistry*, *276*(6), 3798–3804.

Keillor, J. W., Castonguay, R., & Lherbet, C. (2005). Gamma-glutamyl transpeptidase substrate specificity and catalytic mechanism. *Methods in Enzymology*, *401*, 449–467.

King, J. B., West, M. B., Cook, P. F., & Hanigan, M. H. (2009). A novel, species-specific class of uncompetitive inhibitors of gamma-glutamyl transpeptidase. *The Journal of Biological Chemistry*, *284*(14), 9059–9065.
Kinoshita, H., Okabe, H., Beppu, T., Chikamoto, A., Hayashi, H., Imai, K., et al. (2013). Cystine/glutamic acid transporter is a novel marker for predicting poor survival in patients with hepatocellular carcinoma. *Oncology Reports*, *29*(2), 685–689.
Knickelbein, R. G., Ingbar, D. H., Seres, T., Snow, K., Johnston, R. B., Jr., Fayemi, O., et al. (1996). Hyperoxia enhances expression of gamma-glutamyl transpeptidase and increases protein S-glutathiolation in rat lung. *The American Journal of Physiology*, *270*(1 Pt 1), L115–L122.
Kobata, A., & Amano, J. (2005). Altered glycosylation of proteins produced by malignant cells, and application for the diagnosis and immunotherapy of tumours. *Immunology and Cell Biology*, *83*(4), 429–439.
Kojima, J., Kanatani, M., Nakamura, N., Kashiwagi, T., Tohjoh, F., & Akiyama, M. (1980). Electrophoretic fractionation of serum gamma-glutamyl transpeptidase in human hepatic cancer. *Clinica Chimica Acta*, *106*(2), 165–172.
Kwiecien, I., Rokita, H., Lorenc-Koci, E., Sokolowska, M., & Wlodek, L. (2007). The effect of modulation of gamma-glutamyl transpeptidase and nitric oxide synthase activity on GSH homeostasis in HepG2 cells. *Fundamental & Clinical Pharmacology*, *21*(1), 95–103.
Lauterburg, B. H., Adams, J. D., & Mitchell, J. R. (1984). Hepatic glutathione homeostasis in the rat: Efflux accounts for glutathione turnover. *Hepatology*, *4*(4), 586–590.
Lewerenz, J., Hewett, S. J., Huang, Y., Lambros, M., Gout, P. W., Kalivas, P. W., et al. (2013). The cystine/glutamate antiporter system x(c)(−) in health and disease: From molecular mechanisms to novel therapeutic opportunities. *Antioxidants & Redox Signaling*, *18*(5), 522–555.
Lewis, A. D., Hayes, J. D., & Wolf, C. R. (1988). Glutathione and glutathione-dependent enzymes in ovarian adenocarcinoma cell lines derived from a patient before and after the onset of drug resistance: Intrinsic differences and cell cycle effects. *Carcinogenesis*, *9*(7), 1283–1287.
Lieberman, M. W., Barrios, R., Carter, B. Z., Habib, G. M., Lebovitz, R. M., Rajagopalan, S., et al. (1995). gamma-Glutamyl transpeptidase. What does the organization and expression of a multipromoter gene tell us about its functions? *The American Journal of Pathology*, *147*(5), 1175–1185.
Lieberman, M. W., Wiseman, A. L., Shi, Z. Z., Carter, B. Z., Barrios, R., Ou, C. N., et al. (1996). Growth retardation and cysteine deficiency in gamma-glutamyl transpeptidase-deficient mice. *Proceedings of the National Academy of Sciences of the United States of America*, *93*(15), 7923–7926.
Maciag, A. E., Holland, R. J., Robert Cheng, Y. S., Rodriguez, L. G., Saavedra, J. E., Anderson, L. M., et al. (2013). Nitric oxide-releasing prodrug triggers cancer cell death through deregulation of cellular redox balance. *Redox Biology*, *1*(1), 115–124.
Mandel, H. G., Manson, M. M., Judah, D. J., Simpson, J. L., Green, J. A., Forrester, L. M., et al. (1987). Metabolic basis for the protective effect of the antioxidant ethoxyquin on aflatoxin B1 hepatocarcinogenesis in the rat. *Cancer Research*, *47*(19), 5218–5223.
Meister, A. (1973). On the enzymology of amino acid transport. *Science*, *180*(4081), 33–39.
Meister, A. (1991). Glutathione deficiency produced by inhibition of its synthesis, and its reversal; applications in research and therapy. *Pharmacology & Therapeutics*, *51*(2), 155–194.
Meredith, M. J. (1987). Cystathionase activity and glutathione metabolism in redifferentiating rat hepatocyte primary cultures. *Cell Biology and Toxicology*, *3*(4), 361–377.
Mikkelsen, I. M., Mortensen, B., Laperche, Y., & Huseby, N. E. (2002). The expression of gamma-glutamyltransferase in rat colon carcinoma cells is distinctly regulated during differentiation and oxidative stress. *Molecular and Cellular Biochemistry*, *232*(1–2), 87–95.

Mistry, P., & Harrap, K. R. (1991). Historical aspects of glutathione and cancer chemotherapy. *Pharmacology & Therapeutics*, *49*(1–2), 125–132.
Moon, D. O., Kim, B. Y., Jang, J. H., Kim, M. O., Jayasooriya, R. G., Kang, C. H., et al. (2012). K-RAS transformation in prostate epithelial cell overcomes H_2O_2-induced apoptosis via upregulation of gamma-glutamyltransferase-2. *Toxicology In Vitro*, *26*(3), 429–434.
Moriarty-Craige, S. E., & Jones, D. P. (2004). Extracellular thiols and thiol/disulfide redox in metabolism. *Annual Review of Nutrition*, *24*, 481–509.
Morris, C., Courtay, C., Geurts van Kessel, A., ten Hoeve, J., Heisterkamp, N., & Groffen, J. (1993). Localization of a gamma-glutamyl-transferase-related gene family on chromosome 22. *Human Genetics*, *91*(1), 31–36.
Morrow, A. L., Williams, K., Sand, A., Boanca, G., & Barycki, J. J. (2007). Characterization of Helicobacter pylori gamma-glutamyltranspeptidase reveals the molecular basis for substrate specificity and a critical role for the tyrosine 433-containing loop in catalysis. *Biochemistry*, *46*(46), 13407–13414.
Nagano, O., Okazaki, S., & Saya, H. (2013). Redox regulation in stem-like cancer cells by CD44 variant isoforms. *Oncogene*, *32*, 5191–5198.
Nakagawa, Y., Suzuki, T., Nakajima, K., Ishii, H., & Ogata, A. (2009). Biotransformation and cytotoxic effects of hydroxychavicol, an intermediate of safrole metabolism, in isolated rat hepatocytes. *Chemico-Biological Interactions*, *180*(1), 89–97.
Newkirk, K., Heffern, J., Sloman-Moll, E., Sessions, R. B., Rasmussen, A. A., Andrews, P. A., et al. (1997). Glutathione content but not gamma glutamyl cysteine synthetase mRNA expression predicts cisplatin resistance in head and neck cancer cell lines. *Cancer Chemother Pharmacol*, *40*(1), 75–80.
Oguchi, H., Kikkawa, F., Kojima, M., Maeda, O., Mizuno, K., Suganuma, N., et al. (1994). Glutathione related enzymes in cis-diamminedichloroplatinum (II)-sensitive and-resistant human ovarian carcinoma cells. *Anticancer Research*, *14*(1A), 193–200.
Oinonen, C., & Rouvinen, J. (2000). Structural comparison of Ntn-hydrolases. *Protein Science*, *9*(12), 2329–2337.
Okada, T., Suzuki, H., Wada, K., Kumagai, H., & Fukuyama, K. (2006). Crystal structures of gamma-glutamyltranspeptidase from Escherichia coli, a key enzyme in glutathione metabolism, and its reaction intermediate. *Proceedings of the National Academy of Sciences of the United States of America*, *103*(17), 6471–6476.
Orlowski, M., & Meister, A. (1963). gamma-Glutamyl-p-nitroanilide: A new convenient substrate for determination and study of L- and D-gamma-glutamyltranspeptidase activities. *Biochimica et Biophysica Acta*, *73*, 679–681.
Orlowski, M., & Meister, A. (1970). The gamma-glutamyl cycle: A possible transport system for amino acids. *Proceedings of the National Academy of Sciences of the United States of America*, *67*(3), 1248–1255.
Ortega, A. L., Salvador, M., & Estrela, J. M. (2011). Glutathione in cancer cell death. *Cancers*, *3*, 1285–1310.
Pandur, S., Pankiv, S., Johannessen, M., Moens, U., & Huseby, N. E. (2007). Gamma-glutamyltransferase is upregulated after oxidative stress through the Ras signal transduction pathway in rat colon carcinoma cells. *Free Radical Research*, *41*(12), 1376–1384.
Pani, G., Galeotti, T., & Chiarugi, P. (2010). Metastasis: Cancer cell's escape from oxidative stress. *Cancer Metastasis Reviews*, *29*(2), 351–378.
Pankiv, S., Moller, S., Bjorkoy, G., Moens, U., & Huseby, N. E. (2006). Radiation-induced upregulation of gamma-glutamyltransferase in colon carcinoma cells is mediated through the Ras signal transduction pathway. *Biochimica et Biophysica Acta*, *1760*(2), 151–157.
Paolicchi, A., Emdin, M., Passino, C., Lorenzini, E., Titta, F., Marchi, S., et al. (2006). Beta-lipoprotein- and LDL-associated serum gamma-glutamyltransferase in patients with coronary atherosclerosis. *Atherosclerosis*, *186*(1), 80–85.

Pardo, A., Ruiz, V., Arreola, J. L., Ramirez, R., Cisneros-Lira, J., Gaxiola, M., et al. (2003). Bleomycin-induced pulmonary fibrosis is attenuated in gamma-glutamyl transpeptidase-deficient mice. *American Journal of Respiratory and Critical Care Medicine*, *167*(6), 925–932.

PetitClerc, C., Shiele, F., Bagrel, D., Mahassen, A., & Siest, G. (1980). Kinetic properties of gamma-glutamyltransferase from human liver. *Clinical Chemistry*, *26*(12), 1688–1693.

Pham, A. N., Blower, P. E., Alvarado, O., Ravula, R., Gout, P. W., & Huang, Y. (2010). Pharmacogenomic approach reveals a role for the x(c)- cystine/glutamate antiporter in growth and celastrol resistance of glioma cell lines. *The Journal of Pharmacology and Experimental Therapeutics*, *332*(3), 949–958.

Rajpert-De Meyts, E., Heisterkamp, N., & Groffen, J. (1988). Cloning and nucleotide sequence of human gamma-glutamyl transpeptidase. *Proceedings of the National Academy of Sciences of the United States of America*, *85*(23), 8840–8844.

Rajpert-De Meyts, E., Shi, M., Chang, M., Robison, T. W., Groffen, J., Heisterkamp, N., et al. (1992). Transfection with gamma-glutamyl transpeptidase enhances recovery from glutathione depletion using extracellular glutathione. *Toxicology and Applied Pharmacology*, *114*(1), 56–62.

Rao, M. S., Lalwani, N. D., Scarpelli, D. G., & Reddy, J. K. (1982). The absence of gamma-glutamyl transpeptidase activity in putative preneoplastic lesions and in hepatocellular carcinomas induced in rats by the hypolipidemic peroxisome proliferator Wy-14,643. *Carcinogenesis*, *3*(10), 1231–1233.

Ravuri, C., Svineng, G., Pankiv, S., & Huseby, N. E. (2011). Endogenous production of reactive oxygen species by the NADPH oxidase complexes is a determinant of gamma-glutamyltransferase expression. *Free Radical Research*, *45*(5), 600–610.

Reuter, S., Schnekenburger, M., Cristofanon, S., Buck, I., Teiten, M. H., Daubeuf, S., et al. (2009). Tumor necrosis factor alpha induces gamma-glutamyltransferase expression via nuclear factor-kappaB in cooperation with Sp1. *Biochemical Pharmacology*, 77(3), 397–411.

Ripple, M. O., Henry, W. F., Rago, R. P., & Wilding, G. (1997). Prooxidant-antioxidant shift induced by androgen treatment of human prostate carcinoma cells. *Journal of the National Cancer Institute*, *89*(1), 40–48.

Robe, P. A., Martin, D. H., Nguyen-Khac, M. T., Artesi, M., Deprez, M., Albert, A., et al. (2009). Early termination of ISRCTN45828668, a phase 1/2 prospective, randomized study of sulfasalazine for the treatment of progressing malignant gliomas in adults. *BMC Cancer*, *9*, 372.

Rojas, E., Valverde, M., Kala, S. V., Kala, G., & Lieberman, M. W. (2000). Accumulation of DNA damage in the organs of mice deficient in gamma-glutamyltranspeptidase. *Mutation Research*, *447*(2), 305–316.

Roomi, M. W., Gaal, K., Yuan, Q. X., French, B. A., Fu, P., Bardag-Gorce, F., et al. (2006). Preneoplastic liver cell foci expansion induced by thioacetamide toxicity in drug-primed mice. *Experimental and Molecular Pathology*, *81*(1), 8–14.

Rosalki, S. B. (1975). Gamma-glutamyl transpeptidase. *Advances in Clinical Chemistry*, *17*, 53–107.

Ruoso, P., & Hedley, D. W. (2004). Inhibition of gamma-glutamyl transpeptidase activity decreases intracellular cysteine levels in cervical carcinoma. *Cancer Chemotherapy and Pharmacology*, *54*(1), 49–56.

Sawabu, N., Nakagen, M., Ozaki, K., Wakabayashi, T., Toya, D., Hattori, N., et al. (1983). Clinical evaluation of specific gamma-GTP isoenzyme in patients with hepatocellular carcinoma. *Cancer*, *51*(2), 327–331.

Schafer, F. Q., & Buettner, G. R. (2001). Redox environment of the cell as viewed through the redox state of the glutathione disulfide/glutathione couple. *Free Radical Biology & Medicine*, *30*(11), 1191–1212.

Schulman, J. D., Goodman, S. I., Mace, J. W., Patrick, A. D., Tietze, F., & Butler, E. J. (1975). Glutathionuria: Inborn error of metabolism due to tissue deficiency of gamma-glutamyl transpeptidase. *Biochemical and Biophysical Research Communications*, *65*(1), 68–74.

Scopelliti, A. J., Ryan, R. M., & Vandenberg, R. J. (2013). Molecular determinants for functional differences between alanine-serine-cysteine transporter 1 and other glutamate transporter family members. *The Journal of Biological Chemistry*, *288*(12), 8250–8257.

Shi, M., Gozal, E., Choy, H. A., & Forman, H. J. (1993). Extracellular glutathione and gamma-glutamyl transpeptidase prevent H2O2-induced injury by 2,3-dimethoxy-1,4-naphthoquinone. *Free Radical Biology & Medicine*, *15*(1), 57–67.

Shi, Z. Z., Han, B., Habib, G. M., Matzuk, M. M., & Lieberman, M. W. (2001). Disruption of gamma-glutamyl leukotrienase results in disruption of leukotriene D(4) synthesis in vivo and attenuation of the acute inflammatory response. *Molecular and Cellular Biology*, *21*(16), 5389–5395.

Sies, H., Bartoli, G. M., Burk, R. F., & Waydhas, C. (1978). Glutathione efflux from perfused rat liver after phenobarbital treatment, during drug oxidations, and in selenium deficiency. *European Journal of Biochemistry*, *89*(1), 113–118.

Slaga, T. J., Budunova, I. V., Gimenez-Conti, I. B., & Aldaz, C. M. (1996). The mouse skin carcinogenesis model. *The Journal of Investigative Dermatology Symposium Proceedings*, *1*(2), 151–156.

Stark, A. A., Russell, J. J., Langenbach, R., Pagano, D. A., Zeiger, E., & Huberman, E. (1994). Localization of oxidative damage by a glutathione-gamma-glutamyl transpeptidase system in preneoplastic lesions in sections of livers from carcinogen-treated rats. *Carcinogenesis*, *15*(2), 343–348.

Stark, A. A., Zeiger, E., & Pagano, D. A. (1993). Glutathione metabolism by gamma-glutamyltranspeptidase leads to lipid peroxidation: Characterization of the system and relevance to hepatocarcinogenesis. *Carcinogenesis*, *14*(2), 183–189.

Stenius, U., Rubin, K., Gullberg, D., & Hogberg, J. (1990). gamma-Glutamyltranspeptidase-positive rat hepatocytes are protected from GSH depletion, oxidative stress and reversible alterations of collagen receptors. *Carcinogenesis*, *11*(1), 69–73.

Takahashi, Y., Oakes, S. M., Williams, M. C., Takahashi, S., Miura, T., & Joyce-Brady, M. (1997). Nitrogen dioxide exposure activates gamma-glutamyl transferase gene expression in rat lung. *Toxicology and Applied Pharmacology*, *143*(2), 388–396.

Takahashi, Y., Takahashi, S., Yoshimi, T., Miura, T., Mochitate, K., & Kobayashi, T. (1997). Increases in the mRNA levels of gamma-glutamyltransferase and heme oxygenase-1 in the rat lung after ozone exposure. *Biochemical Pharmacology*, *53*(7), 1061–1064.

Takeuchi, S., Wada, K., Toyooka, T., Shinomiya, N., Shimazaki, H., Nakanishi, K., et al. (2013). Increased xCT expression correlates with tumor invasion and outcome in patients with glioblastomas. *Neurosurgery*, *72*(1), 33–41 discussion 41.

Tate, S. S., & Meister, A. (1978). Serine-borate complex as a transition-state inhibitor of gamma-glutamyl transpeptidase. *Proceedings of the National Academy of Sciences of the United States of America*, *75*(10), 4806–4809.

Tateishi, N., Higashi, T., Nomura, T., Naruse, A., & Nakashima, K. (1976). Higher transpeptidation activity and broad acceptor specificity of gamma-glutamyltransferases of tumors. *Gann*, *67*(2), 215–222.

Tew, K. D., & Townsend, D. M. (2012). Glutathione-s-transferases as determinants of cell survival and death. *Antioxidants & Redox Signaling*, *17*(12), 1728–1737.

Townsend, D. M., Deng, M., Zhang, L., Lapus, M. G., & Hanigan, M. H. (2003). Metabolism of Cisplatin to a nephrotoxin in proximal tubule cells. *Journal of the American Society of Nephrology*, *14*(1), 1–10.

Townsend, D. M., & Hanigan, M. H. (2002). Inhibition of gamma-glutamyl transpeptidase or cysteine S-conjugate beta-lyase activity blocks the nephrotoxicity of cisplatin in mice. *The Journal of Pharmacology and Experimental Therapeutics*, *300*(1), 142–148.

Traverso, N., Ricciarelli, R., Nitti, M., Marengo, B., Furfaro, A. L., Pronzato, M. A., et al. (2013). Role of glutathione in cancer progression and chemoresistance. *Oxidative Medicine and Cellular Longevity*, *2013*, 972913.

Vassallo, J. D., Hicks, S. M., Born, S. L., & Daston, G. P. (2004). Roles for epoxidation and detoxification of coumarin in determining species differences in clara cell toxicity. *Toxicological Sciences*, *82*(1), 26–33.

Vassallo, J. D., Kaetzel, R. S., Born, S. L., Lewis, C. L., Lehman-McKeeman, L. D., & Reed, D. J. (2010). Gamma-glutamyl transpeptidase null mice fail to develop tolerance to coumarin-induced Clara cell toxicity. *Food and Chemical Toxicology*, *48*(6), 1612–1618.

Vene, R., Castellani, P., Delfino, L., Lucibello, M., Ciriolo, M. R., & Rubartelli, A. (2011). The cystine/cysteine cycle and GSH are independent and crucial antioxidant systems in malignant melanoma cells and represent druggable targets. *Antioxidants & Redox Signaling*, *15*(9), 2439–2453.

Vina, J., Perez, C., Furukawa, T., Palacin, M., & Vina, J. R. (1989). Effect of oral glutathione on hepatic glutathione levels in rats and mice. *The British Journal of Nutrition*, *62*(3), 683–691.

Visvikis, A., Pawlak, A., Accaoui, M. J., Ichino, K., Leh, H., Guellaen, G., et al. (2001). Structure of the 5' sequences of the human gamma-glutamyltransferase gene. *European Journal of Biochemistry*, *268*(2), 317–325.

Warren, D. L., Brown, D. L., Jr., & Buckpitt, A. R. (1982). Evidence for cytochrome P-450 mediated metabolism in the bronchiolar damage by naphthalene. *Chemico-Biological Interactions*, *40*(3), 287–303.

Warzecha, C. C., Shen, S., Xing, Y., & Carstens, R. P. (2009). The epithelial splicing factors ESRP1 and ESRP2 positively and negatively regulate diverse types of alternative splicing events. *RNA Biology*, *6*(5), 546–562.

West, M. B., Chen, Y., Wickham, S., Heroux, A., Cahill, K., Hanigan, M. H., et al. (2013). Novel insights into eukaryotic gamma-glutamyl transpeptidase 1 from the crystal structure of the glutamate-bound human enzyme. *The Journal of Biological Chemistry*, *288*(44), 31902–31913.

West, M. B., Segu, Z. M., Feasley, C. L., Kang, P., Klouckova, I., Li, C., et al. (2010). Analysis of site-specific glycosylation of renal and hepatic gamma-glutamyl transpeptidase from normal human tissue. *The Journal of Biological Chemistry*, *285*(38), 29511–29524.

West, M. B., Wickham, S., Parks, E. E., Sherry, D. M., & Hanigan, M. H. (2013). Human GGT2 Does Not Autocleave into a Functional Enzyme: A Cautionary Tale for Interpretation of Microarray Data on Redox Signaling. *Antioxidants & Redox Signaling*, *19*(16), 1877–1888.

West, M. B., Wickham, S., Quinalty, L. M., Pavlovicz, R. E., Li, C., & Hanigan, M. H. (2011). Autocatalytic cleavage of human {gamma}-glutamyl transpeptidase is highly dependent on N-glycosylation at asparagine 95. *The Journal of Biological Chemistry*, *286*(33), 28876–28888.

Whitfield, J. B. (2001). Gamma glutamyl transferase. *Critical Reviews in Clinical Laboratory Sciences*, *38*(4), 263–355.

Wickham, S., Regan, N., West, M. B., Kumar, V. P., Thai, J., Li, P. K., et al. (2011). Divergent effects of compounds on the hydrolysis and transpeptidation reactions of gamma-glutamyl transpeptidase. *Journal of Enzyme Inhibition and Medicine Chemistry*, *27*, 476–489.

Wickham, S., Regan, N., West, M. B., Thai, J., Cook, P. F., Terzyan, S. S., et al. (2013). Inhibition of human gamma-glutamyl transpeptidase: Development of more potent, physiologically relevant, uncompetitive inhibitors. *The Biochemical Journal*, *450*(3), 547–557.

Wickham, S., West, M. B., Cook, P. F., & Hanigan, M. H. (2011). Gamma-glutamyl compounds: Substrate specificity of gamma-glutamyl transpeptidase enzymes. *Analytical Biochemistry*, *414*, 208–214.

Williams, G. M., Ohmori, T., Katayama, S., & Rice, J. M. (1980). Alteration by phenobarbital of membrane-associated enzymes including gamma glutamyl transpeptidase in mouse liver neoplasms. *Carcinogenesis*, *1*(10), 813–818.

Woodlock, T. J., Brown, R., Mani, M., Pompeo, L., Hoffman, H., Segel, G. B., et al. (1990). Decreased L system amino acid transport and decreased gamma-glutamyl transpeptidase are independent processes in human chronic lymphocytic leukemia B-lymphocytes. *Journal of Cellular Physiology*, *145*(2), 217–221.

Wright, E. C., Stern, J., Ersser, R., & Patrick, A. D. (1980). Glutathionuria: Gamma-glutamyl transpeptidase deficiency. *Journal of Inherited Metabolic Disease*, *2*(1), 3–7.

Xiong, Y., Uys, J. D., Tew, K. D., & Townsend, D. M. (2011). S-glutathionylation: From molecular mechanisms to health outcomes. *Antioxidants & Redox Signaling*, *15*(1), 233–270.

Xu, K., Meng, X. Y., Wu, J. W., Shen, B., Shi, Y. C., & Wei, Q. (1992). Diagnostic value of serum gamma-glutamyl transferase isoenzyme for hepatocellular carcinoma: A 10-year study. *The American Journal of Gastroenterology*, *87*(8), 991–995.

Yamada, K., Tsuji, T., & Kunieda, T. (2013). Phenotypic characterization of Ggt1(dwg/dwg) mice, a mouse model for hereditary gamma-glutamyltransferase deficiency. *Experimental Animals*, *62*(2), 151–157.

Yamamoto, S., Watanabe, B., Hiratake, J., Tanaka, R., Ohkita, M., & Matsumura, Y. (2011). Preventive effect of GGsTop, a novel and selective gamma-glutamyl transpeptidase inhibitor, on ischemia/reperfusion-induced renal injury in rats. *The Journal of Pharmacology and Experimental Therapeutics*, *339*(3), 945–951.

Yang, L., Fox, S. A., Kirby, J. L., Troan, B. V., & Hinton, B. T. (2006). Putative regulation of expression of members of the Ets variant 4 transcription factor family and their downstream targets in the rat epididymis. *Biology of Reproduction*, *74*(4), 714–720.

Yao, D. F., Huang, Z. W., Chen, S. Z., Huang, J. F., Lu, J. X., Xiao, M. B., et al. (1998). Diagnosis of hepatocellular carcinoma by quantitative detection of hepatoma-specific bands of serum gamma-glutamyltransferase. *American Journal of Clinical Pathology*, *110*(6), 743–749.

Yoshikawa, M., Tsuchihashi, K., Ishimoto, T., Yae, T., Motohara, T., Sugihara, E., et al. (2013). xCT inhibition depletes CD44v-expressing tumor cells that are resistant to EGFR-targeted therapy in head and neck squamous cell carcinoma. *Cancer Research*, *73*(6), 1855–1866.

CHAPTER FOUR

Pleiotropic Functions of Glutathione S-Transferase P

Jie Zhang[*], Christina Grek[*], Zhi-Wei Ye[*], Yefim Manevich[*], Kenneth D. Tew[†], Danyelle M. Townsend[‡,1]

[*]Department of Cell and Molecular Pharmacology and Experimental Therapeutics, Medical University of South Carolina, Charleston, South Carolina, USA
[†]Professor and Chairman, Department of Cell and Molecular Pharmacology, John C. West Chair of Cancer Research, Medical University of South Carolina, Charleston, South Carolina, USA
[‡]Department of Pharmaceutical and Biomedical Sciences, Medical University of South Carolina, Charleston, South Carolina, USA
[1]Corresponding author: e-mail address: townsed@musc.edu

Contents

Abstract

Glutathione S-transferase P (GSTP) is one member of the GST superfamily that is prevalently expressed in mammals. Known to possess catalytic activity through deprotonating glutathione allowing formation of thioether bonds with electrophilic substrates, more recent discoveries have broadened our understanding of the biological roles of this protein. In addition to catalytic detoxification, other properties so far ascribed to GSTP include chaperone functions, regulation of nitric oxide pathways, regulation of a variety of kinase signaling pathways, and participation in the forward reaction of protein S-glutathionylation. The expression of GSTP has been linked with cancer and other human pathologies and more recently even with drug addiction. With respect to human health, polymorphic variants of GSTP may determine individual susceptibility to oxidative stress and/or be critical in the design and development of drugs that have used redox pathways as a discovery platform.

Advances in Cancer Research, Volume 122
ISSN 0065-230X
http://dx.doi.org/10.1016/B978-0-12-420117-0.00004-9

1. INTRODUCTION

The glutathione S-transferase (GST) family of phase II detoxification enzymes has been critically studied for five decades. The cytosolic GST superfamily is composed of at least 7 classes that share >70% homology and have overlapping substrate specificities (Hayes & Pulford, 1995; McIlwain, Townsend, & Tew, 2006). In humans, a single gene on chromosome 11q13 codes for proteins designated in the human p class (GSTP1). The *GSTP1* gene spans 3 kb and encodes 210 amino acids in seven exons (Cowell, Dixon, Pemble, Ketterer, & Taylor, 1988). Polymorphisms at the *GSTP1* locus result in four alleles GSTP1*A–D that differ structurally and functionally (Lo & Ali-Osman, 1998). Initially named as a function of its high levels in human placenta, one characteristic of GSTP is that the homodimeric enzyme can undergo heterodimerization with other GST isoenzymes (Pettigrew & Colman, 2001) or other proteins (Ralat, Manevich, Fisher, & Colman, 2006). Distinctively, all mammalian GSTs bind GSH (at the so-called G-site) and activate it to a thiolate anion (GS^-) that is used to catalyze subsequent conjugation with various electrophiles (Fig. 4.1) (Graminski, Kubo, & Armstrong, 1989; Graminski, Zhang, Sesay, Ammon, & Armstrong, 1989). High levels of GSTP are found in a number of cancers and in cells resistant to anticancer drugs; factors determinate in targeting GSTP in anticancer drug discovery (Tew, 1994). Moreover, in addition to cancer, there are examples where GSTP has been linked to pathologies such as asthma, neurodegenerative diseases, and inflammatory conditions. The identification of human polymorphisms of GSTP now provides a basis for

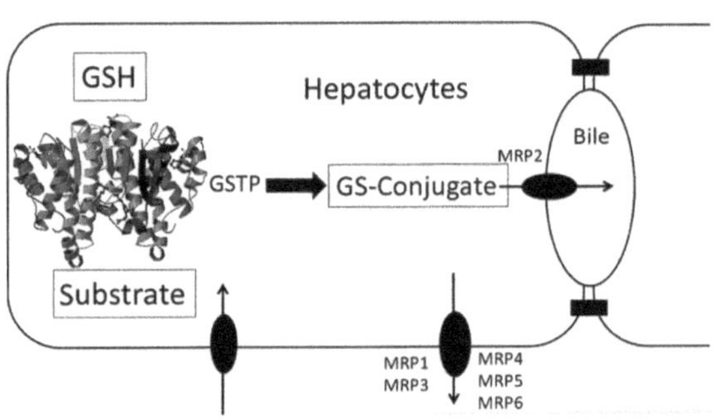

Figure 4.1 Representative example of phase II detoxification of an electrophilic compound via GSTP, with ATP-binding cassette transporter participation in efflux. (See the color plate.)

considering how individual and population differences may be associated with response to oxidative stress (Ali-Osman, Akande, Antoun, Mao, & Buolamwini, 1997; Pal, Hu, Zimniak, & Singh, 2000).

A functional overlap and redundancy invested by the GST isoenzyme family likely explain why the GSTP knockout mouse is viable and healthy, primarily presenting a phenotype with altered sensitivity to certain carcinogens and drugs (Henderson et al., 1998, 2000). Moreover, crosses between this mouse and p53 knockouts only mildly altered susceptibility to the development of spontaneous tumors (Gate, Majumdar, Lunk, & Tew, 2005). Many of the functional attributes of GSTP described in subsequent sections can also be coordinated by other isoenzymes (e.g., GSTA and GSTM), albeit with differing efficiencies. While much of the literature from the 1970s to the 2000s focused upon the detoxification properties of GSTs, more recent evidence has broadened the perspectives on how GSTP contributes to cellular redox homeostasis. The following sections provide information on these advances.

2. SUBCELLULAR DISTRIBUTION OF GSTP

There are many challenges in ascribing enzyme localization to specific organelles. Frequently, the methods used require separation conditions that lead to artifacts or contamination. Immunolocalization techniques also have intrinsic drawbacks. Nevertheless, within the mammalian GST superfamily, three structurally and evolutionarily distinct gene families have tissue- and species-specific distribution and expression patterns, and in some instances, their classifications provide evidence of their subcellular locale and cytosolic (cGST-alpha, cGST-mu, cGST-pi, cGST-omega, cGST-theta, cGST-delta, cGST-sigma, and cGST-zeta), mitochondrial (mGST-alpha, mGST-mu, mGST-pi, and mGST-kappa), and six membrane-bound microsomal (or membrane-associated proteins in eicosanoid and glutathione) transferases (Allocati, Federici, Masulli, & Di Ilio, 2009; Atkinson & Babbitt, 2009; Hayes, Flanagan, & Jowsey, 2005). Recent reports make these designations less rigid, having further characterized GSTs in the plasma membrane, outer mitochondrial membrane, and the nucleus and peroxisomes (Allocati et al., 2009; Gardner & Gallagher, 2001; Goto et al., 2001a; Hayes et al., 2005; Petit et al., 2009; Raza, 2011).

Despite classification as a cytosolic enzyme, GSTP expression has been found in a number of other cellular organelles. GSTP has been detected in human liver mitochondria and the cytoplasm, mitochondria, lysosomes, and nucleus of cancer cells (Ali-Osman, Brunner, Kutluk, & Hess, 1997; Goto et al., 2001b). Mitochondrial GSTP shared the same molecular size as cytoplasmic GSTP, supporting the notion that GSTP mitochondrial targeting does not involve protease processing. Mutation of arginine residues in the N-terminal mitochondrial targeting signal region of full-length GSTP abrogated mitochondrial distribution, suggesting a putative role for positively charged arginine residues in GSTP mitochondrial transport. Mitochondrial GSTP may be involved in protecting the organelle from oxidative stress via the suppression of cardiolipin peroxidation and cytochrome c release from the mitochondrial inner membranes or possibly catalyzing the formation of reactive aldehydes conjugated with GSH.

The presence of nuclear GSTP has been strongly and inversely correlated with patient prognosis in a number of cancers, including ovarian, breast, colon, glioma, lymphomas, and non-small cell lung carcinoma (Ali-Osman, Brunner, et al., 1997; Allen et al., 2007; Huang, Tan, Thiyagarajan, & Bay, 2003; Jankova et al., 2012; Rolland, Raharijaona, Barbarat, Houlgatte, & Thieblemont, 2010; Soh et al., 2005). Nuclear GSTP in Bcl-2-overexpressing breast tumors may gain entry through the nuclear pore and be linked to chemotherapy resistance. Nuclear GSTP immunologic activity is enhanced in cancer cells treated with the chemotherapeutic agents doxorubicin (DOX) and cis-diamminedichloroplatinum (II) (cisplatin) (Goto et al., 2001a; Rolland et al., 2010). Cancer cells lacking nuclear GSTP expression maintained sensitivity to DOX-mediated apoptosis. The authors suggested that subcellular localization of GSTP in these organelles may be important in metabolizing either DOX or cisplatin, catalyzing the formation of DOX–GSH adducts, thereby enhancing drug efflux, protecting against DNA damage, and contributing to the acquisition and maintenance of chemotherapeutic resistance. However, neither DOX nor cisplatin is a direct substrate of GSTP, and as a consequence, if these correlations are predictive, an alternative function for GSTP would be implicated. Maintenance of nuclear GSTP levels was independent of cytoplasmic levels, further implicating some role of specific nuclear protein transport system. Furthermore, treatment with a nuclear transport inhibitor led to increases in drug accumulation in the nucleus and the inhibition of GSTP nuclear translocation and DOX-stimulated GSTP nuclear activity (Goto et al., 2001a; Rolland et al., 2010).

A recent study has shown that GSTP can be detected in the endoplasmic reticulum (ER) of murine alveolar type II epithelial cells (Anathy et al., 2012). GSTP immunoprecipitation revealed the association of GSTP with the tumor necrosis factor (TNF) receptor family member Fas (CD9/Apo-1) in the ER. Stimulation with its ligand, FasL, enhanced GSTP/Fas interaction in a time-dependent manner, followed by complex translocation to the cytosol and the mediation of apoptotic signaling pathways. This suggests a specific role for GSTP in catalyzing the S-glutathionylation of a latent pool of Fas (Fas-SSG) in the ER. The oxidative processing of ER Fas and the interaction and subsequent translocation of GSTP/Fas-SSG were directly linked to apoptotic pathways involving the formation of death-inducing signaling complexes and increases in caspase activity. Inhibition of GSTP with the specific inhibitor TLK199 diminished the thiolase activity of GSTP and decreased Fas-SSG and caspase activation in epithelial cells. Pharmacological and genetic modulation of GSTP showed that GSTP-mediated S-glutathionylation of Fas plays a role in a bleomycin model of acute lung injury (Anathy et al., 2012). Such studies unveil a mechanism through which GSTP participates in regulation of ER proteins.

In general terms, S-glutathionylated proteins are found in a number of distinct subcellular locations indicating that specific organelle localization would facilitate this process. In this context, additional role for GSTP subcellular translocation in mediating protein S-glutathionylation and cell signaling pathways would seem to be likely.

3. GST REGULATION OF KINASE SIGNALING PATHWAYS

GSTP is present at high levels in many solid tumors (particularly ovarian, non-small cell lung, breast, liver, pancreas, and colon cancers and lymphomas) and in a wide range of drug-resistant cell lines and tumors (Tew, 1994). Although the catalytic detoxification function of GSTP is expected to alter drug sensitivity, the physiological significance of increased GSTP in tumor and drug-resistant cells remains unclear because in most instances, the selected drugs were not substrates of GSTP. From this, it can be inferred that the ligand-binding or chaperone properties of GSTs and the capacity of GSTP to regulate the forward reaction of S-glutathionylation (Townsend et al., 2009a) may be of more consequence than its catalytic properties alone. Supporting this conclusion, recent studies have demonstrated the regulatory role of GSTs in the mitogen-activated protein kinase (MAPK) pathways via their noncatalytic, ligand-binding activity. GSTP was the first isoenzyme to emerge as a regulator of redox-mediated kinase signaling (Adler, Yin, Fuchs, et al., 1999; Adler, Yin, Tew, & Ronai, 1999; Gate, Majumdar, Lunk, & Tew, 2004; Wang, Arifoglu, Ronai, & Tew, 2001). Specifically, GSTP binds directly to c-Jun N-terminal kinases (JNKs), which are at the end

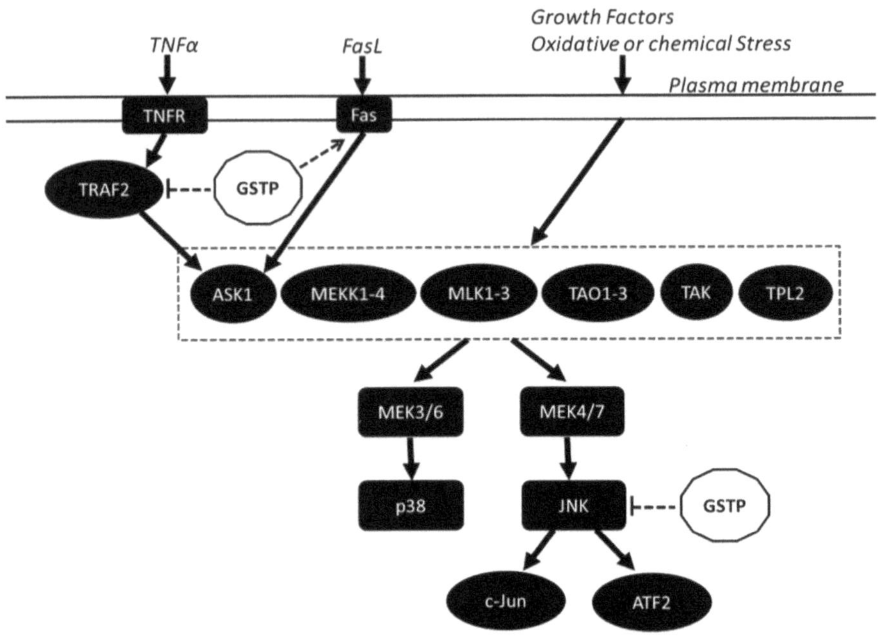

Figure 4.2 Some examples of how GSTP has been shown to participate in regulating various kinase signaling pathways. GSTP is a negative regulator in TNFα-induced MAPK signaling. It interacts physically with TRAF2, which blocks the interaction of TRAF2 and ASK1, attenuates ASK1 autophosphorylation, and in turn suppresses TNFα–ASK1–JNK/p38 signaling pathways. GSTP is also able to directly sequester JNK in a complex, thus preventing it from acting on downstream targets, c-Jun and ATF2. In contrast, GSTP can amplify Fas-induced MAPK signaling. Stimulation of Fas ligand increased the interaction of GSTP with Fas and ERp57 in the ER leading to Fas S-glutathionylation and subsequent mobilization from ER to cytosol, resulting in enhanced Fas–ASK1–JNK/p38 signaling pathways.

of the MAPK pathway, and acts as a negative regulator, thereby modulating proliferation and apoptosis (Fig. 4.2). JNKs are encoded by at least three genes (JNK1, JNK2, and JNK3), which are spliced alternatively into 10 JNK isoforms, including both a shorter C-terminus (JNK1α1, JNK1β1, JNK2α1, JNK2β1, and JNK3α1) and longer C-terminus (JNK1α2, JNK1β2, JNK2α2, JNK2β2, and JNK3α2) isoforms. Both JNK1 and JNK2 are expressed ubiquitously, while JNK3 is localized in the brain, testis, and heart (Davis, 2000; Gupta et al., 1996). The basal activity of JNK is necessarily maintained at low levels through its sequestration within a multiprotein complex that includes at least GSTP and JNK. The rise in ROS levels that occurs under conditions of oxidative or chemical stress results in a dissociation of JNK from GSTP and the subsequent activation of

JNK as indicated by phosphorylation of c-Jun and/or ATF2. This process further impacts downstream events, which in a cell- or tissue-specific manner can lead to events as divergent as proliferation or apoptosis (Adler, Yin, Fuchs, et al., 1999; Gate et al., 2004; Ruscoe et al., 2001). As such, GSTP serves as a sensor of intracellular changes in either endogenous or exogenous redox potential. It became apparent that this type of function is not restricted to GSTP and that other small redox-active proteins provide a level of redundancy in interacting with specific intracellular signaling molecules and regulate their activity in a redox-dependent fashion. Consistent with this notion, thioredoxin has also been found to interact with, and inhibit, apoptosis signal-regulated kinase 1 (ASK1) activity (Saitoh et al., 1998).

The ligand-binding properties of GSTP seem to be independent of its detoxification activity since mutation at critical sites in GSTP (tyrosine 7 required for catalytic proton transfer and cysteine 47 or 101 required for GSTP dimerization), while abrogating its catalytic activity, has been shown to be as effective as wild-type GSTP in inhibiting JNK activity in mouse embryo fibroblast (MEF) cells from $GSTP^{-/-}$ mice (Adler, Yin, Fuchs, et al., 1999). However, *Telintra*, a GSH peptidomimetic drug, or TLK199 (γ-glutamyl-*S*-(benzyl)cysteinyl-*R*-phenyl glycine diethyl ester), a select GSTP inhibitor, does inhibit the kinase regulatory activity of GSTP in mouse fibroblast 3T3/4A cells (Adler, Yin, Fuchs, et al., 1999) and human myeloid leukemia HL60 cells (Ruscoe et al., 2001). Moreover, NBDHEX (6-(7-nitro-2, 1, 3-benzoxadiazol-4-ylthio) hexanol), another GST inhibitor that forms a σ-complex with GSH in the active site of GSTP, also abolished the interaction between GSTP and JNK1α2 (De Luca, Federici, De Canio, Stella, & Caccuri, 2012). In this context, NBDHEX induces the dissociation of GSTP1-1/JNK1 heterocomplex and the subsequent phosphorylation of JNK1, leading to cell cycle arrest and apoptosis in several tumor cell lines (Turella et al., 2005). Inhibitor binding likely induces changes in the conformation of one or more domains of GSTP and therefore interferes with the GSTP interaction with JNK. Indeed, evidence suggests that the C-terminal domain (residues 194–201) of GSTP, one of the most flexible regions after inhibitor binding, is critical for GSTP and JNK interaction (Monaco et al., 1999). Moreover, the extended C-terminal region (residues 200–424) of JNK was also shown to be critical in this interaction (9), suggesting that the longer JNK isoforms might preferentially interact with GSTs and that the capacity of GSTP to suppress JNK activity could be through an allosteric inhibition mechanism, since the catalytic kinase domain is localized in the N-terminal of JNK. However, similar dissociation constants were reported in a recent study using JNK1Δ or JNK1α2, suggesting that the long C-terminal tail that distinguishes JNK1α2 from the shorter splicing variant JNK1α1 is not implicated in GSTP recognition (De Luca et al., 2012). Evidence has recently been presented to indicate that GSTP binds strongly with the activated phosphorylated form of JNK1 than with JNK2, while the interaction with the unphosphorylated, inactive JNK occurs only in the presence of the substrate ATF2. A direct interaction between GSTP and ATF2 was also demonstrated and explained the ability of GSTP to inhibit JNK catalytic activity as a consequence of the competition between GSTP and the activated JNK for the substrate ATF2 (Thevenin, Zony, Bahnson, & Colman, 2011). Contrary to this, De Luca et al. (2012) found that GSTP is able to form a complex with the unphosphorylated and inactive JNK1α2 isoform, even in the absence of the substrate. The complex strongly reduced the extent of activation of JNK1α2 and preserved GSTP from inactivation. They also showed that GSH exerted a negative effect on the affinity of GSTP for JNK1α2, suggesting that the intracellular levels of this thiol may participate in fine-tuning of the MAPK signaling pathways. Moreover, there are indications that GSTP plays an anti-inflammatory role by preventing lipopolysaccharide-induced production of proinflammatory factors, suggested to occur through the ERK, JNK and p38, and NfkB axes of regulation (Xue et al., 2005). GSTP is also important in the regulation of the transcriptional activity of Stat3 (signal transduction and activator of transcription) and a regulator of the cell cycle via its influence on EGF signaling (Kou, Chen, Feng, Luo, & Yin, 2013).

The stoichiometry of the GSTP–JNK interaction has been difficult to assess by standard procedures. In spite of the catalytic function of GSTP being dependent upon dimerization, in some instances, it was assumed that the monomeric form of GSTP would interact with JNK (Adler, Yin, Fuchs, et al., 1999; Bernardini et al., 2000). A recent study (Gildenhuys et al., 2010) supports the principle that the JNK–GSTP interaction occurs via the dimer of the latter. Using equilibrium folding and unfolding kinetics experiments, as well as molecular modeling, the thermodynamic instability of the GSTP monomer was shown, with the implication that the dimeric GSTP binds to JNK. At this stage, further studies will be required to determine the precise details underlying those mechanisms involved in these reactions. It is worth reflecting that understanding these processes will also be complicated by the fact that different polymorphic variants of GSTP, with differences in residues 105 and 114, triggered different effects. GSTP*C (Val105/Val114) was shown to be the only GSTP haplotype that was able to inhibit JNK apoptotic activity *in vivo* (Holley et al., 2007). More recently, it was demonstrated that GSTP*C is a more potent inhibitor of JNK activity than the wild-type GSTP*A (Ile105/Ala114) (Thevenin et al., 2011). Such factors when considered in the context of human populations will affect susceptibility traits to stress response through these JNK-mediated pathways.

GSTP also has a negative regulatory role in regulating tumor necrosis factor-alpha (TNFα)-induced MAPK signaling. This occurs through the formation of ligand-binding or chaperone interactions with tumor necrosis factor receptor-associated factor 2 (TRAF2), which blocks the interaction of TRAF2 and ASK1, attenuates ASK1 autophosphorylation, and in turn suppresses ASK1–MEK–JNK/p38 signaling pathways, inhibiting apoptosis. Obverse to this situation, reduced levels of GSTP increased TNFα-dependent TRAF2–ASK1 associations, activating both ASK1 and JNK/p38. Similar to the GSTP–JNK interaction, the GSH conjugation activity of GSTP is not necessary for GSTP–TRAF2 binding since mutant GSTP lacking the catalytic tyrosine (Y7F) also represses ASK1 and the interaction between GSTP and TRAF2 is observed only in unstimulated cells. The C-terminal TRAF domain of TRAF2 and the C-terminal TRAF2-binding motif (TWQE amino acids 38–41) in GSTP seem necessary for the binding of TRAF2 to GSTP. A mutant GSTP lacking the TRAF domain-binding motif exhibited a significant decline of capacity to bind TRAF2 and block TRAF2–ASK1 signaling compared with the wild-type GSTP (Wu et al., 2006).

A further indication of functional redundancy is provided by the fact that the direct interaction of JNK with GSTs is not limited to just GSTP. Indeed, in addition to GSTP, GSTA and GSTM were also capable of associating with JNK complex *in vitro* (Adler, Yin, Fuchs, et al., 1999). GSTP exhibited greater JNK inhibitory activity than did GSTM1-1, which was more potent than GSTA1-1. Furthermore, Romero et al. (2006) had shown that GSTA1-1 interacts physically with JNK to suppress activation of JNK signaling by a proinflammatory cytokine and oxidative stress in Caco-2 cells. They showed that GSTA1-1 levels were higher in postconfluent than in preconfluent cells and that JNK activation was significantly reduced in the former when exposed to stress conditions including interleukin-1β, H_2O_2, and UV irradiation, implying a protective role for GSTA1-1 in JNK-associated apoptosis. In a different study, Desmots et al. (Desmots, Loyer, Rissel, Guillouzo, & Morel, 2005) demonstrated that GSTA4-4 and JNK coimmunoprecipitate in mouse liver tissue and GSTA4-4 expression was inversely correlated with JNK activities in primary hepatocyte cultures, implying that GSTA4-4 might be an endogenous regulator of JNK activity by direct binding. However, they showed a correlation between increased expression of GSTA4-4 and JNK phosphorylation during hepatocyte isolation and TNFα treatment, suggesting a proapoptotic role for GSTA4-4 in JNK-associated apoptosis. It seems likely that GSTs may be ligand-bound with other upstream receptors, kinases, and/or phosphatases, thus regulating their activities and leading to phosphorylation of JNK. As mentioned earlier (Anathy et al., 2012), GSTP can amplify Fas ligand (FasL)-induced apoptosis in epithelial cells. The stimulation of FasL increased the interaction of GSTP with Fas and ERp57, a member of the TNF receptor superfamily and a protein disulfide isomerase, in the ER leading to Fas S-glutathionylation and subsequent mobilization from the ER to the cytosol, resulting in enhanced apoptosis. Knockdown or inhibition of ERp57 and GSTP enhanced survival. As such, the pleiotropic functions of GSTP are extended to the Fas–ASK1–JNK/p38 signaling pathways as well. These findings indicate complex mechanisms of kinase signaling pathway regulation by GSTs that may serve as checks and balances for commitment to apoptosis versus survival under different stimuli.

4. GSTP IN REDOX REGULATION AND S-GLUTATHIONYLATION

Posttranslational modification of proteins is an elegant means by which cells have evolved a high degree of functional redundancy through reversible cycling reactions, enhancing protein function in response to varying stimuli in a cell- or tissue-specific manner. In general, kinase signaling pathways have been accepted as one of the cornerstones of signal transduction. The phosphorylation cycle is regulated by kinases and phosphatases that can function as either generalist or specialist enzyme. Redox signaling is gaining traction and is now recognized as physiologically important, and its dysregulation is attributed to a variety of pathologies (Grek, Zhang, Manevich, Townsend, & Tew, 2013; Tew & Townsend, 2011b; Townsend, 2007; Townsend, Tew, & Tapiero, 2003; Xiong, Uys, Tew, & Townsend, 2011). S-glutathionylation is a posttranslational modification where GSH is conjugated to cysteine residues (P-SSG). Regulation through S-glutathionylation has been ascribed to a large number of proteins that were designated into the following clusters: cytoskeletal, glycolysis/energy metabolism, kinase signaling pathways, calcium homeostasis, antioxidant enzymes, and protein folding (Townsend, 2007) (Fig. 4.3). Here, we will draw comparisons

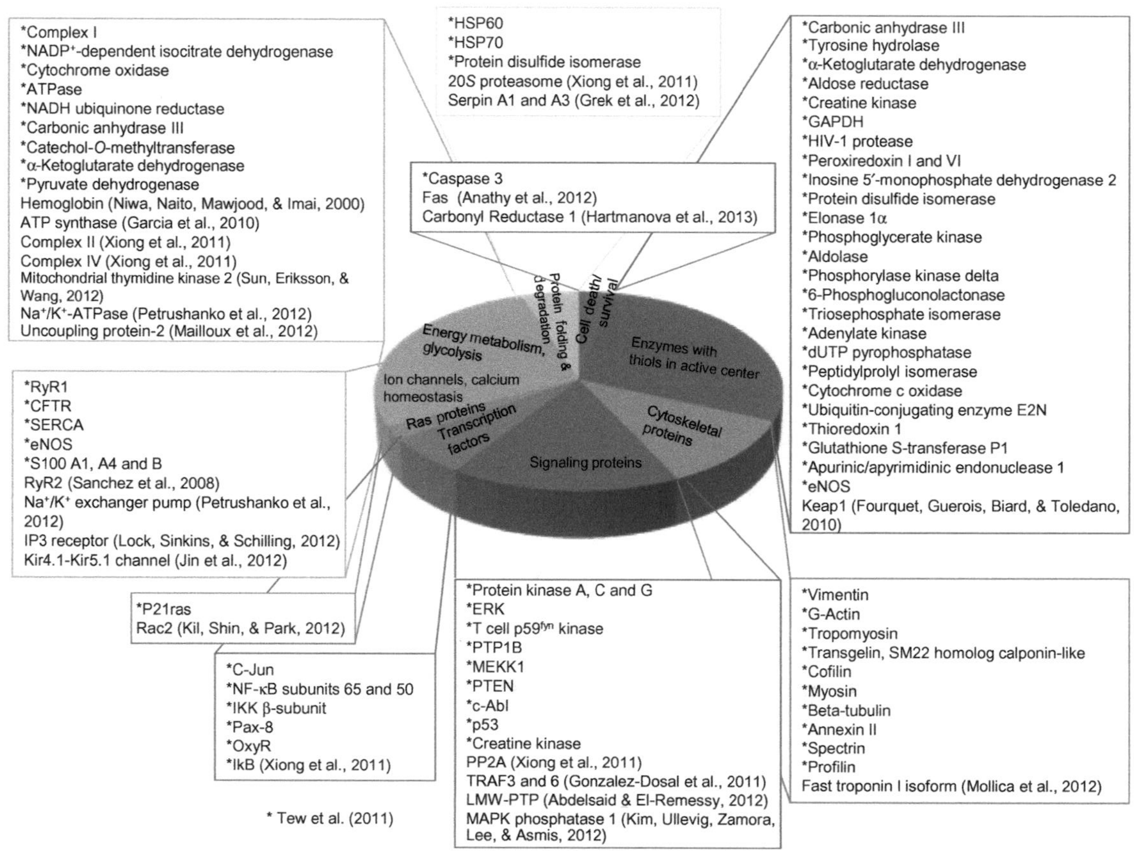

Figure 4.3 Representation from the literature of proteins clustered into functional groups that are known to be susceptible to S-glutathionylation. (See the color plate.)

of the S-glutathionylation cycle with examples of thiolases and dethiolases having both generalist and specialist functions.

5. S-GLUTATHIONYLATION REACTIONS

Glutathione is the most prevalent thiol in the cell, and the GSH:GSSG ratio has been used to describe the redox potential. Although recently, Flohe has revisited much of the earlier material that defines GSH:GSSG ratios and posited that small rate constants of related spontaneous reactions when compared with enzyme catalysis define the superiority of kinetic parameters over electrochemical or thermodynamic ones for an in-depth understanding of GSH-dependent biological phenomena. He concludes that at best, the GSH:GSSG potential might be useful analytically to disclose disturbances in redox metabolism, but more importantly, the enzyme-catalyzed reactions are more relevant to the understanding cellular redox homeostasis (Flohe, 2012).

The emerging role of S-glutathionylation underscores the importance of including the protein-associated pool of GSH in evaluating redox homeostasis. The redox potential is decreased following oxidative or nitrosative stress induced by physiological or pathological conditions. It is not surprising that cells have evolved a redox-sensitive subproteome to trigger signaling cascades following exposure to reactive oxygen (ROS) and nitrogen (RNS) species. Both ROS and RNS lead to oxidation of redox-sensitive proteins and subsequent S-glutathionylation (Fig. 4.4). Not all cysteine residues are targets for S-glutathionylation. Rather, cysteines with a low p*Ka* have the nucleophilicity that under oxidative/nitrosative conditions can be oxidized to a thiyl radical (RS) with strong reactivity toward oxygen followed by addition of GSH. S-glutathionylation leads to an increase of 305 MW and a net negative charge the consequences of which impact protein structure, function, and subcellular localization.

In vitro studies have shown that S-glutathionylation reactions can occur spontaneously with ROS/RNS generating compounds in the presence of GSH. In cells, several enzymes have been shown to promote the forward reaction of S-glutathionylation, for example, thiolase activity. At present, the details regarding specificity of thiolase enzymes are limited. GSTP is among those enzymes identified to date that have been shown to promote S-glutathionylation reactions.

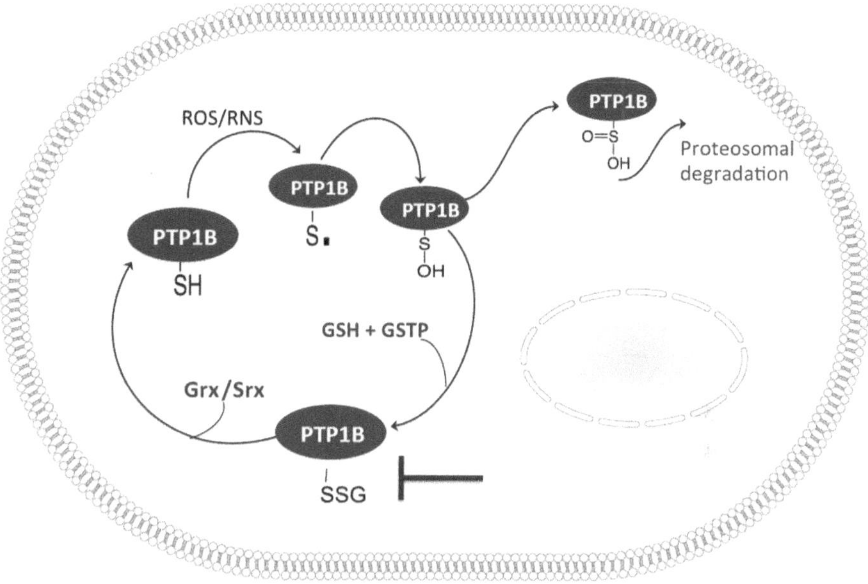

Figure 4.4 S-glutathionylation cycle of redox sensors. PTP1B is a representative protein with a low-p*K*a residue (redox sensor) that is a target for oxidative or nitrosative stress. Cysteine residues within redox sensors can be oxidized to form protein sulfenic (P-OH) and sulfinic (P-OOH) acids. Protein S-glutathionylation (P-SSG) reactions can be spontaneous or are mediated by GGT, Grx, or GSTP. P-SSG proteins have a wide variety of functions in cellular physiology/pathology.

6. S-GLUTATHIONYLASE ACTIVE PROTEINS

A large proportion of S-glutathionylated proteins that have been identified are compartmentalized intracellularly (Townsend, 2007). However, evidence suggests that extracellular and cell surface proteins can be S-glutathionylated and impact the proliferative/apoptotic balance. Gamma-glutamyl transpeptidase (GGT) is a cell surface enzyme that hydrolyzes GSH to yield glutamate and cysteinylglycine (CG) that can more readily enter cells. As such, GGT impacts intracellular redox homeostasis through enhancing intracellular GSH content. However, GGT plays an active role in the S-glutathionylation cycle (Corti et al., 2005; Pompella, 2005). Both GSH and CG have been shown to form mixed disulfides in proteins in the presence of activated GGT; however, nearly 85% are GSH-bound (Corti et al., 2005; Pompella, 2005). Acivicin, a GGT inhibitor, decreased the formation of protein mixed disulfides. The specificity and targets of GGT-mediated S-glutathionylation remain poorly understood despite the key role GGT plays in the liver and at the blood–brain barrier. With regard to pathophysiology, GGT is overexpressed in a wide variety of cancer cells and is associated with multidrug resistance. Further exploration into the role of GGT in promoting S-glutathionylase reactions of extracellular and cell surface proteins will be insightful to normal physiology and pathophysiology.

Glutaredoxin (Grx) is a member of the thioredoxin superfamily and has thiol disulfide oxidoreductase activity. Grx is widely recognized in the S-glutathionylation cycle as a deglutathionylating enzyme and will be discussed as such in the following section (for review see Mieyal, Starke, Gravina, Dothey, & Chung, 1991; Mieyal, Starke, Gravina, & Hocevar, 1991; Shelton, Chock, & Mieyal, 2005). There are, however, limited conditions in which Grx1 and Grx2 have demonstrated S-glutathionylase activity. Specifically, Grx-mediated S-glutathionylation utilized GS·as the proximal donor with limited redox-sensitive targets, including GAPDH (Gallogly, Starke, Leonberg, Ospina, & Mieyal, 2008). GAPDH is among the few proteins that are S-glutathionylated on cysteine residues not in a basic environment and consequently lacks the characteristic low p*K*a (Peskin & Winterbourn, 2006). Grx appears to be a "specialist" in terms of glutathionylase activity, and it is plausible that the unique p*K*a property in GAPDH may be important to the limited substrate specificity of Grx thiolase activity, whereas other characteristic low-p*K*a cysteines that are modified serve as the generalist/promiscuous substrate for deglutathionylase activity.

GSTP was shown to promote S-glutathionylation reactions and may function as a promiscuous/generalist S-glutathionylase (Anathy et al., 2012; Hutchens, Manevich, He, Tew, & Townsend, 2011; Townsend et al., 2009b). Using MEF from *Gstp1p2* wild-type and *Gstp1p2*-deficient mice, a range of redox-sensitive proteins were subject to S-glutathionylation following treatments with either ROS- or RNS-mediating agents (Townsend et al., 2009b). HEK293 cells transfected with GSTP-WT and a catalytic inactive mutant, Y7F, validated that the catalytic activity is required for S-glutathionylation. Polymorphisms within GSTP impact enzyme activity and may have important implications in an individual's ability to respond to ROS/RNS, a causative factor in multiple disease states (Townsend et al., 2003). The S-glutathionylation activity of GSTP is autoregulated, much like kinases. Specifically, downregulation of thiolase activity is attributed to S-glutathionylation of GSTP on cys47 and cys101, which acts as an oligomer switch and can cause multimerization of the protein (Townsend et al., 2009b). While GSTP has a wide variety of substrates for S-glutathionylase activity, the cellular consequences of many have not been vetted. However, S-glutathionylation of peroxiredoxin (Prdx) and Fas ligand is attributed to GSTP thiolase activity (Anathy et al., 2012; Manevich, Feinstein, & Fisher, 2004; Manevich, Hutchens, Tew, & Townsend, 2012; Ralat, Misquitta, Manevich, Fisher, & Colman, 2008).

Peroxiredoxins are non-seleno peroxidases with two cysteine-containing active sites that function to reduce hydrogen peroxide through redox reactions with Trx (Rhee, Kang, Chang, Jeong, & Kim, 2001). Peroxiredoxin VI (Prdx6) is a singular catalytic cysteine-containing peroxiredoxin (Choi, Kang, Yang, Rhee, & Ryu, 1998). It is a bifunctional enzyme with phospholipid hydroperoxide peroxidase and acidic Ca^{2+}-independent phospholipase A_2 (PLA_2) activities (Manevich & Fisher, 2005b). Inactivation of Prdx6 occurs following oxidative stress via formation of sulfinic acid on the catalytic cysteine residue. Catalytic cysteines of multiple enzymes are redox-sensitive and often are buried inside of protein globule to provide a unique environment for catalysis. Commonly, these crucial cysteines are inaccessible to the bulk and polarity of GSH. Recently, it was shown that GSTP could deliver a reducing equivalent (GS^-) to the catalytic Cys47 sulfenate (buried inside of protein globule) through its heterodimerization with Prdx6 (Manevich et al., 2004). The evidence of Prdx6–GSTP heterodimerization has been proven by its chromatographic purification and N-terminal sequencing showing equimolar amounts of two proteins in this complex (Ralat et al., 2008). The Prdx6–GSTP heterodimer has been further characterized by determining its molecular mass (52,200 Da, pH 8.0) by dynamic light scattering and by demonstration (using the deletion mutants) that two sites of GSTP, 41–85 and 115–124, are critical for its formation (Ralat et al., 2006, 2008). The delivery of reducing equivalents resulted in transient S-glutathionylation of Prdx6, which affects its folding and brings together disulfide bonds of Prdx6, S-glutathionylated Cys47, and catalytic Cys47 of GSTP (Manevich & Fisher, 2005a; Ralat et al., 2006). As a result of the close proximity of these sites, a disulfide-based heterodimer of Prdx6 and GSTP was formed (Ralat et al., 2006). This influences folding of both proteins and results in exposure of disulfide bond to GSH and, consequently, in sequential reduction of both Prdx6 and GSTP catalytic cysteines. The overall result of this heterodimer-originated and GSTP-catalyzed

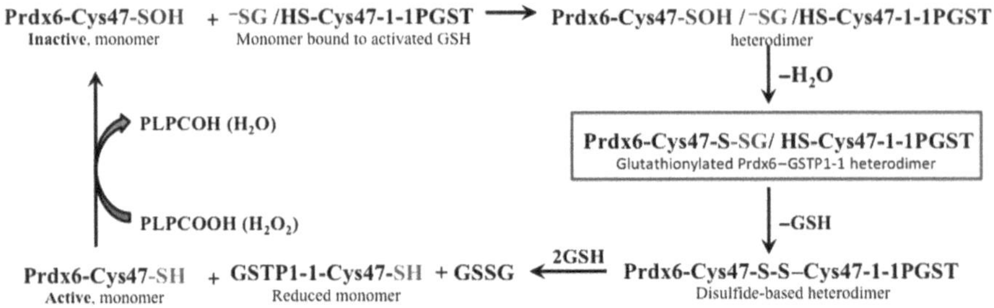

Figure 4.5 Catalytic cycle of Prdx6 activation by GSTP1. Reduction of phospholipid hydroperoxide (PLPCOOH) or H_2O_2 by Prdx6 results in oxidation of its catalytic Cys47 to sulfenic acid (shown in red). This oxidized monomer of Prdx6 forms a heterodimer with thiolate anion-bearing (shown in blue) GSTP. The spontaneous reaction of the thiolate anion with the catalytic Cys47 sulfenate of Prdx6 results in the S-glutathionylation of the latter (shown in box; the critical step of Prdx6 activation). Alignment of the S-glutathionylated catalytic Cys47 of Prdx6 with the catalytic Cys47 of GSTP results in the formation of a disulfide-based heterodimer. GSH access to the disulfide bond results in catalytic cysteine reduction/activation (shown in green) and heterodimer dissociation. (See the color plate.)

glutathionylation-driven reduction is activation of both Prdx6 and GSTP. The catalytic cycle of GSTP-mediated activation of Prdx6 is presented in Fig. 4.5.

7. DEGLUTATHIONYLASE ACTIVE PROTEINS

Redox signaling events are triggered under physiological levels of oxidative and nitrosative stresses, and the temporal fluxes in homeostasis are restored through deglutathionylation of redox sensors and increased GSH/GSSG levels. In the forward reaction of the S-glutathionylation cycle, GSTP plays a role as a generalist with a multitude of target proteins, while Grx and GGT may have more specialized roles. This paradigm follows through in the reverse reaction, deglutathionylation. Grx mediates the removal of GSH through direct thiol disulfide exchange reactions that lead to the formation of GSSG that is subsequently reduced by glutathione reductase. Grx is a generalist deglutathionylase with a broad spectrum of substrates, dysregulation of some of which can lead to different pathologies (Shelton et al., 2005).

Sulfiredoxin (Srx) is an antioxidant enzyme with oxidoreductase activity. The primary function of Srx is in the metabolism and reactivation of Prdx's (Baek et al., 2012; Jeong, Bae, Toledano, & Rhee, 2012). Overexpression of Srx in mammalian cells demonstrated that the enzyme has a dual function, including deglutathionylase activity (Bowers, Manevich, Townsend, & Tew, 2012; Findlay, Tapiero, & Townsend, 2005; Findlay et al., 2006; Lei, Townsend, & Tew, 2008; Park, Mieyal, Rhee, & Chock, 2009). Substrates for Srx-mediated deglutathionylation are limited but include Prdx, PTP1B, and actin (Findlay et al., 2006; Park et al., 2009). Srx contains a single cysteine residue in the active site, whereas Grx has a CXXC motif, suggesting distinct molecular mechanism of deglutathionylation. Park et al. evaluated the specificity of the two enzymes toward Prdx1 and determined that Srx has a greater binding affinity than Grx toward Prdx-SSG (Park et al., 2009). Collectively, these data suggest that Grx and Srx play distinct roles in the S-glutathionylation cycle that can be regulated through substrate specificity and/or by tissue and subcellular distribution.

8. GSTP, NITRIC OXIDE SYNTHASES, AND NO HOMEOSTASIS

Under both normal and pathological conditions, nitric oxide (NO) is a critical secondary messenger in controlling cellular signaling events (Isenberg, Martin-Manso, Maxhimer, & Roberts, 2009). NO is generated intracellularly by nitric oxide synthases (NOS) (Stuehr, 1997). It is a reactive molecule (radical) and may either directly or through reaction with molecular oxygen (generating N_2O_3 with high affinity for thiols; Stamler, Singel, & Loscalzo, 1992) or superoxide anion radicals (generating peroxynitrite; Beckman, Beckman, Chen, Marshall, & Freeman, 1990) lead to nitrosylation or S-glutathionylation of protein thiols (West, Hill, Xuan, & Bhatnagar, 2006). The thiol reaction with peroxynitrite may also cause generation of thiyl radicals or sulfenates (Filipovska & Murphy, 2006), creating a situation where protein glutathionylation status may be independent of the global GSH:GSSG ratio. Dinitrosyl iron complexes (DNIC) ($[(RS^-)_2Fe^+(NO^+)_2\ldots(^-SR)_2]^-$) incorporate NO, increasing its stability and enhancing its target specificity (Martinez & Andriantsitohaina, 2009; Vanin, 2009). Generally, nitrosative stress (RNS) is a composite of nitric oxide (NO) and nitric oxide-derived compounds, including nitroxyl anion (NO^- and HNO), nitrosonium cation (NO^+), higher oxides of nitrogen (N_2O, NO_2, and N_2O_3), peroxynitrite ($ONOO^-$/ONOOH), *S*-nitrosothiols (RSNO), and DNIC.

In the presence of O_2, nitric oxide can react with glutathione sulfhydryls resulting (most likely through intermediate(s) of higher nitrogen oxides) in formation of GSNO, which serves as a physiological scavenger of the NO_x species (Wink et al., 1994). GSNO, in turn, can react with GSH yielding

GSSG, nitrite, nitrous oxide, and ammonia as end products (Singh, Wishnok, Keshive, Deen, & Tannenbaum, 1996). High intracellular concentrations of GSH can shift the equilibrium of the reaction with NO to GSSG, but site-specifically, some GSNO can still be present. Recent developments with stabilized DNIC–glutathione preparations allow retention of the protein S-nitrosylation for at least 1 year, and these may be precursors for a variety of broad-spectrum NO-containing drug moieties (Vanin, 2009).

9. GSTP BINDING OF NITRIC OXIDE CARRIERS

Using crystallization, isothermal titration, and differential scanning calorimetry techniques, GSNO was shown to bind strongly to the G-site of homodimeric human GSTP. The nitrosyl moiety of bound GSNO is involved in multiple interactions with the protein. The binding of GSNO to wild-type GSTP induces negative cooperativity with the kinetics concomitant with the binding at physiological temperatures. GSNO inhibits the wild-type GSTP enzyme competitively at low temperatures, but covalently at higher temperatures. Because the C47S mutation prevents this covalent modification, S-nitrosylation of the Cys47 sulfhydryl has been implicated. This is compatible with the crucial role of the Cys47 in intersubunit substrate-binding cooperativity and with the physiological role of Cys47 S-nitrosylation by GSNO in certain pathologies (Tellez-Sanz et al., 2006). Both GSNO and DNIC can participate in storage and transport of NO in biological systems. *In vitro*, without GSH, GSNO induces rapid and stable S-nitrosylation of GSTP Cys47 (101). MS analysis failed to show any S-glutathionylation under those conditions, which presumably indicates an NO (N_xO_y) release from GSNO. S-nitrosylation of GSTP at Cys47 decreases ($\sim 10\times$) its affinity for GSH. Partial protection of one Cys47 residue of homodimeric GSTP from S-nitrosylation minimizes loss of its enzymatic activity. In living cells, intracellular GSH is generally present at millimolar levels, and another type of NO sequestration—dinitrosyl–diglutathionyl–iron complex (DNDGIC, formed by GSNO decomposition in the presence of traces of ferrous ions) becomes predominant (Lo Bello et al., 2001). *In vivo*, DNDGIC can serve as an NO donor and iron/GSH transporter (Vasieva, 2011). Through one of its glutathione moieties, DNDGIC binds tightly ($K_i < 10^{-12}$ M) to the G-site of GSTP. The other GSH moiety is displaced by the active site Tyr7, stabilizing the enzyme-inhibitor complex. The NO moieties in this complex are stabilized by interaction with

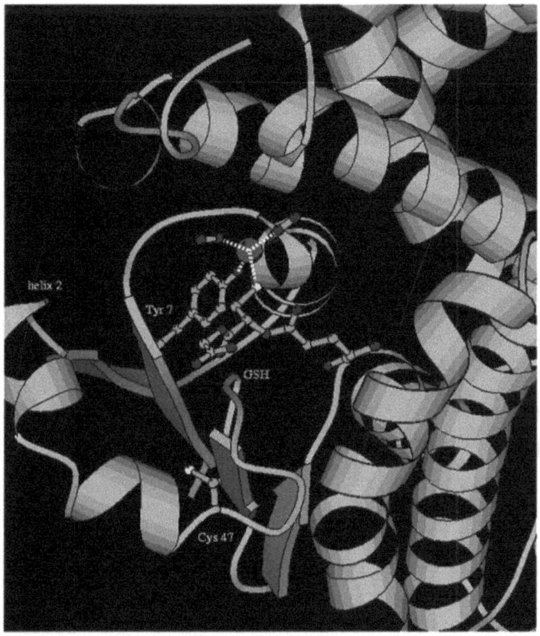

Figure 4.6 Ribbon depiction of the DNGIC/GSTP1-1 complex. The close-up view of the active site of the covalent DNGIC (as obtained after removal of the excess GSH) is shown. The model shows that (i) one of the GSH ligands of DNDGIC can dock into the G-site and adopt the canonical extended conformation seen in crystal structures of GST–GSH complexes, (ii) Tyr7 is close enough to displace the other GSH ligand to generate a stable enzyme-inhibitor complex, and (iii) the NO moieties of the complex form van der Waal's interactions with Ile104 and Tyr108. In addition, there are possible polar interactions with Tyr108 and the main chain nitrogen of Gly205. The iron atom is depicted as an *orange sphere*, oxygen atoms are colored *red*, nitrogen atoms are *blue*, sulfur atoms are *yellow*, and carbon atoms are *green*. This figure was drawn using MOLSCRIPT. *Reproduced from Lo Bello et al. (2001) with permission.* (See the color plate.)

Ile104 and Tyr108 (Fig. 4.6). The half-life of DNDGIC is ~1 min, but in the complex with GSTP, this is substantially increased (~3 h). This binding induces a negative cooperativity in other unoccupied monomer G-sites of the GSTP homodimer, downregulating its basal activity (Lo Bello et al., 2001). In addition, DNDGIC is an irreversible inhibitor of glutathione reductase (GR), although this inhibition was not associated with DNDGIC–GR complexes and/or nitrosation of the active site cysteines 58 and 63. Simultaneously, one NO moiety with acid lability was present per molecule of inhibited monomeric enzyme (Boese, Keese, Becker, Busse, & Mulsch, 1997). Such results indicate that GSTP can serve as a cellular NO donor and/or carrier potentially regulating the activities of other (GR) thiol redox-modifying enzymes.

10. GSTP-MEDIATED SITE-SPECIFIC PROTEIN NITROSYLATION/GLUTATHIONYLATION

When GSTP forms heterodimers with target proteins (Prdx6; Ralat et al., 2006, 2008), it can enable the delivery of GS^- to cysteines that may otherwise be inaccessible to GSH. GSNO (DNDGIC) bound to GSTP monomers could also modify target protein affinity and impact posttranslational modifications of cysteines. For example, with GSNO bound to the G-site of GSTP, the NO group of the complex is on the surface, in proximity to the Cys47 of GSTP. One could anticipate that NO donor–GSTP complexes may overcome steric constraints and provide specificity for fine-tuning regulation of signal transduction or target enzyme activities. The translational importance of such transnitrosylation reactions has been discussed elsewhere (Marino & Gladyshev, 2010).

Two cysteines (Cys689 and Cys908) that are highly conserved within the NOS family are susceptible to S-glutathionylation. These are critical for normal eNOS functions, and when modified, the enzyme produces $O_2^{\bullet-}$, a process that cannot be inhibited by either *N*-nitro-l-arginine methylester (L-NAME) or removal of calcium. S-glutathionylation of eNOS uncouples eNOS in endothelial cells impairing endothelium-dependent vasodilation (Crabtree, Brixey, Batchelor, Hale, & Channon, 2013). Induction of eNOS uncoupling by S-glutathionylation switches eNOS from NO production to $O_2^{\bullet-}$ generation, with concomitant loss of endothelial-mediated vasodilation contributing to vascular constriction, which can contribute to hypertensive cardiovascular disease. The involvement of GSTP in mediating this S-glutathionylation seems a viable possibility but is not yet established.

11. GSTP POLYMORPHISMS AND PHARMACOGENETICS

Translational relevance is a desired component of medical research today. One area of GST biology that is likely to provide such an outlet is the identification of human polymorphic variants of GSTP, which arise from nucleotide transitions that change codon 105 from Ile to Val and codon 114 from Ala to Val, generating four GSTP1 alleles: wild-type GSTP1-1A (Ile105/Ala114), GSTP1-1B (Val105/Ala114), GSTP1-1C (Val105/

Val114), and GSTP1-1D (Ile105/Val114) (Ali-Osman, Akande, et al., 1997; Watson, Stewart, Smith, Massey, & Bell, 1998). The Ile105 → Val105 and Ala114 → Val114 substitutions do not alter affinity for GSH, but sterically alter the substrate-binding site of the enzyme (Watson et al., 1998; Zimniak et al., 1994). The allele frequencies for *A–C in Caucasian populations are 0.685, 0.262, and 0.068, respectively (Garte et al., 2001). Altered conformation of the substrate-binding site(s) may contribute to different catalytic rates for the formation of thioether conjugates between GSH and some electrophiles (Ali-Osman, Akande, et al., 1997; Zimniak et al., 1994). The hydrophobicity and size of residue 114 could serve as a determinant of the substrate specificity of each isoenzyme (Hu, Herzog, Zimniak, & Singh, 1999). Since GSTP1-1D has enzyme activity toward CDNB comparable to GSTP1-1A, Val105 may circumvent the influence of Val114 (Watson et al., 1998). There are a few examples where polymorphisms influence response to certain anticancer drugs: GSTP1-1A reportedly enhances the formation of cisplatin–glutathione conjugates (Goto et al., 1999). There are also some early indications of epidemiological correlations of isotype expression with etiological aspects of endometrial (Chan et al., 2005), bladder, and testicular cancers (Harries, Stubbins, Forman, Howard, & Wolf, 1997). Perhaps most directly, these polymorphisms have the potential to influence disease susceptibility and response to ROS/RNS (McIlwain et al., 2006). For example, they have been shown to impact S-glutathionylation and lipid peroxidation (Manevich et al., 2012), altering Prdx6 peroxidase function through their effect on the affinity of GSTP for Prdx6 during heterodimerization (Manevich, Hutchens, Tew, & Townsend, 2013). Specifically, GSTP1*A and GSTP1*C have a lower binding affinity (Kd) for Prdx6 and consequently altered lipid peroxidation. The clinical implication of polymorphisms of GSTP may be attributable to its role in phase II metabolism/detoxification. However, the role of GSTP in the forward reaction of the S-glutathionylation cycle remains a work in progress and clinically could supersede its role in metabolism.

12. GSH PATHWAYS AND GSTP AS DRUG PLATFORMS

Premised upon the elevated levels of GSTP expression in tumors and the induction of expression underlying the drug resistance phenotype, the GSTP platform was used as a target area for anticancer drug discovery. Further, human GST polymorphisms serve to link possible pharmacogenetic variations in catalytic activity and drug metabolism, as well as cancer susceptibility and prognosis (Townsend & Tew, 2003). Early discovery identified GSH analogs, GSH conjugates, small organic molecules, and various natural products as putative GSTP inhibitors (Laborde, 2010; Mahajan & Atkins, 2005; Tew, 1994).

Telintra was designed as a selective inhibitor of GSTP. It is a peptidomimetic of GSH, substituted with aromatic moieties and esterified to enhance membrane permeability. An extensive preclinical and toxicological work established its safety profile and confirmed that inhibition of GSTP led to activation of Jun kinase, a key regulator of cellular growth and differentiation of blood precursor cells. In normal mice, TLK199 treatment increased circulating white blood cells and stimulated bone marrow progenitor proliferation. Early studies using *Telintra*-treated GSTP-deficient mice indicated that myeloproliferative events were allied with downstream activation of STAT proteins (Gate et al., 2004). *Telintra* is currently in clinical development (Ezatiostat HCl) for the treatment of low- to intermediate-risk myelodysplastic syndrome (MDS) patients. MDS affects approx. 300,000 people worldwide with etiology as a heterogeneous group of clonal hematopoietic stem cell disorders typically associated with aberrant hematopoiesis of one or more cell lineages (red blood cells, white blood cells, or platelets) and a variable risk of progression to acute myeloid leukemia. Two recent reports (Lyons, Wilks, Young, & Brown, 2011; Raza et al., 2009) have suggested that *Telintra* has therapeutic benefit in this disease, including durable red blood cell transfusion independence and multilineage responses. Recent trends in drug development have emphasized the value of stratifying patient selection on the basis of biomarkers that might positively influence response. One such study used a gene marker analysis from 30 patients enrolled in a phase II trial (Galili et al., 2012). The *Telintra* response profile identified two miRNAs that regulate genes implicated in MDS pathology and genes controlling the JNK–c-Jun pathways. This latter correlation is linked with *Telintra*-mediated activation, and interestingly, JNK–c-Jun were underexpressed in responders and overexpressed in nonresponders, further supporting the mechanism of action of *Telintra*. Additional attempts to optimize inhibitor allelic specificity, stability, and/or potency have included testing a number of peptidomimetic conjugate analogs with additional modifications of the GSH peptide backbone. Examples of these include compounds that are resistant to cleavage by GGT (Burg, Riepsaame, Pont, Mulder, & van de Water, 2006).

Other concepts have included the design of specific GSH conjugates, where the thiol group is modified and these have met with varying degrees of preclinical success. Allosteric inhibition of GSTP was successfully accomplished by conjugating GSH to DOX. Interestingly, activation of JNK and apoptotic pathways following GSH–DOX–GSTP binding appeared to occur without complex dissociation and be mediated by mitochondrial cytochrome c release (Asakura et al., 2007).

Several small molecule inhibitors and natural products have been reported to successfully target GSTP (see Laborde (2010) for a recent review). However, broad isoenzyme inhibition (i.e., lack of specificity) makes the majority of these less than ideal pharmacological agents. Ethacrynic acid (EA) is a potent, reversible small molecule inhibitor that nonspecifically inhibits GSTP activity by acting both as a GSH conjugate and as a noncompetitive GSH inhibitor for GST binding (Ploemen, Van Schanke, Van Ommen, & Van Bladeren, 1994). While preclinical studies indicated that EA administration effectively sensitized chemotherapy-resistant tumors, EA is also a potent diuretic and this negatively impacted a phase I clinical trial with this agent in combination with thiotepa (O'Dwyer et al., 1991). Similarly, 7-nitro-2, 1, 3-benzoxadiazole derivatives are potent, irreversible suicide GSTP inhibitors that not only induce activation of the c-Jun–JNK-mediated signal transduction pathway but also bind other GSTs and have not found clinical application (Federici et al., 2009a; Ricci et al., 2005).

An alternative approach to targeting GSTP has been tried. Inactive prodrugs that take advantage of the high GSTP levels in cancer cells have been synthesized and tested. For example, Telcyta (canfosfamide HCl, TLK286) is a conjugate of a GSH analog and an N, N, N', N'-tetrakis(2-chloroethyl) phosphorodiamidate that is metabolically activated via GSTP cleavage, thus therapeutically targeting GSTP overexpressing cells (Satyam et al., 1996). This drug has been extensively tested preclinically and also in both phase II and phase III clinical trials (Kavanagh et al., 2010; Vergote et al., 2009). Although the phase III results were not positive enough to advance the drug to an NDA application, further clinical trials continue.

GSTP was shown to catalyze NO release from a diazeniumdiolate, O^2-{2,4-dinitro-5-[4-(N-methylamino)benzoyloxy]phenyl}1-(N,N-dimethylamino)diazen-1-ium-1,2-diolate (PABA/NO), through addition of GSH and subsequent Meisenheimer complex formation (Saavedra et al., 2006). Antitumor activity and mechanism of action studies confirmed the importance of GSTP in the pharmacology of PABA/NO and related diazenium diolates (Findlay et al., 2004; Townsend et al., 2006). Additionally, it was shown, by crystallization of the complexes, that 6-(7-nitro-2, 1, 3-benzoxadiazol-4-ylthio)hexanol (NBDHEX) is specifically binding to the H-site of GSTM2-2 and resulting in its glutathionylation as a stable σ-complex. Interestingly, NBDHEX binds to the H-site of GSTP, but with lower affinity as compared to GSTM2-2 and without conjugation with GSH in the active center. Apparently, affinity of NBDHEX binding to GSTP was increased ~4× by $Ile^{104}Val$ or $Ile^{104}Ala$ mutation (Federici et al., 2009b). This GSTP property provides a plausible platform for development of novel therapeutics site-specifically releasing NO and other active glutathionylated nitroaromatic compounds (Hutchens et al., 2011; Manevich, Townsend, Hutchens, & Tew, 2010; Tew & Townsend, 2011a).

13. CONCLUSIONS AND PERSPECTIVES

The prevalence and ubiquitous biology of GSTP imply functional properties that extend beyond detoxification. Consistent with this, acting as chaperone-like proteins, GST isoenzymes (and GSTP in particular) have been found to regulate kinase signaling pathways. As a carrier of NO, GSTP also participates in regulating certain aspects of nitrosative stress, important in normal physiology, where imbalance is causally related to specific human disease pathologies (Tew & Townsend, 2011b; Townsend, 2007; Townsend et al., 2003; Xiong et al., 2011). Oxidation/reduction reactions control the activities of a number of proteins that can be functionally clustered. S-glutathionylation of such proteins can occur spontaneously; however, GSTP is documented to facilitate the delivery of GS to acceptor cysteines, catalyzing reactions in parts of the protein that might otherwise be inaccessible. Viability of the GSTP knockout mouse implies functional redundancy within the GST isoenzyme family. Nevertheless, the high level of GSTP in some human cancers and in drug-resistant tumor cells has provided a platform for discovery/development of therapeutic agents and diagnostic tools. Some of these are in clinical trials, and while it remains too early to predict their value, this area of research has translational relevance. Polymorphic expression of GSTP also provides a platform to understand individual differences in response to ROS/RNS (whether through drugs or environmental exposures) and susceptibility of drug addiction. While these developments remain in their early stages of development, they hold great promise for the future.

ACKNOWLEDGMENTS

This work was supported by grants from the National Center for Research Resources (5P20RR024485-02) and the National Institute of General Medical Sciences (8 P20 GM103542-02) from the National Institutes of Health and by CA08660, CA117259, and R56 ES017453 and support from the South Carolina Centers of Economic Excellence program. J. Z. was financially supported by the Swedish Research Council (No. 524-2011-6998).

REFERENCES

Abdelsaid, M. A., & El-Remessy, A. B. (2012). S-Glutathionylation of LMW-PTP regulates VEGF-mediated FAK activation and endothelial cell migration. *Journal of Cell Science, 125*(Pt 20), 4751–4760.

Adler, V., Yin, Z., Fuchs, S. Y., Benezra, M., Rosario, L., Tew, K. D., et al. (1999). Regulation of JNK signaling by GSTp. *The EMBO Journal, 18*, 1321–1334.

Adler, V., Yin, Z., Tew, K. D., & Ronai, Z. (1999). Role of redox potential and reactive oxygen species in stress signaling. *Oncogene, 18*, 6104–6111.

Ali-Osman, F., Akande, O., Antoun, G., Mao, J. X., & Buolamwini, J. (1997). Molecular cloning, characterization, and expression in *Escherichia coli* of full-length cDNAs of three human glutathione S-transferase Pi gene variants. Evidence for differential catalytic activity of the encoded proteins. *The Journal of Biological Chemistry, 272*, 10004–10012.

Ali-Osman, F., Brunner, J. M., Kutluk, T. M., & Hess, K. (1997). Prognostic significance of glutathione S-transferase pi expression and subcellular localization in human gliomas. *Clinical Cancer Research, 3*, 2253–2261.

Allen, T. C., Granville, L. A., Cagle, P. T., Haque, A., Zander, D. S., & Barrios, R. (2007). Expression of glutathione S-transferase pi and glutathione synthase correlates with survival in early stage non-small cell carcinomas of the lung. *Human Pathology, 38*, 220–227.

Allocati, N., Federici, L., Masulli, M., & Di Ilio, C. (2009). Glutathione transferases in bacteria. *The FEBS Journal, 276*, 58–75.

Anathy, V., Roberson, E., Cunniff, B., Nolin, J. D., Hoffman, S., Spiess, P., et al. (2012). Oxidative processing of latent Fas in the endoplasmic reticulum controls the strength of apoptosis. *Molecular and Cellular Biology, 32*, 3464–3478.

Asakura, T., Sasagawa, A., Takeuchi, H., Shibata, S., Marushima, H., Mamori, S., et al. (2007). Conformational change in the active center region of GST P1-1, due to binding of a synthetic conjugate of DXR with GSH, enhanced JNK-mediated apoptosis. *Apoptosis, 12*, 1269–1280.

Atkinson, H. J., & Babbitt, P. C. (2009). Glutathione transferases are structural and functional outliers in the thioredoxin fold. *Biochemistry, 48*, 11108–11116.

Baek, J. Y., Han, S. H., Sung, S. H., Lee, H. E., Kim, Y. M., Noh, Y. H., et al. (2012). Sulfiredoxin protein is critical for redox balance and survival of cells exposed to low steady-state levels of H_2O_2. *The Journal of Biological Chemistry, 287*, 81–89.

Beckman, J. S., Beckman, T. W., Chen, J., Marshall, P. A., & Freeman, B. A. (1990). Apparent hydroxyl radical production by peroxynitrite: Implications for endothelial injury from nitric oxide and superoxide. *Proceedings of the National Academy of Sciences of the United States of America, 87*, 1620–1624.

Bernardini, S., Bernassola, F., Cortese, C., Ballerini, S., Melino, G., Motti, C., et al. (2000). Modulation of GST P1-1 activity by polymerization during apoptosis. *Journal of Cellular Biochemistry, 77*, 645–653.

Boese, M., Keese, M. A., Becker, K., Busse, R., & Mulsch, A. (1997). Inhibition of glutathione reductase by dinitrosyl-iron-dithiolate complex. *The Journal of Biological Chemistry, 272*, 21767–21773.

Bowers, R. R., Manevich, Y., Townsend, D. M., & Tew, K. D. (2012). Sulfiredoxin redox-sensitive interaction with S100A4 and non-muscle myosin IIA regulates cancer cell motility. *Biochemistry, 51*, 7740–7754.

Burg, D., Riepsaame, J., Pont, C., Mulder, G., & van de Water, B. (2006). Peptide-bond modified glutathione conjugate analogs modulate GSTpi function in GSH-conjugation, drug sensitivity and JNK signaling. *Biochemical Pharmacology, 71*, 268–277.

Chan, Q. K., Khoo, U. S., Ngan, H. Y., Yang, C. Q., Xue, W. C., Chan, K. Y., et al. (2005). Single nucleotide polymorphism of pi-class glutathione s-transferase and susceptibility to endometrial carcinoma. *Clinical Cancer Research, 11*, 2981–2985.

Choi, H. J., Kang, S. W., Yang, C. H., Rhee, S. G., & Ryu, S. E. (1998). Crystal structure of a novel human peroxidase enzyme at 2.0 A resolution. *Nature Structural Biology*, *5*, 400–406.
Corti, A., Paolicchi, A., Franzini, M., Dominici, S., Casini, A. F., & Pompella, A. (2005). The S-thiolating activity of membrane gamma-glutamyltransferase: Formation of cysteinyl-glycine mixed disulfides with cellular proteins and in the cell microenvironment. *Antioxidants & Redox Signaling*, 7, 911–918.
Cowell, I. G., Dixon, K. H., Pemble, S. E., Ketterer, B., & Taylor, J. B. (1988). The structure of the human glutathione S-transferase pi gene. *The Biochemical Journal*, *255*, 79–83.
Crabtree, M. J., Brixey, R., Batchelor, H., Hale, A. B., & Channon, K. M. (2013). Integrated redox sensor and effector functions for tetrahydrobiopterin- and glutathionylation-dependent endothelial nitric-oxide synthase uncoupling. *The Journal of Biological Chemistry*, *288*, 561–569.
Davis, R. J. (2000). Signal transduction by the JNK group of MAP kinases. *Cell*, *103*, 239–252.
De Luca, A., Federici, L., De Canio, M., Stella, L., & Caccuri, A. M. (2012). New insights into the mechanism ofJNK1 inhibition by glutathione transferase p1-1. *Biochemistry*, *51*, 7304–7312.
Desmots, F., Loyer, P., Rissel, M., Guillouzo, A., & Morel, F. (2005). Activation of C-Jun N-terminal kinase is required for glutathione transferase A4 induction during oxidative stress, not during cell proliferation, in mouse hepatocytes. *FEBS Letters*, *579*, 5691–5696.
Federici, L., Lo Sterzo, C., Pezzola, S., Di Matteo, A., Scaloni, F., Federici, G., et al. (2009a). Structural basis for the binding of the anticancer compound 6-(7-nitro-2,1,3-benzoxadiazol-4-ylthio)hexanol to human glutathione s-transferases. *Cancer Research*, *69*, 8025–8034.
Federici, L., Lo Sterzo, C., Pezzola, S., Di Matteo, A., Scaloni, F., Federici, G., et al. (2009b). Structural basis for the binding of the anticancer compound 6-(7-nitro-2,1,3-benzoxadiazol-4-ylthio)hexanol to human glutathione s-transferases. *Cancer Research*, *69*, 8025–8034.
Filipovska, A., & Murphy, M. P. (2006). Overview of protein glutathionylation. *Current Protocols in Toxico*, Chapter 6, Unit6 10.
Findlay, V. J., Tapiero, H., & Townsend, D. M. (2005). Sulfiredoxin: A potential therapeutic agent? *Biomedicine & Pharmacotherapy*, *59*, 374–379.
Findlay, V. J., Townsend, D. M., Morris, T. E., Fraser, J. P., He, L., & Tew, K. D. (2006). A novel role for human sulfiredoxin in the reversal of glutathionylation. *Cancer Research*, *66*, 6800–6806.
Findlay, V. J., Townsend, D. M., Saavedra, J. E., Buzard, G. S., Citro, M. L., Keefer, L. K., et al. (2004). Tumor cell responses to a novel glutathione S-transferase-activated nitric oxide-releasing prodrug. *Molecular Pharmacology*, *65*, 1070–1079.
Flohe, L. (2012). The fairytale of the GSSG/GSH redox potential. *Biochimica et Biophysica Acta*, *1830*(5), 3139–3142.
Fourquet, S., Guerois, R., Biard, D., & Toledano, M. B. (2010). Activation of NRF2 by nitrosative agents and H2O2 involves KEAP1 disulfide formation. *The Journal of Biological Chemistry*, *285*, 8463–8471.
Galili, N., Tamayo, P., Botvinnik, O. B., Mesirov, J. P., Brooks, M. R., Brown, G., et al. (2012). Prediction of response to therapy with ezatiostat in lower risk myelodysplastic syndrome. *Journal of Hematology & Oncology*, *5*, 20.
Gallogly, M. M., Starke, D. W., Leonberg, A. K., Ospina, S. M., & Mieyal, J. J. (2008). Kinetic and mechanistic characterization and versatile catalytic properties of mammalian glutaredoxin 2: Implications for intracellular roles. *Biochemistry*, *47*, 11144–11157.
Garcia, J., Han, D., Sancheti, H., Yap, L. P., Kaplowitz, N., & Cadenas, E. (2010). Regulation of mitochondrial glutathione redox status and protein glutathionylation by respiratory substrates. *The Journal of Biological Chemistry*, *285*, 39646–39654.

Gardner, J. L., & Gallagher, E. P. (2001). Development of a peptide antibody specific to human glutathione S-transferase alpha 4-4 (hGSTA4-4) reveals preferential localization in human liver mitochondria. *Archives of Biochemistry and Biophysics, 390*, 19–27.

Garte, S., Gaspari, L., Alexandrie, A. K., Ambrosone, C., Autrup, H., Autrup, J. L., et al. (2001). Metabolic gene polymorphism frequencies in control populations. *Cancer Epidemiology, Biomarkers & Prevention, 10*, 1239–1248.

Gate, L., Majumdar, R. S., Lunk, A., & Tew, K. D. (2004). Increased myeloproliferation in glutathione S-transferase pi-deficient mice is associated with a deregulation of JNK and Janus kinase/STAT pathways. *The Journal of Biological Chemistry, 279*, 8608–8616.

Gate, L., Majumdar, R. S., Lunk, A., & Tew, K. D. (2005). Influence of glutathione S-transferase pi and p53 expression on tumor frequency and spectrum in mice. *International Journal of Cancer, 113*, 29–35.

Gildenhuys, S., Wallace, L. A., Burke, J. P., Balchin, D., Sayed, Y., & Dirr, H. W. (2010). Class Pi glutathione transferase unfolds via a dimeric and not monomeric intermediate: Functional implications for an unstable monomer. *Biochemistry, 49*, 5074–5081.

Gonzalez-Dosal, R., Horan, K. A., Rahbek, S. H., Ichijo, H., Chen, Z. J., Mieyal, J. J., et al. (2011). HSV infection induces production of ROS, which potentiate signaling from pattern recognition receptors: Role for S-glutathionylation of TRAF3 and 6. *PLoS Pathogens*, 7, e1002250.

Goto, S., Ihara, Y., Urata, Y., Izumi, S., Abe, K., Koji, T., et al. (2001a). Doxorubicin-induced DNA intercalation and scavenging by nuclear glutathione S-transferase pi. *The FASEB Journal, 15*, 2702–2714.

Goto, S., Ihara, Y., Urata, Y., Izumi, S., Abe, K., Koji, T., et al. (2001b). Doxorubicin-induced DNA intercalation and scavenging by nuclear glutathione S-transferase pi. *The FASEB Journal, 15*, 2702–2714.

Goto, S., Iida, T., Cho, S., Oka, M., Kohno, S., & Kondo, T. (1999). Overexpression of glutathione S-transferase pi enhances the adduct formation of cisplatin with glutathione in human cancer cells. *Free Radical Research, 31*, 549–558.

Graminski, G. F., Kubo, Y., & Armstrong, R. N. (1989). Spectroscopic and kinetic evidence for the thiolate anion of glutathione at the active site of glutathione S-transferase. *Biochemistry, 28*, 3562–3568.

Graminski, G. F., Zhang, P. H., Sesay, M. A., Ammon, H. L., & Armstrong, R. N. (1989). Formation of the 1-(S-glutathionyl)-2,4,6-trinitrocyclohexadienate anion at the active site of glutathione S-transferase: Evidence for enzymic stabilization of sigma-complex intermediates in nucleophilic aromatic substitution reactions. *Biochemistry, 28*, 6252–6258.

Grek, C. L., Townsend, D. M., Uys, J. D., Manevich, Y., Coker, W. J., 3rd., Pazoles, C. J., et al. (2012). S-glutathionylated serine proteinase inhibitors as plasma biomarkers in assessing response to redox-modulating drugs. *Cancer Research, 72*, 2383–2393.

Grek, C. L., Zhang, J., Manevich, Y., Townsend, D. M., & Tew, K. D. (2013). Causes and consequences of cysteine S-glutathionylation. *The Journal of Biological Chemistry, 288*, 26497–26504.

Gupta, S., Barrett, T., Whitmarsh, A. J., Cavanagh, J., Sluss, H. K., Derijard, B., et al. (1996). Selective interaction of JNK protein kinase isoforms with transcription factors. *The EMBO Journal, 15*, 2760–2770.

Harries, L. W., Stubbins, M. J., Forman, D., Howard, G. C., & Wolf, C. R. (1997). Identification of genetic polymorphisms at the glutathione S-transferase Pi locus and association with susceptibility to bladder, testicular and prostate cancer. *Carcinogenesis, 18*, 641–644.

Hartmanova, T., Tambor, V., Lenco, J., Staab-Weijnitz, C. A., Maser, E., & Wsol, V. (2013). S-Nitrosoglutathione covalently modifies cysteine residues of human carbonyl reductase 1 and affects its activity. *Chemico-Biological Interactions, 202*, 136–145.

Hayes, J. D., Flanagan, J. U., & Jowsey, I. R. (2005). Glutathione transferases. *Annual Review of Pharmacology and Toxicology*, *45*, 51–88.

Hayes, J. D., & Pulford, D. J. (1995). The glutathione S-transferase supergene family: Regulation of GST and the contribution of the isoenzymes to cancer chemoprotection and drug resistance. *Critical Reviews in Biochemistry and Molecular Biology*, *30*, 445–600.

Henderson, C. J., Smith, A. G., Ure, J., Brown, K., Bacon, E. J., & Wolf, C. R. (1998). Increased skin tumorigenesis in mice lacking pi class glutathione S-transferases. *Proceedings of the National Academy of Sciences of the United States of America*, *95*, 5275–5280.

Henderson, C. J., Wolf, C. R., Kitteringham, N., Powell, H., Otto, D., & Park, B. K. (2000). Increased resistance to acetaminophen hepatotoxicity in mice lacking glutathione S-transferase Pi. *Proceedings of the National Academy of Sciences of the United States of America*, *97*, 12741–12745.

Holley, S. L., Fryer, A. A., Haycock, J. W., Grubb, S. E., Strange, R. C., & Hoban, P. R. (2007). Differential effects of glutathione S-transferase pi (GSTP1) haplotypes on cell proliferation and apoptosis. *Carcinogenesis*, *28*, 2268–2273.

Hu, X., Herzog, C., Zimniak, P., & Singh, S. V. (1999). Differential protection against benzo [a]pyrene-7,8-dihydrodiol-9,10-epoxide-induced DNA damage in HepG2 cells stably transfected with allelic variants of pi class human glutathione S-transferase. *Cancer Research*, *59*, 2358–2362.

Huang, J., Tan, P. H., Thiyagarajan, J., & Bay, B. H. (2003). Prognostic significance of glutathione S-transferase-pi in invasive breast cancer. *Modern Pathology*, *16*, 558–565.

Hutchens, S., Manevich, Y., He, L., Tew, K. D., & Townsend, D. M. (2011). Cellular resistance to a nitric oxide releasing glutathione S-transferase P-activated prodrug, PABA/NO. *Investigational New Drugs*, *29*, 719–729.

Isenberg, J. S., Martin-Manso, G., Maxhimer, J. B., & Roberts, D. D. (2009). Regulation of nitric oxide signalling by thrombospondin 1: Implications for anti-angiogenic therapies. *Nature Reviews Cancer*, *9*, 182–194.

Jankova, L., Robertson, G., Chan, C., Tan, K. L., Kohonen-Corish, M., Fung, C. L., et al. (2012). Glutathione S-transferase Pi expression predicts response to adjuvant chemotherapy for stage C colon cancer: A matched historical control study. *BMC Cancer*, *12*, 196.

Jeong, W., Bae, S. H., Toledano, M. B., & Rhee, S. G. (2012). Role of sulfiredoxin as a regulator of peroxiredoxin function and regulation of its expression. *Free Radical Biology & Medicine*, *53*, 447–456.

Jin, X., Yu, L., Wu, Y., Zhang, S., Shi, Z., Chen, X., et al. (2012). S-Glutathionylation underscores the modulation of the heteromeric Kir4.1-Kir5.1 channel in oxidative stress. *The Journal of Physiology*, *590*, 5335–5348.

Kavanagh, J. J., Levenback, C. F., Ramirez, P. T., Wolf, J. L., Moore, C. L., Jones, M. R., et al. (2010). Phase 2 study of canfosfamide in combination with pegylated liposomal doxorubicin in platinum and paclitaxel refractory or resistant epithelial ovarian cancer. *Journal of Hematology & Oncology*, *3*, 9.

Kil, I. S., Shin, S. W., & Park, J. W. (2012). S-glutathionylation regulates GTP-binding of Rac2. *Biochemical and Biophysical Research Communications*, *425*, 892–896.

Kim, H. S., Ullevig, S. L., Zamora, D., Lee, C. F., & Asmis, R. (2012). Redox regulation of MAPK phosphatase 1 controls monocyte migration and macrophage recruitment. *Proceedings of the National Academy of Sciences of the United States of America*, *109*, E2803–E2812.

Kou, X., Chen, N., Feng, Z., Luo, L., & Yin, Z. (2013). GSTP1 negatively regulates Stat3 activation in epidermal growth factor signaling. *Oncology Letters*, *5*, 1053–1057.

Laborde, E. (2010). Glutathione transferases as mediators of signaling pathways involved in cell proliferation and cell death. *Cell Death and Differentiation*, *17*, 1373–1380.

Lei, K., Townsend, D. M., & Tew, K. D. (2008). Protein cysteine sulfinic acid reductase (sulfiredoxin) as a regulator of cell proliferation and drug response. *Oncogene*, *27*, 4877–4887.

Lo Bello, M., Nuccetelli, M., Caccuri, A. M., Stella, L., Parker, M. W., Rossjohn, J., et al. (2001). Human glutathione transferase P1-1 and nitric oxide carriers; a new role for an old enzyme. *The Journal of Biological Chemistry*, *276*, 42138–42145.

Lo, H. W., & Ali-Osman, F. (1998). Structure of the human allelic glutathione S-transferase-pi gene variant, hGSTP1 C, cloned from a glioblastoma multiforme cell line. *Chemico-Biological Interactions*, *111–112*, 91–102.

Lock, J. T., Sinkins, W. G., & Schilling, W. P. (2012). Protein S-glutathionylation enhances Ca2+−induced Ca2+ release via the IP3 receptor in cultured aortic endothelial cells. *The Journal of Physiology*, *590*, 3431–3447.

Lyons, R. M., Wilks, S. T., Young, S., & Brown, G. L. (2011). Oral ezatiostat HCl (Telintra(R), TLK199) and Idiopathic Chronic Neutropenia (ICN): A case report of complete response of a patient with G-CSF resistant ICN following treatment with ezatiostat, a glutathione S-transferase P1-1 (GSTP1-1) inhibitor. *Journal of Hematology & Oncology*, *4*, 43.

Mahajan, S., & Atkins, W. M. (2005). The chemistry and biology of inhibitors and pro-drugs targeted to glutathione S-transferases. *Cellular and Molecular Life Sciences*, *62*, 1221–1233.

Mailloux, R. J., Fu, A., Robson-Doucette, C., Allister, E. M., Wheeler, M. B., Screaton, R., et al. (2012). Glutathionylation state of uncoupling protein-2 and the control of glucose-stimulated insulin secretion. *The Journal of Biological Chemistry*, *287*, 39673–39685.

Manevich, Y., Feinstein, S. I., & Fisher, A. B. (2004). Activation of the antioxidant enzyme 1-CYS peroxiredoxin requires glutathionylation mediated by heterodimerization with pi GST. *Proceedings of the National Academy of Sciences of the United States of America*, *101*, 3780–3785.

Manevich, Y., & Fisher, A. B. (2005a). Peroxiredoxin 6, a 1-Cys peroxiredoxin, functions in antioxidant defense and lung phospholipid metabolism. *Free Radical Biology & Medicine*, *38*, 1422–1432.

Manevich, Y., & Fisher, A. B. (2005b). Peroxiredoxin 6, a 1-Cys peroxiredoxin, functions in antioxidant defense and lung phospholipid metabolism. *Free Radical Biology & Medicine*, *38*, 1422–1432.

Manevich, Y., Hutchens, S., Tew, K. D., & Townsend, D. M. (2012). Allelic variants of glutathione S-transferase P1-1 differentially mediate the peroxidase function of peroxiredoxin VI and alter membrane lipid peroxidation. *Free Radical Biology & Medicine*, *54C*, 62–70.

Manevich, Y., Hutchens, S., Tew, K. D., & Townsend, D. M. (2013). Allelic variants of glutathione S-transferase P1-1 differentially mediate the peroxidase function of peroxiredoxin VI and alter membrane lipid peroxidation. *Free Radical Biology & Medicine*, *54*, 62–70.

Manevich, Y., Townsend, D. M., Hutchens, S., & Tew, K. D. (2010). Diazeniumdiolate mediated nitrosative stress alters nitric oxide homeostasis through intracellular calcium and S-glutathionylation of nitric oxide synthetase. *PloS One*, *5*, e14151.

Marino, S. M., & Gladyshev, V. N. (2010). Structural analysis of cysteine S-nitrosylation: A modified acid-based motif and the emerging role of trans-nitrosylation. *Journal of Molecular Biology*, *395*, 844–859.

Martinez, M. C., & Andriantsitohaina, R. (2009). Reactive nitrogen species: Molecular mechanisms and potential significance in health and disease. *Antioxidants & Redox Signaling*, *11*, 669–702.

McIlwain, C. C., Townsend, D. M., & Tew, K. D. (2006). Glutathione S-transferase polymorphisms: Cancer incidence and therapy. *Oncogene*, *25*, 1639–1648.

Mieyal, J. J., Starke, D. W., Gravina, S. A., Dothey, C., & Chung, J. S. (1991). Thioltransferase in human red blood cells: Purification and properties. *Biochemistry*, *30*, 6088–6097.

Mieyal, J. J., Starke, D. W., Gravina, S. A., & Hocevar, B. A. (1991). Thioltransferase in human red blood cells: Kinetics and equilibrium. *Biochemistry, 30*, 8883–8891.

Mollica, J. P., Dutka, T. L., Merry, T. L., Lamboley, C. R., McConell, G. K., McKenna, M. J., et al. (2012). S-glutathionylation of troponin I (fast) increases contractile apparatus Ca2+ sensitivity in fast-twitch muscle fibres of rats and humans. *The Journal of Physiology, 590*, 1443–1463.

Monaco, R., Friedman, F. K., Hyde, M. J., Chen, J. M., Manolatus, S., Adler, V., et al. (1999). Identification of a glutathione-S-transferase effector domain for inhibition of jun kinase, by molecular dynamics. *Journal of Protein Chemistry, 18*, 859–866.

Niwa, T., Naito, C., Mawjood, A. H., & Imai, K. (2000). Increased glutathionyl hemoglobin in diabetes mellitus and hyperlipidemia demonstrated by liquid chromatography/electrospray ionization-mass spectrometry. *Clinical Chemistry, 46*, 82–88.

O'Dwyer, P. J., LaCreta, F., Nash, S., Tinsley, P. W., Schilder, R., Clapper, M. L., et al. (1991). Phase I study of thiotepa in combination with the glutathione transferase inhibitor ethacrynic acid. *Cancer Research, 51*, 6059–6065.

Pal, A., Hu, X., Zimniak, P., & Singh, S. V. (2000). Catalytic efficiencies of allelic variants of human glutathione S-transferase Pi in the glutathione conjugation of alpha, beta-unsaturated aldehydes. *Cancer Letters, 154*, 39–43.

Park, J. W., Mieyal, J. J., Rhee, S. G., & Chock, P. B. (2009). Deglutathionylation of 2-Cys peroxiredoxin is specifically catalyzed by sulfiredoxin. *The Journal of Biological Chemistry, 284*, 23364–23374.

Peskin, A. V., & Winterbourn, C. C. (2006). Taurine chloramine is more selective than hypochlorous acid at targeting critical cysteines and inactivating creatine kinase and glyceraldehyde-3-phosphate dehydrogenase. *Free Radical Biology & Medicine, 40*, 45–53.

Petit, E., Michelet, X., Rauch, C., Bertrand-Michel, J., Terce, F., Legouis, R., et al. (2009). Glutathione transferases kappa 1 and kappa 2 localize in peroxisomes and mitochondria, respectively, and are involved in lipid metabolism and respiration in Caenorhabditis elegans. *The FEBS Journal, 276*, 5030–5040.

Petrushanko, I. Y., Yakushev, S., Mitkevich, V. A., Kamanina, Y. V., Ziganshin, R. H., Meng, X., et al. (2012). S-glutathionylation of the Na, K-ATPase catalytic alpha subunit is a determinant of the enzyme redox sensitivity. *The Journal of Biological Chemistry, 287*, 32195–32205.

Pettigrew, N. E., & Colman, R. F. (2001). Heterodimers of glutathione S-transferase can form between isoenzyme classes pi and mu. *Archives of Biochemistry and Biophysics, 396*, 225–230.

Ploemen, J. H., Van Schanke, A., Van Ommen, B., & Van Bladeren, P. J. (1994). Reversible conjugation of ethacrynic acid with glutathione and human glutathione S-transferase p11. *Cancer Research, 54*, 915–919.

Pompella, A. (2005). Redox events and cell membrane proteins. *Archives of Biochemistry and Biophysics, 434*, 1–2.

Ralat, L. A., Manevich, Y., Fisher, A. B., & Colman, R. F. (2006). Direct evidence for the formation of a complex between 1-cysteine peroxiredoxin and glutathione S-transferase pi with activity changes in both enzymes. *Biochemistry, 45*, 360–372.

Ralat, L. A., Misquitta, S. A., Manevich, Y., Fisher, A. B., & Colman, R. F. (2008). Characterization of the complex of glutathione S-transferase pi and 1-cysteine peroxiredoxin. *Archives of Biochemistry and Biophysics, 474*, 109–118.

Raza, H. (2011). Dual localization of glutathione S-transferase in the cytosol and mitochondria: Implications in oxidative stress, toxicity and disease. *The FEBS Journal, 278*, 4243–4251.

Raza, A., Galili, N., Callander, N., Ochoa, L., Piro, L., Emanuel, P., et al. (2009). Phase 1-2a multicenter dose-escalation study of ezatiostat hydrochloride liposomes for injection

(Telintra, TLK199), a novel glutathione analog prodrug in patients with myelodysplastic syndrome. *Journal of Hematology & Oncology*, *2*, 20.
Rhee, S. G., Kang, S. W., Chang, T. S., Jeong, W., & Kim, K. (2001). Peroxiredoxin, a novel family of peroxidases. *IUBMB Life*, *52*, 35–41.
Ricci, G., De Maria, F., Antonini, G., Turella, P., Bullo, A., Stella, L., et al. (2005). 7-Nitro-2,1,3-benzoxadiazole derivatives, a new class of suicide inhibitors for glutathione S-transferases. Mechanism of action of potential anticancer drugs. *The Journal of Biological Chemistry*, *280*, 26397–26405.
Rolland, D., Raharijaona, M., Barbarat, A., Houlgatte, R., & Thieblemont, C. (2010). Inhibition of GST-pi nuclear transfer increases mantle cell lymphoma sensitivity to cisplatin, cytarabine, gemcitabine, bortezomib and doxorubicin. *Anticancer Research*, *30*, 3951–3957.
Romero, L., Andrews, K., Ng, L., O'Rourke, K., Maslen, A., & Kirby, G. (2006). Human GSTA1-1 reduces c-Jun N-terminal kinase signalling and apoptosis in Caco-2 cells. *The Biochemical Journal*, *400*, 135–141.
Ruscoe, J. E., Rosario, L. A., Wang, T., Gate, L., Arifoglu, P., Wolf, C. R., et al. (2001). Pharmacologic or genetic manipulation of glutathione S-transferase P1-1 (GSTpi) influences cell proliferation pathways. *The Journal of Pharmacology and Experimental Therapeutics*, *298*, 339–345.
Saavedra, J. E., Srinivasan, A., Buzard, G. S., Davies, K. M., Waterhouse, D. J., Inami, K., et al. (2006). PABA/NO as an anticancer lead: Analogue synthesis, structure revision, solution chemistry, reactivity toward glutathione, and in vitro activity. *Journal of Medicinal Chemistry*, *49*, 1157–1164.
Saitoh, M., Nishitoh, H., Fujii, M., Takeda, K., Tobiume, K., Sawada, Y., et al. (1998). Mammalian thioredoxin is a direct inhibitor of apoptosis signal-regulating kinase (ASK) 1. *The EMBO Journal*, *17*, 2596–2606.
Sanchez, G., Escobar, M., Pedrozo, Z., Macho, P., Domenech, R., Hartel, S., et al. (2008). Exercise and tachycardia increase NADPH oxidase and ryanodine receptor-2 activity: Possible role in cardioprotection. *Cardiovascular Research*, *77*, 380–386.
Satyam, A., Hocker, M. D., Kane-Maguire, K. A., Morgan, A. S., Villar, H. O., & Lyttle, M. H. (1996). Design, synthesis, and evaluation of latent alkylating agents activated by glutathione S-transferase. *Journal of Medicinal Chemistry*, *39*, 1736–1747.
Shelton, M. D., Chock, P. B., & Mieyal, J. J. (2005). Glutaredoxin: Role in reversible protein s-glutathionylation and regulation of redox signal transduction and protein translocation. *Antioxidants & Redox Signaling*, *7*, 348–366.
Singh, S. P., Wishnok, J. S., Keshive, M., Deen, W. M., & Tannenbaum, S. R. (1996). The chemistry of the S-nitrosoglutathione/glutathione system. *Proceedings of the National Academy of Sciences of the United States of America*, *93*, 14428–14433.
Soh, Y., Goto, S., Kitajima, M., Moriyama, S., Kotera, K., Nakayama, T., et al. (2005). Nuclear localisation of glutathione S-transferase pi is an evaluation factor for drug resistance in gynaecological cancers. *Clinical Oncology (Royal College of Radiologists)*, *17*, 264–270.
Stamler, J. S., Singel, D. J., & Loscalzo, J. (1992). Biochemistry of nitric oxide and its redox-activated forms. *Science*, *258*, 1898–1902.
Stuehr, D. J. (1997). Structure-function aspects in the nitric oxide synthases. *Annual Review of Pharmacology and Toxicology*, *37*, 339–359.
Sun, R., Eriksson, S., & Wang, L. (2012). Oxidative stress induced S-glutathionylation and proteolytic degradation of mitochondrial thymidine kinase 2. *The Journal of Biological Chemistry*, *287*, 24304–24312.
Tellez-Sanz, R., Cesareo, E., Nuccetelli, M., Aguilera, A. M., Baron, C., Parker, L. J., et al. (2006). Calorimetric and structural studies of the nitric oxide carrier S-nitrosoglutathione bound to human glutathione transferase p1-1. *Protein Science*, *15*, 1093–1105.

Tew, K. D. (1994). Glutathione-associated enzymes in anticancer drug resistance. *Cancer Research, 54*, 4313–4320.

Tew, K. D., Manevich, Y., Grek, C., Xiong, Y., Uys, J., & Townsend, D. M. (2011). The role of glutathione S-transferase P in signaling pathways and S-glutathionylation in cancer. *Free Radical Biology & Medicine, 51*, 299–313.

Tew, K. D., & Townsend, D. M. (2011a). Redox platforms in cancer drug discovery and development. *Current Opinion in Chemical Biology, 15*, 156–161.

Tew, K. D., & Townsend, D. M. (2011b). Regulatory functions of glutathione S-transferase P1-1 unrelated to detoxification. *Drug Metabolism Reviews, 43*, 179–193.

Thevenin, A. F., Zony, C. L., Bahnson, B. J., & Colman, R. F. (2011). GST pi modulates JNK activity through a direct interaction with JNK substrate, ATF2. *Protein Science, 20*, 834–848.

Townsend, D. M. (2007). S-glutathionylation: Indicator of cell stress and regulator of the unfolded protein response. *Molecular Interventions, 7*, 313–324.

Townsend, D. M., Findlay, V. J., Fazilev, F., Ogle, M., Fraser, J., Saavedra, J. E., et al. (2006). A glutathione S-transferase pi-activated prodrug causes kinase activation concurrent with S-glutathionylation of proteins. *Molecular Pharmacology, 69*, 501–508.

Townsend, D. M., Manevich, Y., He, L., Hutchens, S., Pazoles, C. J., & Tew, K. D. (2009a). Novel role for glutathione S-transferase pi. Regulator of protein S-Glutathionylation following oxidative and nitrosative stress. *The Journal of Biological Chemistry, 284*, 436–445.

Townsend, D. M., Manevich, Y., He, L., Hutchens, S., Pazoles, C. J., & Tew, K. D. (2009b). Novel role for glutathione S-transferase pi. Regulator of protein S-Glutathionylation following oxidative and nitrosative stress. *The Journal of Biological Chemistry, 284*, 436–445.

Townsend, D., & Tew, K. (2003). Cancer drugs, genetic variation and the glutathione-S-transferase gene family. *American Journal of Pharmacogenomics, 3*, 157–172.

Townsend, D. M., Tew, K. D., & Tapiero, H. (2003). The importance of glutathione in human disease. *Biomedicine & Pharmacotherapy, 57*, 145–155.

Turella, P., Cerella, C., Filomeni, G., Bullo, A., De Maria, F., Ghibelli, L., et al. (2005). Proapoptotic activity of new glutathione S-transferase inhibitors. *Cancer Research, 65*, 3751–3761.

Vanin, A. F. (2009). Dinitrosyl iron complexes with thiolate ligands: Physico-chemistry, biochemistry and physiology. *Nitric Oxide, 21*, 1–13.

Vasieva, O. (2011). The many faces of glutathione transferase pi. *Current Molecular Medicine, 11*, 129–139.

Vergote, I., Finkler, N., del Campo, J., Lohr, A., Hunter, J., Matei, D., et al. (2009). Phase 3 randomised study of canfosfamide (Telcyta, TLK286) versus pegylated liposomal doxorubicin or topotecan as third-line therapy in patients with platinum-refractory or -resistant ovarian cancer. *European Journal of Cancer, 45*, 2324–2332.

Wang, T., Arifoglu, P., Ronai, Z., & Tew, K. D. (2001). Glutathione S-transferase P1-1 (GSTP1-1) inhibits c-Jun N-terminal kinase (JNK1) signaling through interaction with the C terminus. *The Journal of Biological Chemistry, 276*, 20999–21003.

Watson, M. A., Stewart, R. K., Smith, G. B., Massey, T. E., & Bell, D. A. (1998). Human glutathione S-transferase P1 polymorphisms: Relationship to lung tissue enzyme activity and population frequency distribution. *Carcinogenesis, 19*, 275–280.

West, M. B., Hill, B. G., Xuan, Y. T., & Bhatnagar, A. (2006). Protein glutathiolation by nitric oxide: An intracellular mechanism regulating redox protein modification. *The FASEB Journal, 20*, 1715–1717.

Wink, D. A., Nims, R. W., Darbyshire, J. F., Christodoulou, D., Hanbauer, I., Cox, G. W., et al. (1994). Reaction kinetics for nitrosation of cysteine and glutathione in aerobic nitric oxide solutions at neutral pH. Insights into the fate and physiological effects of intermediates generated in the NO/O_2 reaction. *Chemical Research in Toxicology*, *7*, 519–525.

Wu, Y., Fan, Y., Xue, B., Luo, L., Shen, J., Zhang, S., et al. (2006). Human glutathione S-transferase P1-1 interacts with TRAF2 and regulates TRAF2-ASK1 signals. *Oncogene*, *25*, 5787–5800.

Xiong, Y., Uys, J. D., Tew, K. D., & Townsend, D. M. (2011). S-glutathionylation: From molecular mechanisms to health outcomes. *Antioxidants & Redox Signaling*, *15*, 233–270.

Xue, B., Wu, Y., Yin, Z., Zhang, H., Sun, S., Yi, T., et al. (2005). Regulation of lipopolysaccharide-induced inflammatory response by glutathione S-transferase P1 in RAW264.7 cells. *FEBS Letters*, *579*, 4081–4087.

Zimniak, P., Nanduri, B., Pikula, S., Bandorowicz-Pikula, J., Singhal, S. S., Srivastava, S. K., et al. (1994). Naturally occurring human glutathione S-transferase GSTP1-1 isoforms with isoleucine and valine in position 104 differ in enzymic properties. *European Journal of Biochemistry*, *224*, 893–899.

CHAPTER FIVE

A Comparison of Reversible Versus Irreversible Protein Glutathionylation

Danyelle M. Townsend*, **Volodymyr I. Lushchak**†, **Arthur J.L. Cooper**‡,1

*Department of Pharmaceutical and Biomedical Sciences, Medical University of South Carolina, Charleston, South Carolina, USA
†Department of Biochemistry and Biotechnology, Vassyl Stefanyk Precarpathian National University, Ivano-Frankivsk, Ukraine
‡Department of Biochemistry and Molecular Biology, New York Medical College, Valhalla, New York, USA
[1]Corresponding author: e-mail address: arthur_cooper@nymc.edu

Contents

Abstract

Glutathionylation is generally a reversible posttranslational modification that occurs to cysteine residues that have been exposed to reactive oxygen species (P-SSG). This cyclical process can regulate various clusters of proteins, including those involved in critical cellular signaling functions. However, certain conditions can favor the formation of dehydroamino acids, such as 2,3-didehydroalanine (2,3-dehydroalanine, DHA) and 2,3-didehydrobutyrine (2,3-dehydrobutyrine), which can act as Michael acceptors. In turn, these can form Michael adducts with glutathione (GSH), resulting in the formation of a

Advances in Cancer Research, Volume 122
ISSN 0065-230X
http://dx.doi.org/10.1016/B978-0-12-420117-0.00005-0

stable thioether conjugate, an irreversible process referred to as nonreducible glutathionylation. This is predicted to be prevalent in nature, particularly in more slowly turning over proteins. Such nonreducible glutathionylation can be distinguished from the more facile cycling signaling processes and is predicted to be of gerontological, toxicological, pharmacological, and oncological relevance. Here, we compare reversible and irreversible glutathionylation.

1. INTRODUCTION

Glutathione is a tripeptide (L-γ-glutamyl-L-cysteinyl-glycine, Fig. 5.1) with multiple biological functions (Lushchak, 2012; Meister & Anderson, 1983; Sies, 1999). It is an abundant low-molecular-mass thiol antioxidant, which either interacts directly with reactive oxygen and nitrogen species (ROS and RNS, respectively) or serves as a cofactor for many antioxidant and associated enzymes such as peroxidases and transferases (Foster, Hess, & Stamler, 2009). In addition, glutathione is (1) a storage form of cysteine; (2) a

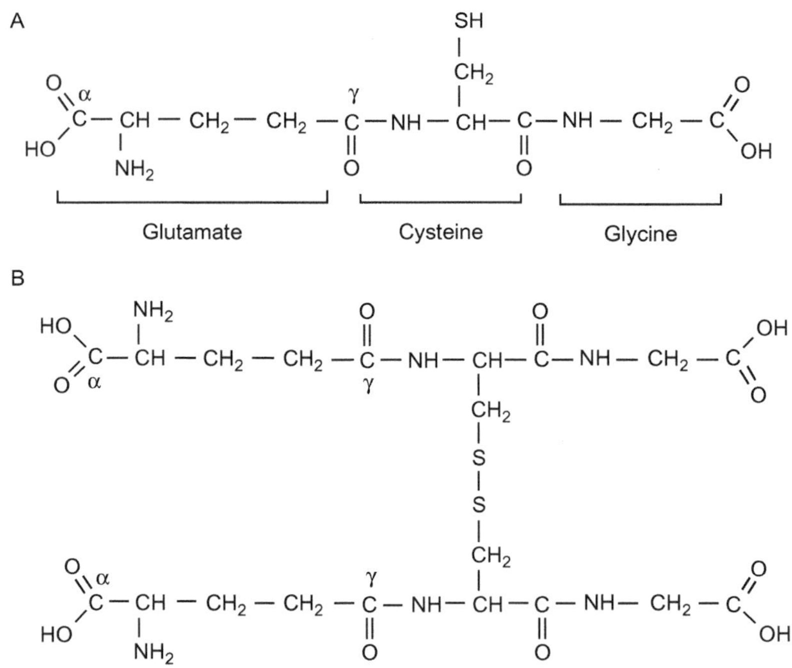

Figure 5.1 Chemical structure of glutathione in reduced (A) and oxidized (disulfide) forms (B).

storage form and transporter of nitric oxide (as GSNO); (3) involved in the metabolism of estrogens, leukotrienes, and prostaglandins, reduction of ribonucleotides to deoxyribonucleotides, and maturation of iron–sulfur clusters of proteins; (4) involved in the regulation of certain transcription factors; and (5) involved in the detoxification of many endogenous compounds and xenobiotics (the mercapturate pathway). Glutathione also can be used even for the detoxification of ions of transition metals such as chromium (Giustarini et al., 2005; Holland & Avery, 2011; Lushchak, Kubrak, Nykorak, Storey, & Lushchak, 2008). Free glutathione exists *in vivo* mostly as two forms—reduced (GSH) and oxidized (glutathione disulfide; GSSG). Its biological activity is primarily related to the active thiol group of the cysteine residue. In the intracellular milieu, glutathione is relatively stable due to the presence of an unusual γ-peptide bond between glutamate and cysteine residues. Intracellular peptidases specifically cleave peptide bonds formed from the α-carboxyl group, but not from the γ-carboxyl group. Recent attention has been drawn to the importance of the glutathione pool that is utilized in the posttranslational modification of cysteine residues, S-glutathionylation.

Glutathione is synthesized in a two-step process catalyzed by the consecutive action of γ-glutamyl-L-cysteine ligase (γGLCL, EC 6.3.2.2) and glutathione synthetase (GLS, EC 6.3.2.3). The first enzyme in the pathway is generally considered to be a regulatory enzyme in the overall synthesis and is feedback-inhibited by glutathione (Richman & Meister, 1975). Glutathione is consumed through reactions involving oxidation, conjugation, and hydrolysis. Oxidation can take place nonenzymatically through direct interaction with ROS and RNS and via enzymatic reactions catalyzed by glutathione-dependent peroxidases (Fig. 5.2). Diverse glutathione S-transferases (GSTs) catalyze conjugation of glutathione to endogenous and exogenous electrophiles. Finally, a portion of the intracellular glutathione pool may be released to the extracellular environment in either reduced or oxidized forms (Fig. 5.2). Extracellular glutathione may be hydrolyzed by the ectoenzyme γ-L-glutamyl transpeptidase (GGT, EC 2.3.2.2) to cysteinylglycine, which in turn may be hydrolyzed by dipeptidases to cysteine and glycine (Meister, 1983). Cells can take up the products liberated by glutathione hydrolysis as individual amino acids or dipeptides. Thus, a balance between production, consumption, hydrolysis, and transport determines the concentrations of intra- and extracellular glutathione pools. These processes are finely regulated and, under normal conditions, are well balanced. Regulation of glutathione levels occurs at the levels of transcription

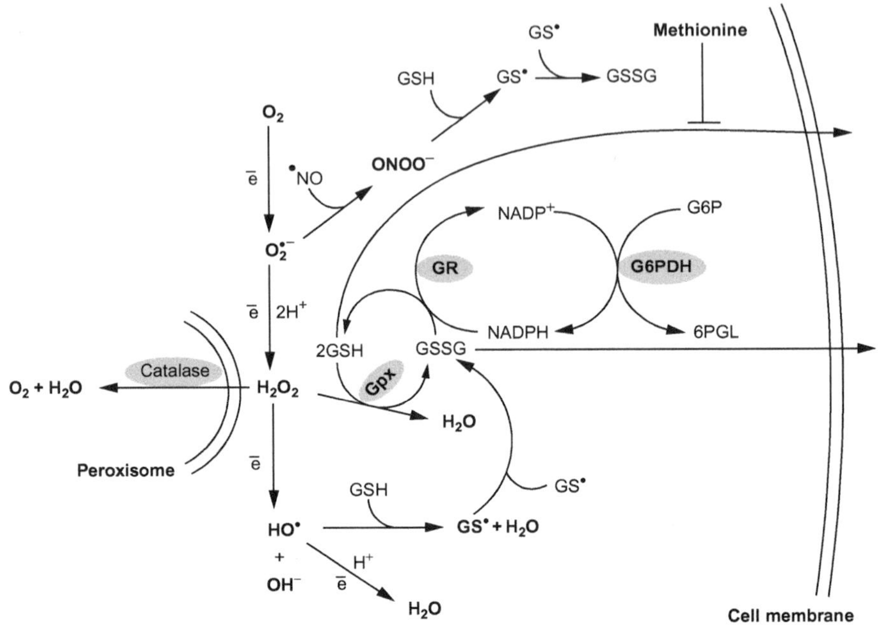

Figure 5.2 Involvement of glutathione in the elimination of reactive oxygen and nitrogen species. Hydroxyl radical and nitric oxide (after oxidation to the NO^+ form (nitrosyl cation)) or peroxynitrite ($ONOO^-$) may interact directly with GSH leading to GSSG formation. Hydrogen peroxide may be removed by catalase or by glutathione peroxidase (GPx). The latter requires glutathione to reduce peroxide. GR, glutathione reductase; G6PDH, glucose-6-phosphate dehydrogenase; G6P, glucose-6-phosphate; 6PGL, 6-phosphogluconolactone.

and translation and by posttranslational modifications of the enzymes involved in its synthesis (Lushchak, 2012).

Since glutathione plays a pivotal role as an antioxidant and participates in many regulatory and metabolic processes, the glutathione biosynthetic pathway has attracted attention from pharmacologists and biomedical scientists as a possible target for medical interventions. These strategies are directed toward decreasing or increasing glutathione levels either at the whole body level or in specific tissues. One mechanism for depleting glutathione reserves is to use inhibitors of the first step in the pathway (i.e., γGLCL), such as L-buthionine sulfoximine (BSO) (Griffith & Meister, 1979). Another strategy is to use externally added electrophiles that will react with glutathione either nonenzymatically or via reactions catalyzed by GSTs (Tew, 2007). A common strategy for increasing glutathione concentrations is to use compounds such as glutathione ethyl esters and *N*-acetyl L-cysteine each readily converted to glutathione or incorporated into glutathione *in vivo*. Dietary supplementation with cysteine-enriched peptides (some whey proteins) although less commonly considered does increase tissue glutathione levels (Droge, 2005). Glutathione levels can also be enhanced through increased expression of the enzymes involved in its biosynthesis. The upregulation of γGLCL and glutathione synthetase and increased activities of these enzymes can be induced by low-intensity ROS via redox-sensitive transcription factors, such as SoxR in bacteria, Yap1 in yeasts, and Nrf2/Keap1 in mammals (Lushchak, 2011).

Cancer cells differ from their normal counterparts in processes related to ROS metabolism because (i) ROS are frequently seen as potential inducers of transformation of normal cells to malignant cells and (ii) cancer cells possess substantially imbalanced free radical processes (Ralph, Rodriguez-Enriquez, Neuzil, Saavedra, & Moreno-Sanchez, 2010). Therefore, antioxidants are considered as potential prophylactic tools and frequently are included in diverse therapeutic procedures. Antioxidants may also be useful in preventing side effects from anticancer therapies such as radiation and drugs (Nakayama, Alladin, Igbokwe, & White, 2011). Such approaches are dependently linked with the utility of glutathione in modifying thiol groups of proteins, that is, glutathionylation. This posttranslational modification affects the operation of many proteins involved in intermediary metabolism (Cooper, Pinto, & Callery, 2011; Townsend, 2007). However, prominent effects of protein glutathionylation may be through its influence on different regulatory signaling pathways (Mieyal & Chock, 2012; Tew & Townsend, 2011).

GSH efficiently interacts with $HO^{\bullet}$, HOCl, $RO^{\bullet}$, $RO_2^{\bullet}$, $^1O_2$, peroxynitrite ($ONOO^{\bullet-}$), and numerous other ROS, leading to the formation of a thiyl radical ($GS^{\bullet}$) (Fig. 5.2). Glutathione also helps the cell to neutralize many products of ROS-promoted oxidation of lipids, including malondialdehyde and 4-hydroxy-2-nonenal. The thiyl radical can also combine with another glutathione thiyl radical leading to the production of GSSG. GSSG is also generated in the reactions catalyzed by glutathione peroxidases (GPx, EC1.11.1.9) (Eq. 5.1):

$$\mathrm{ROOH + 2GSH \rightarrow ROH + GSSG + H_2O} \qquad (5.1)$$

and by glutaredoxins (GRX, EC 1.20.4.1) (Eq. 5.2):

$$\mathrm{Oxidized glutaredoxin + 2GSH \rightarrow reduced glutaredoxin + GSSG} \qquad (5.2)$$

Clearly, as a result of neutralization of reactive species and formation of GSSG, the potential exists for (reduced) glutathione levels to be decreased. There are two major mechanisms for maintaining the glutathione pool in a reduced state: *de novo* synthesis (briefly covered in the preceding text) and reduction of GSSG. Reduction of GSSG in most organisms is carried out by glutathione reductase using NADPH as reductant. NADPH is maintained in the reduced state by several enzymatic reactions/pathways. Especially important is the pentose phosphate shunt (PPS), particularly the first and limiting step of this pathway catalyzed by glucose-6-phosphate dehydrogenase (G6PDH, EC 1.1.1.49) (Eq. 5.3):

$$\begin{aligned}&\text{Glucose-6-phosphate} + \text{NADP}^+ + \text{H}_2\text{O}\\&\quad\rightarrow \text{6-phosphoglucolactone} + \text{NADPH} + \text{H}^+\end{aligned} \tag{5.3}$$

Another important source of NADPH is the next enzymatic step of the PPS, catalyzed by 6-phosphogluconate dehydrogenase (6PGDH, EC 1.1.1.43) (Eq. 5.4):

$$\begin{aligned}&\text{NADP}^+ + \text{6-phosphogluconate} \rightarrow \text{NADPH}\\&\qquad + \text{ribulose-5-phosphate} + \text{CO}_2 + \text{H}^+\end{aligned} \tag{5.4}$$

In some tissues, particularly in the brain, malic enzyme (malate dehydrogenase (oxaloacetate-decarboxylating) ($NADP^+$); EC 1.1.1.40) is thought to be an important source of NADPH. The enzyme catalyzes the following reaction (Eq. 5.5):

$$\begin{aligned}&(S)\text{-Malate} + \text{NADP}^+ + \text{oxaloacetate}\\&\quad\rightarrow \text{pyruvate} + \text{CO}_2 + \text{NADPH} + \text{H}^+\end{aligned} \tag{5.5}$$

Finally, $NADP^+$-dependent isocitrate dehydrogenase (isocitrate dehydrogenase ($NADP^+$); EC 1.1.1.42) may also be an efficient supplier of NADPH:

$$\text{Isocitrate} + \text{NADP}^+ \rightarrow \text{2-oxoglutarate} + \text{CO}_2 + \text{NADPH} + \text{H}^+ \tag{5.6}$$

Recently, substantial attention has been paid to detoxification of reactive carbonyl species, such as glyoxal (ethane-1,2-dione) and methylglyoxal (2-oxopropanal), and their detoxification by a pathway that involves glutathione (Inoue, Maeta, & Nomura, 2011; Li, Maloney, Circu, Alexander, & Aw, 2013). Cellular methylglyoxal is produced *in vivo* mainly by the spontaneous decomposition of glyceraldehyde-3-phosphate and possibly other triosephosphates. The toxicity of methylglyoxal is based on its covalent interaction with arginine, lysine, or cysteine residues in proteins and guanine bases of nucleic acids. Methylglyoxal is detoxified by the combined action of two enzymes collectively known as the glyoxalase pathway. The first enzyme in this pathway, glyoxalase I (Glo I, EC 4.4.1.5), catalyzes the following conjugation reaction (Eq. 5.7):

$$\text{Methylglyoxal} + \text{GSH} \rightarrow (R)\text{-}S\text{-lactoyl-GSH} \quad (5.7)$$

In the second reaction, glyoxalase II (Glo II, EC 3.12.6) hydrolyzes the product of the glyoxalase I reaction:

$$(R)\text{-}S\text{-lactoyl-GSH} + H_2O \rightarrow \text{d(-)lactic acid} + \text{GSH} \quad (5.8)$$

The glyoxalase pathway represents the main route of methylglyoxal elimination in yeast (Penninckx, Jaspers, & Legrain, 1983) and mammals (Xue et al., 2012).

2. REVERSIBLE PROTEIN GLUTATHIONYLATION REACTIONS

Most frequently, protein glutathionylation is a reversible (oxidative) posttranslational modification where GSH forms disulfide linkage with cysteine residues (P-SSG), affecting certain groups of proteins (Fig. 5.3; Grek, Zhang, Manevich, Townsend, & Tew, 2013; Townsend, 2007). P-SSG increases the net negative charge of proteins and has the potential to impact structure and function of redox-sensitive targets. Redox signal transduction is influenced by temporal fluxes in ROS/RNS homeostasis and is transmitted through forward and reverse glutathionylation of redox sensors accompanied by altered GSH/GSSG ratios. GSTP1 can catalyze the forward reaction of the glutathionylation cycle, while glutaredoxin (Grx) mediates the removal of GSH through direct thiol disulfide exchange reactions leading to the formation of GSSG, which is subsequently reduced by glutathione reductase. Sulfiredoxin (Srx) is an antioxidant protein with oxidoreductase activity and a primary function in the metabolism and reactivation of Prdx. Overexpression of Srx in mammalian cells demonstrated that the enzyme has additional functions, including deglutathionylase activity (Abbas, Riquier, & Drapier, 2013; Moon et al., 2013). Srx contains a single cysteine residue in the active site, whereas Grx has a CXXC motif, suggesting distinct molecular mechanism of deglutathionylation. Park, Mieyal, Rhee, and Chock (2009) evaluated the specificity of the two enzymes toward Prdx1 and determined that Srx has a greater binding affinity toward Prdx-SSG than does

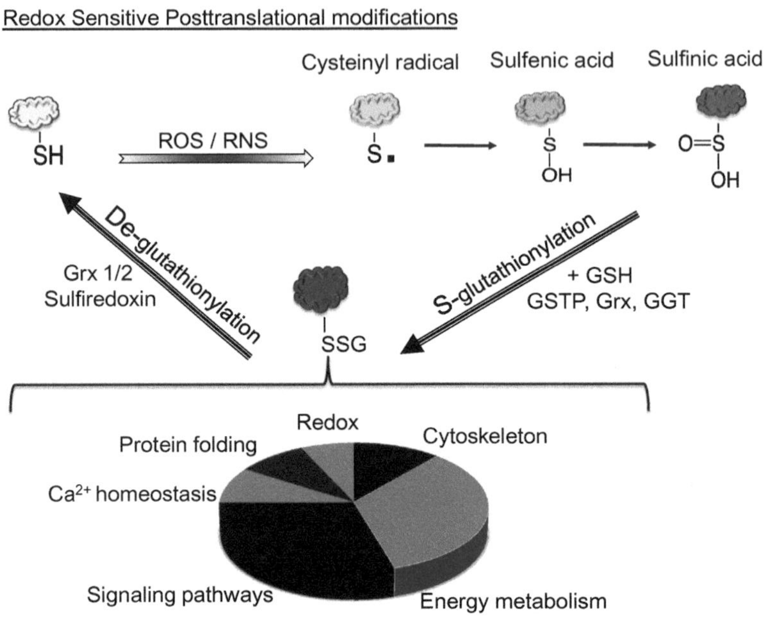

Figure 5.3 *S*-Glutathionylation cycle. Proteins in which cysteine residues possess unusually low p*Ka* values (redox sensors) are targets for oxidative or nitrosative stress. Cysteine residues within redox sensors can be oxidized to form protein sulfenic (P-SOH) and sulfinic (P-SOOH) acids. Some protein glutathionylation (P-SSG) reactions are mediated by GGT, Grx, or GSTP. P-SSG proteins have a wide variety of functions in cellular physiology/pathology, summaries of which can be found in Townsend (2007); the categories are depicted in the pie chart. (See the color plate.)

Grx. Grx and Srx may have distinct roles in the S-glutathionylation cycle perhaps influenced by substrate specificity and/or by tissue and subcellular distribution. Grx is a generalist deglutathionylase with a broad spectrum of substrates that may be dysregulated in some human pathologies (Gallogly, Starke, Leonberg, Ospina, & Mieyal, 2008).

3. IRREVERSIBLE GLUTATHIONYLATION

Here, we first briefly discuss the formation of dehydroamino acids catalyzed by enzymatic and nonenzymatic procedures. This background sets the stage for a discussion of the formation of dehydroamino acids in peptides and proteins. Dehydroamino acids, such as 2,3-didehydroalanine (2,3-dehydroalanine, DHA) and 2,3-didehydrobutyrine (2,3-dehydrobutyrine), are excellent Michael acceptors. Thus, when present in proteins, dehydroamino acid residues are predicted to form Michael adducts with glutathione (GSH), resulting in the formation of a stable thioether (a process known as nonreducible glutathionylation). Thus far, this phenomenon has been observed in mammals *in vivo* only in human cataractous lens but is predicted to be prevalent in nature, particularly in more slowly turning over proteins. We also discuss recent findings showing that proteins can be irreversibly glutathionylated by means of a double electrophile that reacts covalently with both a protein residue and the cysteine residue of GSH. Nonreducible glutathionylation is predicted to be of gerontological, toxicological, pharmacological, and oncological relevance.

4. 2,3-DEHYDROALANINE AND 2,3-DIDEHYDROBUTYRINE FORMATION IN ENZYME-CATALYZED REACTIONS

2,3-Dehydroalanine (aminoacrylate) is the end product of enzyme-catalyzed β-elimination reactions with amino acids containing a leaving group in the β position (Eq. 5.9). These reactions are catalyzed by pyridoxal 5′-phosphate (PLP)-containing enzymes. The aminoacrylate tautomerizes to the corresponding α-iminopropionate (Eq. 5.10), which is then hydrolyzed to pyruvate and ammonia (Eq. 5.11). The net reaction is shown in Eq. (5.12):

$$XCH_2CH(NH_3{}^+)CO_2^- \rightarrow CH_2{=}C(NH_3{}^+)CO_2^- + XH \quad (5.9)$$

$$CH_2{=}C(NH_3{}^+)CO_2^- \rightarrow CH_3C(=NH_2{}^+)CO_2^- \quad (5.10)$$

$$CH_3C(=NH_2{}^+)CO_2^- + H_2O \rightarrow CH_3C(=O)CO_2{}^- + NH_4{}^+ \quad (5.11)$$

$$\text{Net}: XCH_2CH(NH_3{}^+)CO_2^- + H_2O \rightarrow CH_3C(=O)CO_2{}^- + XH + NH_4{}^+ \quad (5.12)$$

An example of this type of reaction is that catalyzed by mammalian serine/threonine dehydratase. When serine is the substrate, X=OH, the enamine formed is aminoacrylate and the eliminated fragment (XH) in Eq. (5.9) is H_2O. A similar sequence of reactions occurs when threonine is the substrate (Eqs. 5.13–5.16). In this case, the eliminated fragment is also H_2O, but the enamine formed is 2,3-dehydrobutyrine. For a detailed analysis of the reaction mechanism of rat serine/threonine dehydratase, see Zhao and Liu (2008). Another example of a PLP-dependent enzyme that catalyzes a β-elimination reaction is cystathionine β-lyase. In this example, the eliminated molecule is homocysteine:

$$HOCH(CH_3)CH(NH_3^+)CO_2^- \rightarrow CH_3CH{=}C(NH_3^+)CO_2^- + H_2O \tag{5.13}$$

$$CH_3CH{=}C(NH_3^+)CO_2^- \rightarrow CH_3CH_2C({=}NH_2^+)CO_2^- \tag{5.14}$$

$$CH_3C({=}NH_2^+)CO_2^- + H_2O \rightarrow CH_3CH_2C({=}O)CO_2^- + NH_4^+ \tag{5.15}$$

$$\text{Net}: HOCH_2CH_2CH(NH_3^+)CO_2^- + H_2O \rightarrow CH_3CH_2C({=}O)CO_2^- + H_2O + NH_4^+ \tag{5.16}$$

When X (Eq. 5.9) is an exceptionally good leaving group, PLP-containing enzymes that do not normally catalyze a β-elimination reaction may catalyze a nonphysiological β-elimination (β-lyase) reaction. Such nonphysiological β-elimination reactions may be catalyzed, for example, by the PLP-dependent cytosolic and mitochondrial isozymes of aspartate aminotransferase. Thus, both enzymes catalyze the β-elimination of chloride from β-chloroalanine with the concomitant formation of pyruvate and ammonia. In another example, both enzymes catalyze a β-elimination reaction with the cysteine conjugates derived from halogenated alkenes. For a recent review, see Cooper, Krasnikov, et al. (2011).

Aminoacrylate released from the active site during enzyme-catalyzed β-elimination reactions involving β-substituted alanines has a half-life of up to a few minutes *in vitro*. This time is long enough for it to be trapped if a suitable nucleophile is included in the reaction mixture. For example, aminoacrylate formed as a result of β-elimination reactions catalyzed by aspartate aminotransferase can be trapped with thiosulfate (Cavallini, Federici, Bossa, & Granata, 1973) or β-mercaptoethanol (Adams, Lowpetch, Thorndycroft, Whyte, & Young, 2005). Because aminoacrylate (an enamine) released from β-substituted alanines and the corresponding enamine derived from threonine (2,3-dehydrobutyrine) are very reactive, they have the potential to interact with and damage nearby macromolecules. Additionally, many PLP-containing enzymes that catalyze nonphysiological β-elimination reactions are syncatalytically inactivated (i.e., the damage is self-inflicted) by covalent adduct formation of the enamine/imine with PLP cofactor and/or with a susceptible residue (Cooper, Krasnikov, et al., 2011).

It used to be thought that tautomerization of the enamine released from the active site of a β-lyase to the imine followed by hydrolysis occurs strictly by a nonenzymatic route. However, it has recently been shown that in the case of a bacterial threonine/serine deaminase, an associated protein known as RidA (*r*eactive *i*ntermediate *d*eiminase A) greatly accelerates the rate at which the released enamine/imine is converted to pyruvate and ammonia (Lambrecht, Flynn, & Downs, 2012; Lambrecht, Schmitz, & Downs, 2013), thus minimizing the potential for self-inflicted damage and damage to nearby macromolecules. The RidA proteins are highly conserved and present in all domains of life (Lambrecht et al., 2013). It is possible that this protective mechanism is less effective with aging and that an accumulation of damaged proteins may contribute to cancer progression.

The reactions mentioned in this section serve as a prelude to the following discussion of the unsaturated amino acids in peptides and proteins.

5. HISTORICAL CHARACTERIZATION OF DEHYDROPEPTIDES

Greenstein and colleagues in the 1940s prepared a number of acylated and peptide-linked dehydroamino acids and described some of their physical and enzymatic properties (Greenstein & Leuthardt, 1947; Levintow, Fu, Price, & Greenstein, 1950). They found that although 2,3-didehydroamino acids rapidly decompose to α-keto acids, 2,3-didehydroamino acids are relatively stable if the amino group is acylated or if the 2,3-didehydroamino acid is in peptide linkage. This work appears to have been largely forgotten but is important in any discussion of the biological importance of dehydroamino acids.

6. EXAMPLES OF ENZYME-CATALYZED FORMATION OF DEHYDROAMINO ACIDS IN PEPTIDE LINKAGE

A widely studied example of peptide-linked dehydroamino acids relates to the formation of a class of ribosome-directed polypeptide-based antibiotics known as lantibiotics (Asaduzzaman & Sonomoto, 2009) The synthesis of lantibiotics requires intramolecular addition of cysteine residues to DHA residues in the lantibiotic peptide precursor. DHA residues are formed by the enzymatic dehydration of serine residues. Michael attack on a DHA residue by the side group of a cysteine residue gives rise to a peptide covalently cross-linked by a thioether-containing lanthionine residue $[-O_2CCH(NH_3^+)CH_2SCH_2CH(NH_3^+)CO_2^-]$. In contrast to the easily reduced disulfide in glutathione disulfide (GSSG), the sulfide in lanthionine is stable to reduction under physiological conditions. Another interesting example of the formation of a DHA residue occurs during thyroid hormone biosynthesis. A di-iodinated aromatic ring in thyroglobulin is transferred from part of a di-iodinated tyrosine residue to an adjacent di-iodinated tyrosine residue forming a DHA residue and an adjacent thyroxine (T4) residue, respectively; upon proteolysis of the mature protein, T4 is released (Gavaret, Cahnmann, & Nunez, 1981; Gavaret, Nunez, & Cahnmann, 1980).

An example of a dehydrobutyrine intermediate is that involved in the formation of the *N*-acyl terminus of the complex bioactive polypeptides polytheonamides A and B derived from the marine sponge *Theonella swinhoei* (Freeman et al., 2012). Another example is the intramolecular addition of a cysteine residue to a 2,3-didehydrobutyrine residue generating a 3-methyl lanthionine link during the synthesis of the antimicrobial peptide nisin, a peptide used extensively in the food industry as a preservative (Li & van der Donk, 2007).

DHA, 2,3-didehydrobutyrine, lanthionine, and methyl lanthionine residues in proteins can be detected by automated Edman degradation coupled to direct detection by electrospray-ionization mass spectrometry (Walk et al., 1999). The formation of lanthionine (and methyl lanthionine) residues is discussed here because similar Michael chemistry is involved in the irreversible (nonreducible) glutathionylation of proteins/peptides discussed in the succeeding text.

7. NONENZYMATIC METHODS FOR THE INTRODUCTION OF DHA RESIDUES INTO GLUTATHIONE AND PROTEINS

Under certain conditions, the cysteine residue of glutathione (nucleophile) can be converted to a DHA residue (electrophile)—an umpolung conversion. An interesting example involves busulfan. Busulfan is a bifunctional alkylating agent used for the treatment of hematologic and other malignancies prior to stem cell transplantation (Iwamoto et al., 2004). The compound is converted *in vivo* to a glutathione S-conjugate (L-γ-glutamyl-β-(*S*-tetrahydrothiophenium)-L-alanylglycine; $GS^{+}THT$) by direct interaction with GSH and enzymatic catalysis by GSTs, especially GST A1-1 (Ritter, Bohnenstengel, Hofmann, Kroemer, & Sperker, 1999). $GS^{+}THT$ undergoes a base-catalyzed β-elimination reaction to yield γ-glutamyldehydroalanylglycine (EdAG) and tetrahydrothiophene (Cooper et al., 2008). This reaction, however, occurs readily *in vitro* at physiological pH values and temperature, and EdAG was identified as a metabolite of busulfan in a human liver cytosol fraction (Younis et al., 2008). EdAG condenses with glutathione in a Michael addition reaction to produce a lanthionine-containing thioether (GSG) (Fig. 5.4), which is a nonreducible analog of GSSG (Younis et al., 2008). EdAG has recently been shown to be an excellent hydroxyl radical scavenging agent (Peer et al., 2012) as noted previously for several antioxidant *N*-acyl dehydrolanines (Buc-Calderon, Sipe, Flitter, Mason, & Roberfroid, 1990). The authors noted that "observation of the hydroxyl trapping properties of EdAG suggests that the busulfan metabolite EdAG may contribute to or mitigate redox-related

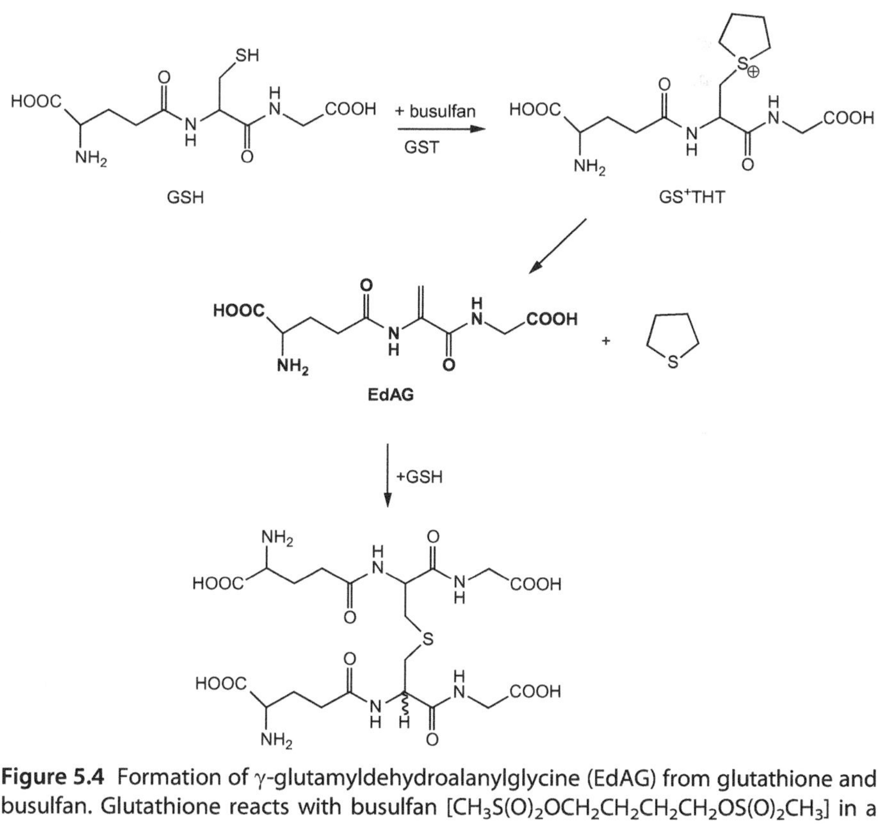

Figure 5.4 Formation of γ-glutamyldehydroalanylglycine (EdAG) from glutathione and busulfan. Glutathione reacts with busulfan [$CH_3S(O)_2OCH_2CH_2CH_2CH_2OS(O)_2CH_3$] in a reaction accelerated by GSTs. Two equivalents of $CH_3S(O)_2H$ are released with the formation of the corresponding glutathione conjugate (GS^+THT). This conjugate contains a cyclic sulfonium moiety, which is an excellent leaving group. The elimination of thiophene results in the formation of a glutathione analog (EdAG) in which the cysteine residue is converted to a dehydroalanine residue. EdAG can then participate in a Michael addition reaction with glutathione to generate the Michael adduct GSG that contains a stable thioether bond. *From Younis et al. (2008) with permission.*

cytotoxicity associated with the therapeutic use of busulfan, and reaffirms indicators that support a role in free radical biology for dehydroalanine-containing peptides and proteins" (Buc-Calderon et al., 1990).

Cysteine (and serine) residues in proteins can also be converted to DHA residues. Koshland's group carried out some of the earliest work in this area in the 1960s (Strumeyer, White, & Koshland, 1963; Weiner, White, Hoare, & Koshland, 1966). The authors tosylated the active site serine of chymotrypsin. Under alkaline conditions, the tosylated serine residue was converted to a DHA residue. Since that time, a number of procedures have been described to convert a serine or a cysteine residue to a dehydroalanine residue. These procedures are discussed in Chalker, Bernardes, and Davis (2011), Cooper, Pinto, et al. (2011), Cooper et al. (2008), Griffin, Srinivasan, Zheng, Huang, and Coughlin (2001), Chalker, Gunnoo, et al. (2011), and Younis et al. (2008). As discussed by Chalker, Bernardes, et al. (2011), the DHA residue can readily react with a number of RSH compounds to generate S-substituted cysteine residues, where R may be, for example, $-PO_3^{2-}$, glycosyl group, alkyl group, aminoalkyl group, or lipid side chain.

8. REDUCIBLE GLUTATHIONYLATION OF LENS PROTEINS

GSH is present at high concentrations in the human lens (~6 mM) where, as an antioxidant, it is essential for maintaining transparency (Craghill, Cronshaw, & Harding, 2004). The size of the GSH pool diminishes with age, and loss of GSH is associated with cataract formation. Cataractous lenses exhibit a decreased GSH/GSSG ratio compared with clear lenses (Zhang, Chai, Yan, Guo, & Harding, 2008). As discussed by Harding and colleagues (Craghill et al., 2004 and references cited therein), the unique development and structure of the lens may explain the need for high levels of GSH. During cataractogenesis, especially in the nucleus part of the lens, the lens proteins unfold and thiols that were buried become exposed and reactive. Some of these thiols become oxidized and react to form (a) mixed disulfides with GSH and cysteine and (b) disulfide cross-linked aggregates. With increasing severity of the cataract, total protein thiol decreases with a concurrent increase in protein disulfide content, including disulfide linkages to GSH. A major glutathionylated protein in the lens was found to have a molecular mass of about 47 kDa, which was shown to be composed of βB1-, βB2-, and γS-crystallins (Craghill et al., 2004). In theory, the disulfides, including glutathionylated cysteine residues, in the lens are reducible. Thus, the possibility has been considered that maintaining the GSH status may be helpful in preventing damaging protein disulfide formation and in ameliorating cataract formation (Craghill et al., 2004; Zhang et al., 2008). In a streptozotocin model of diabetes in rats, the administration of eye drops containing *N*-acetyl-L-cysteine and glutathione ethyl ester (compounds readily converted to GSH *in vivo*) showed significant inhibition of the progression of diabetic cataracts at early (but not late) stages of the diabetes-associated cataractogenesis (Zhang et al., 2008).

9. IRREVERSIBLE, NONREDUCIBLE GLUTATHIONYLATION OF LENS PROTEINS

The lens nucleus contains proteins that are present from birth. Moreover, the outer fibrous cells no longer make proteins after the lens has been fully shaped (Zhang et al., 2008 and references therein). Thus, it is not surprising that after many decades, some nonenzymatic posttranslational protein modifications occur in the lens. One such modification is the loss of H_2S or H_2O from cysteine or serine residues, respectively, resulting in the formation of DHA residues (Fig. 5.5). Michael addition by cysteine, histidine, or lysine residues will result in proteins cross-linked internally (or to

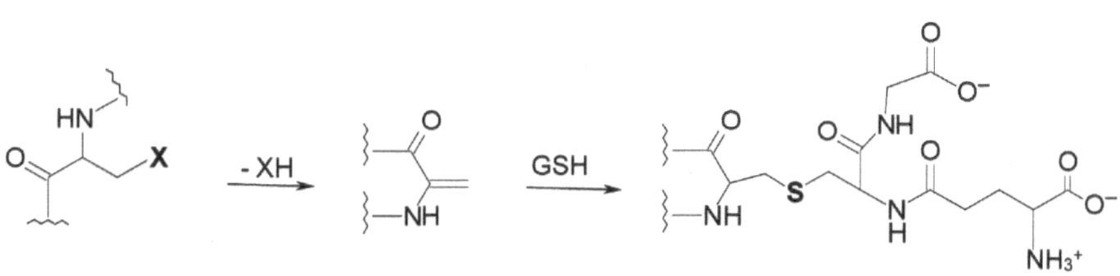

Figure 5.5 Formation of a dehydroalanyl (DHA) residue in proteins and peptides followed by irreversible glutathionylation. A protein/peptide containing a leaving group (X) in the β position of an amino acid residue may undergo spontaneous β-elimination to generate a protein/peptide containing a DHA residue (center). Michael addition of the sulfhydryl group of glutathione will generate an adduct containing a nonreducible thioether bond, resulting in irreversible glutathionylation. Residues such as cysteine (X = –SH) and serine (X = –OH) are expected to be relatively stable. However, for proteins with very long turnover times, such as those found in the lens, β-elimination reactions may occur as a consequence of aging. Moreover, other residues such as selenocysteine, serine *O*-phosphate, and serine *O*-sulfate are expected to be more labile and amenable to β-elimination. *Modified from Cooper, Pinto, et al. (2011).*

other proteins) by lanthionine bridges, histidinoalanine linkages, or lysylalanine cross-links, respectively (Linetsky, Hill, LeGrand, & Hu, 2004; Linetsky & LeGrand, 2005). Linetsky and colleagues showed that lanthionine and histidinoalanine residues in hydrolyzed proteins are present at higher concentrations in cataractous lenses than in normal lenses; lysylalanine could only be detected in cataractous lenses (Linetsky et al., 2004).

Lanthionine residues may be formed in lens tissues not only via Michael addition of a protein cysteine residue to a DHA residue but also by attack of either free cysteine or the cysteine moiety of GSH. Linetsky and LeGrand (Linetsky & LeGrand, 2005) showed that the increased lanthionine residues occurring in the aged and cataractous human lens result mainly from the addition of GSH to DHA residues, resulting in irreversible, nonreducible glutathionylation.

EdAG may be present at low concentrations as a result of naturally occurring umpolung transformations involving GSH. Thus, we suggest that another possibility for irreversible glutathionylation of proteins is the addition of EdAG to cysteine residues. This process is expected to be of minor consequence for most proteins but may become more prevalent with time especially with slowly turning over proteins or when individuals are exposed to xenobiotics that can promote the formation of EdAG from GSH. In this context, it is interesting to note that patients systemically treated with busulfan are at increased risk of cataractous changes to the posterior cortex of the lens (Li, Tripathi, & Tripathi, 2008). It is possible that increased formation of EdAG contributes to the damage to the lens.

10. NONREDUCIBLE GLUTATHIONYLATION INVOLVING COVALENT TETHERING

The polyunsaturated lipid peroxidation product 4-oxo-2-nonenal (ONE) is a highly reactive protein cross-linking agent (Zhu, Gallogly, Mieyal, Anderson, & Sayre, 2009). The compound also adds to GSH, resulting in the formation of a reactive 4-ketoaldehyde that has the potential to covalently add to proteins (Cooper, Pinto, et al., 2011). Zhu et al. (2009) showed that almost every lysine residue was modified when 0.25 mM bovine β-lactoglobulin was incubated in 100 mM potassium phosphate buffer containing 0.25–2 mM ONE and 1 mM GSH (an antielectrophile) at 37 °C for 24 h. The GSH is effectively bound to the protein via a thioether tether. The authors suggested that "stable antielectrophile-ONE-protein cross-links may serve as biomarkers of oxidative stress and may represent a novel mechanism of irreversible protein glutathionylation" (Zhu et al., 2009).

In another recent article, the arylated monodiazeniumdiolate JS-K was compared to the related bis-diazeniumdiolate (Double JS-K) for their ability to conjugate GSH to proteins (Holland et al., 2012). The results revealed a previously unrecognized mechanism of protein modification displayed by certain members of the arylated diazeniumdiolate family. GSH was shown to be irreversibly tethered to proteins in lung cancer cells and leukemia cells via reaction of the –SH of GSH and a protein –SH via a bivalent electrophile spacer, and the authors suggested that it is likely that other bivalent electrophiles, including some current chemotherapeutic agents, may support a similar type of cross-linking glutathionylation (Holland et al., 2012).

Finally, the anticancer drug piperlongumine has recently been shown to covalently tether GSH to proteins (Adams et al., 2012). The authors suggested a sequence of events involving first a Michael addition of GSH to the more electrophilic C2–C3 olefin in piperlongumine, followed by the formation of a noncovalent complex between the piperlongumine–GSH adduct and a GSH-binding protein, and finally a Michael addition of a nucleophilic residue of the GSH-binding protein to the less electrophilic C7–C8 olefin of piperlongumine that is accelerated by the formation of the complex (Adams et al., 2012).

11. CONCLUSIONS

In contrast to the cyclical reactions that are characterized by the addition/removal of GSH to reactive cysteines (Townsend, 2007; Townsend et al., 2009), irreversible glutathionylation of proteins in which the covalent linkage is via a lanthionine residue has thus far been detected *in vivo* only in lens proteins. The irreversible glutathionylation is more prevalent in cataractous lenses. We suggest, however, that irreversible glutathionylation may be more widespread in body tissues than is currently appreciated. Slowly turning over proteins throughout the body may become "damaged" in such a way that unfolding exposes and activates cysteine and serine residues making them more prone to conversion to DHA residues followed by Michael addition of GSH. Such processes are expected to increase with oxidative/nitrosative stress and with increasing age. Irreversible glutathionylation may lead to loss of function and/or enhanced removal of the protein. In addition to a cysteine residue of GSH reacting with a DHA residue, it is possible that irreversible glutathionylation may occur by reaction of a protein cysteine residue with EdAG. To our knowledge, irreversible protein glutathionylation via the addition of EdAG to a cysteine residue in a protein has not yet been demonstrated directly. However, our finding that GSH can add to EdAG suggests that if EdAG is generated *in vivo*, the possibility exists that it will react with protein cysteine residues. Certain drugs (and other xenobiotics) that can interact with GSH *in vivo* to generate EdAG (e.g., busulfan) may be toxic in part through irreversible glutathionylation. Finally, several examples have recently been described of bivalent xenobiotic electrophiles that can act as a tether covalently anchoring GSH to proteins.

In summary, there is currently only limited information on irreversible protein glutathionylation via lanthionine linkages or via bivalent electrophile linkages. However, the scope of these reactions, particularly in regard to disease, aging, drug metabolism, and disposition of anticancer drugs, is relatively large. Irreversible protein glutathionylation is a fruitful area for future research.

ACKNOWLEDGMENTS

This work was supported by grants from the National Center for Research Resources (5P20RR024485-02) and the National Institute of General Medical Sciences (8 P20 GM103542-02) from the National Institutes of Health and by CA08660, CA117259, and R56 ES017453 and support from the South Carolina Centers of Excellence program.

REFERENCES

Abbas, K., Riquier, S., & Drapier, J. C. (2013). Peroxiredoxins and sulfiredoxin at the crossroads of the NO and H2O2 signaling pathways. *Methods in Enzymology*, *527*, 113–128.

Adams, D. J., Dai, M., Pellegrino, G., Wagner, B. K., Stern, A. M., Shamji, A. F., et al. (2012). Synthesis, cellular evaluation, and mechanism of action of piperlongumine analogs. *Proceedings of the National Academy of Sciences of the United States of America*, *109*, 15115–15120.

Adams, B., Lowpetch, K., Thorndycroft, F., Whyte, S. M., & Young, D. W. (2005). Stereochemistry of reactions of the inhibitor/substrates L- and D-beta-chloroalanine with beta-mercaptoethanol catalysed by L-aspartate aminotransferase and D-amino acid aminotransferase respectively. *Organic and Biomolecular Chemistry*, *3*, 3357–3364.

Asaduzzaman, S. M., & Sonomoto, K. (2009). Lantibiotics: Diverse activities and unique modes of action. *Journal of Bioscience and Bioengineering*, *107*, 475–487.

Buc-Calderon, P., Sipe, H. J., Jr., Flitter, W., Mason, R. P., & Roberfroid, M. (1990). N-acyl dehydroalanines scavenge oxygen radicals and inhibit in vitro free radical mediated processes. *Chemico-Biological Interactions*, *73*, 77–88.

Cavallini, D., Federici, G., Bossa, F., & Granata, F. (1973). The protective effect of thiosulfate upon the inactivation of aspartate aminotransferase by aminoacrylic-acid-producing substrates. *European Journal of Biochemistry*, *39*, 301–304.

Chalker, J. M., Bernardes, G. J., & Davis, B. G. (2011). A "tag-and-modify" approach to site-selective protein modification. *Accounts of Chemical Research*, *44*, 730–741.

Chalker, J. M., Gunnoo, S. B., Boutureira, O., Gerstberger, S. C., Fernández-González, M., Bernardes, G. J. L., et al. (2011). Methods for converting cysteine to dehydroalanine on peptides and proteins. *Chemical Science*, *2*, 1666–1676.

Cooper, A. J. L., Krasnikov, B. F., Niatsetskaya, Z. V., Pinto, J. T., Callery, P. S., Villar, M. T., et al. (2011). Cysteine S-conjugate beta-lyases: Important roles in the metabolism of naturally occurring sulfur and selenium-containing compounds, xenobiotics and anticancer agents. *Amino Acids*, *41*, 7–27.

Cooper, A. J. L., Pinto, J. T., & Callery, P. S. (2011). Reversible and irreversible protein glutathionylation: Biological and clinical aspects. *Expert Opinion on Drug Metabolism & Toxicology*, 7, 891–910.

Cooper, A. J. L., Younis, I. R., Niatsetskaya, Z. V., Krasnikov, B. F., Pinto, J. T., Petros, W. P., et al. (2008). Metabolism of the cysteine S-conjugate of busulfan involves a beta-lyase reaction. *Drug Metabolism and Disposition*, *36*, 1546–1552.

Craghill, J., Cronshaw, A. D., & Harding, J. J. (2004). The identification of a reaction site of glutathione mixed-disulphide formation on gammaS-crystallin in human lens. *The Biochemical Journal*, *379*, 595–600.

Droge, W. (2005). Oxidative stress and ageing: Is ageing a cysteine deficiency syndrome? *Philosophical Transactions of the Royal Society of London. Series B, Biological Sciences*, *360*, 2355–2372.

Foster, M. W., Hess, D. T., & Stamler, J. S. (2009). Protein S-nitrosylation in health and disease: A current perspective. *Trends in Molecular Medicine*, *15*, 391–404.

Freeman, M. F., Gurgui, C., Helf, M. J., Morinaka, B. I., Uria, A. R., Oldham, N. J., et al. (2012). Metagenome mining reveals polytheonamides as posttranslationally modified ribosomal peptides. *Science*, *338*, 387–390.

Gallogly, M. M., Starke, D. W., Leonberg, A. K., Ospina, S. M., & Mieyal, J. J. (2008). Kinetic and mechanistic characterization and versatile catalytic properties of mammalian glutaredoxin 2: Implications for intracellular roles. *Biochemistry*, *47*, 11144–11157.

Gavaret, J. M., Cahnmann, H. J., & Nunez, J. (1981). Thyroid hormone synthesis in thyroglobulin. The mechanism of the coupling reaction. *The Journal of Biological Chemistry*, *256*, 9167–9173.

Gavaret, J. M., Nunez, J., & Cahnmann, H. J. (1980). Formation of dehydroalanine residues during thyroid hormone synthesis in thyroglobulin. *The Journal of Biological Chemistry*, *255*, 5281–5285.

Giustarini, D., Milzani, A., Aldini, G., Carini, M., Rossi, R., & Dalle-Donne, I. (2005). S-nitrosation versus S-glutathionylation of protein sulfhydryl groups by S-nitrosoglutathione. *Antioxidants and Redox Signaling*, 7, 930–939.

Greenstein, J. P., & Leuthardt, F. M. (1947). Presence of dehydropeptidases I and II activity in plant materials. *Journal of the National Cancer Institute*, *8*, 35–37.

Grek, C. L., Zhang, J., Manevich, Y., Townsend, D. M., & Tew, K. D. (2013). Causes and consequences of cysteine S-glutathionylation. *The Journal of Biological Chemistry*, *288*, 26497–26504.

Griffin, C. T., Srinivasan, Y., Zheng, Y. W., Huang, W., & Coughlin, S. R. (2001). A role for thrombin receptor signaling in endothelial cells during embryonic development. *Science*, *293*, 1666–1670.

Griffith, O. W., & Meister, A. (1979). Glutathione: Interorgan translocation, turnover, and metabolism. *Proceedings of the National Academy of Sciences of the United States of America*, *76*, 5606–5610.

Holland, S. L., & Avery, S. V. (2011). Chromate toxicity and the role of sulfur. *Metallomics, 3*, 1119–1123.
Holland, R. J., Maciag, A. E., Kumar, V., Shi, L., Saavedra, J. E., Prud'homme, R. K., et al. (2012). Cross-linking protein glutathionylation mediated by O2-arylated bis-diazeniumdiolate "Double JS-K". *Chemical Research in Toxicology, 25*, 2670–2677.
Inoue, Y., Maeta, K., & Nomura, W. (2011). Glyoxalase system in yeasts: Structure, function, and physiology. *Seminars in Cell & Developmental Biology, 22*, 278–284.
Iwamoto, T., Hiraku, Y., Oikawa, S., Mizutani, H., Kojima, M., & Kawanishi, S. (2004). DNA intrastrand cross-link at the 5'-GA-3' sequence formed by busulfan and its role in the cytotoxic effect. *Cancer Science, 95*, 454–458.
Lambrecht, J. A., Flynn, J. M., & Downs, D. M. (2012). Conserved YjgF protein family deaminates reactive enamine/imine intermediates of pyridoxal 5'-phosphate (PLP)-dependent enzyme reactions. *The Journal of Biological Chemistry, 287*, 3454–3461.
Lambrecht, J. A., Schmitz, G. E., & Downs, D. M. (2013). RidA proteins prevent metabolic damage inflicted by PLP-dependent dehydratases in all domains of life. *MBio, 4*, e00033-00013.
Levintow, L., Fu, S. C., Price, V. E., & Greenstein, J. P. (1950). Preparation and enzymatic hydrolysis of three new homologous dehydropeptides. *The Journal of Biological Chemistry, 184*, 633–640.
Li, W., Maloney, R. E., Circu, M. L., Alexander, J. S., & Aw, T. Y. (2013). Acute carbonyl stress induces occludin glycation and brain microvascular endothelial barrier dysfunction: Role for glutathione-dependent metabolism of methylglyoxal. *Free Radical Biology and Medicine, 54*, 51–61.
Li, J., Tripathi, R. C., & Tripathi, B. J. (2008). Drug-induced ocular disorders. *Drug Safety, 31*, 127–141.
Li, B., & van der Donk, W. A. (2007). Identification of essential catalytic residues of the cyclase NisC involved in the biosynthesis of nisin. *The Journal of Biological Chemistry, 282*, 21169–21175.
Linetsky, M., Hill, J. M., LeGrand, R. D., & Hu, F. (2004). Dehydroalanine crosslinks in human lens. *Experimental Eye Research, 79*, 499–512.
Linetsky, M., & LeGrand, R. D. (2005). Glutathionylation of lens proteins through the formation of thioether bond. *Molecular and Cellular Biochemistry, 272*, 133–144.
Lushchak, O. V., Kubrak, O. I., Nykorak, M. Z., Storey, K. B., & Lushchak, V. I. (2008). The effect of potassium dichromate on free radical processes in goldfish: Possible protective role of glutathione. *Aquatic Toxicology, 87*, 108–114.
Lushchak, V. I. (2011). Adaptive response to oxidative stress: Bacteria, fungi, plants and animals. *Comparative Biochemistry and Physiology. Toxicology & Pharmacology, 153*, 175–190.
Lushchak, V. I. (2012). Glutathione homeostasis and functions: Potential targets for medical interventions. *Journal of Amino Acids, 2012*, 736837.
Meister, A. (1983). Selective modification of glutathione metabolism. *Science, 220*, 472–477.
Meister, A., & Anderson, M. E. (1983). Glutathione. *Annual Review of Biochemistry, 52*, 711–760.
Mieyal, J. J., & Chock, P. B. (2012). Posttranslational modification of cysteine in redox signaling and oxidative stress: Focus on s-glutathionylation. *Antioxidants and Redox Signaling, 16*, 471–475.
Moon, J. C., Kim, G. M., Kim, E. K., Lee, H. N., Ha, B., Lee, S. Y., et al. (2013). Reversal of 2-Cys peroxiredoxin oligomerization by sulfiredoxin. *Biochemical and Biophysical Research Communications, 432*, 291–295.
Nakayama, A., Alladin, K. P., Igbokwe, O., & White, J. D. (2011). Systematic review: Generating evidence-based guidelines on the concurrent use of dietary antioxidants and chemotherapy or radiotherapy. *Cancer Investigation, 29*, 655–667.

Park, J. W., Mieyal, J. J., Rhee, S. G., & Chock, P. B. (2009). Deglutathionylation of 2-Cys peroxiredoxin is specifically catalyzed by sulfiredoxin. *The Journal of Biological Chemistry, 284*, 23364–23374.

Peer, C. J., Younis, I. R., Leonard, S. S., Gannett, P. M., Minarchick, V. C., Kenyon, A. J., et al. (2012). Glutathione conjugation of busulfan produces a hydroxyl radical-trapping dehydroalanine metabolite. *Xenobiotica, 42*, 1170–1177.

Penninckx, M. J., Jaspers, C. J., & Legrain, M. J. (1983). The glutathione-dependent glyoxalase pathway in the yeast Saccharomyces cerevisiae. *The Journal of Biological Chemistry, 258*, 6030–6036.

Ralph, S. J., Rodriguez-Enriquez, S., Neuzil, J., Saavedra, E., & Moreno-Sanchez, R. (2010). The causes of cancer revisited: "Mitochondrial malignancy" and ROS-induced oncogenic transformation—Why mitochondria are targets for cancer therapy. *Molecular Aspects of Medicine, 31*, 145–170.

Richman, P. G., & Meister, A. (1975). Regulation of gamma-glutamyl-cysteine synthetase by nonallosteric feedback inhibition by glutathione. *The Journal of Biological Chemistry, 250*, 1422–1426.

Ritter, C. A., Bohnenstengel, F., Hofmann, U., Kroemer, H. K., & Sperker, B. (1999). Determination of tetrahydrothiophene formation as a probe of in vitro busulfan metabolism by human glutathione S-transferase A1-1: Use of a highly sensitive gas chromatographic-mass spectrometric method. *Journal of Chromatography. B, Biomedical Sciences and Applications, 730*, 25–31.

Sies, H. (1999). Glutathione and its role in cellular functions. *Free Radical Biology and Medicine, 27*, 916–921.

Strumeyer, D. H., White, W. N., & Koshland, D. E., Jr. (1963). Role of serine in chymotrypsin action. Conversion of the active serine to dehydroalanine. *Proceedings of the National Academy of Sciences of the United States of America, 50*, 931–935.

Tew, K. D. (2007). Redox in redux: Emergent roles for glutathione S-transferase P (GSTP) in regulation of cell signaling and S-glutathionylation. *Biochemical Pharmacology, 73*, 1257–1269.

Tew, K. D., & Townsend, D. M. (2011). Regulatory functions of glutathione S-transferase P1-1 unrelated to detoxification. *Drug Metabolism Reviews, 43*, 179–193.

Townsend, D. M. (2007). S-glutathionylation: Indicator of cell stress and regulator of the unfolded protein response. *Molecular Interventions, 7*, 313–324.

Townsend, D. M., Manevich, Y., He, L., Hutchens, S., Pazoles, C. J., & Tew, K. D. (2009). Novel role for glutathione S-transferase pi. Regulator of protein S-Glutathionylation following oxidative and nitrosative stress. *The Journal of Biological Chemistry, 284*, 436–445.

Walk, T. B., Sussmuth, R., Kempter, C., Gnau, V., Jack, R. W., & Jung, G. (1999). Identification of unusual amino acids in peptides using automated sequential Edman degradation coupled to direct detection by electrospray-ionization mass spectrometry. *Biopolymers, 49*, 329–340.

Weiner, H., White, W. N., Hoare, D. G., & Koshland, D. E., Jr. (1966). The formation of anhydrochymotrypsin by removing the elements of water from the serine at the active site. *Journal of the American Chemical Society, 88*, 3851–3859.

Xue, M., Rabbani, N., Momiji, H., Imbasi, P., Anwar, M. M., Kitteringham, N., et al. (2012). Transcriptional control of glyoxalase 1 by Nrf2 provides a stress-responsive defence against dicarbonyl glycation. *The Biochemical Journal, 443*, 213–222.

Younis, I. R., Elliott, M., Peer, C. J., Cooper, A. J., Pinto, J. T., Konat, G. W., et al. (2008). Dehydroalanine analog of glutathione: An electrophilic busulfan metabolite that binds to human glutathione S-transferase A1-1. *The Journal of Pharmacology and Experimental Therapeutics, 327*, 770–776.

Zhang, S., Chai, F. Y., Yan, H., Guo, Y., & Harding, J. J. (2008). Effects of N-acetylcysteine and glutathione ethyl ester drops on streptozotocin-induced diabetic cataract in rats. *Molecular Vision, 14*, 862–870.

Zhao, Z., & Liu, H. (2008). A quantum mechanical/molecular mechanical study on the catalysis of the pyridoxal 5'-phosphate-dependent enzyme L-serine dehydratase. *The Journal of Physical Chemistry. B, 112*, 13091–13100.

Zhu, X., Gallogly, M. M., Mieyal, J. J., Anderson, V. E., & Sayre, L. M. (2009). Covalent cross-linking of glutathione and carnosine to proteins by 4-oxo-2-nonenal. *Chemical Research in Toxicology, 22*, 1050–1059.

CHAPTER SIX

Glutathione Transferases in the Bioactivation of Azathioprine

Olof Modén*, Bengt Mannervik*,†,1
*Department of Chemistry-BMC, Uppsala University, Uppsala, Sweden
†Department of Neurochemistry, Stockholm University, Stockholm, Sweden
[1]Corresponding author: e-mail address: bengt.mannervik@neurochem.su.se

Contents

Advances in Cancer Research, Volume 122
ISSN 0065-230X
http://dx.doi.org/10.1016/B978-0-12-420117-0.00006-2

Abstract

The prodrug azathioprine is primarily used for maintaining remission in inflammatory bowel disease, but approximately 30% of the patients suffer adverse side effects. The prodrug is activated by glutathione conjugation and release of 6-mercaptopurine, a reaction most efficiently catalyzed by glutathione transferase (GST) A2-2. Among five genotypes of GST A2-2, the variant A2*E has threefold–fourfold higher catalytic efficiency with azathioprine, suggesting that the expression of A2*E could boost 6-mercaptopurine release and adverse side effects in treated patients.

Structure–activity studies of the GST A2-2 variants and homologous alpha class GSTs were made to delineate the determinants of high catalytic efficiency compared to other alpha class GSTs. Engineered chimeras identified GST peptide segments of importance, and replacing the corresponding regions in low-activity GSTs by these short segments produced chimeras with higher azathioprine activity. By contrast, H-site mutagenesis led to decreased azathioprine activity when active-site positions 208 and 213 in these favored segments were mutagenized. Alternative substitutions indicated that hydrophobic residues were favored.

A pertinent question is whether variant A2*E represents the highest azathioprine activity achievable within the GST structural framework. This issue was addressed by mutagenesis of H-site residues assumed to interact with the substrate based on molecular modeling. The mutants with notably enhanced activities had small or polar residues in the mutated positions. The most active mutant L107G/L108D/F222H displayed a 70-fold enhanced catalytic efficiency with azathioprine. The determination of its structure by X-ray crystallography showed an expanded H-site, suggesting improved accommodation of the transition state for catalysis.

1. PREAMBLE

Azathioprine was originally developed as an anticancer agent based on its biotransformation that releases the antimetabolite 6-mercaptopurine (6-MP). However, both azathioprine and 6-MP have been used more widely as immunosuppressive and anti-inflammatory agents in organ transplantation and in treatment of various autoimmune and chronic inflammatory diseases, such as multiple sclerosis, rheumatoid arthritis, and inflammatory bowel disease (IBD) (Tiede et al., 2003). Regarding organ transplantation, there are today drugs that not only act faster than the thiopurines but also have more adverse effects (Barba et al., 2012), while the results for organ function and cancer incidence are similar in the longer perspective (Clayton, McDonald, Chapman, & Chadban, 2012). In spite of the widespread clinical use of azathioprine for half a century, the enzymology of the first step of the activation of azathioprine has remained obscure until recently (Eklund, Moberg, Bergquist, & Mannervik, 2006). In this chapter, the role of glutathione transferases (GSTs) will be placed in the context of azathioprine metabolism with possible implications for clinical side effects. Particular focus will be placed on human glutathione transferase A2-2 and molecular features that govern relationships between protein structure and catalytic activity with azathioprine.

2. BACKGROUND

2.1. Azathioprine

The thiopurine drugs 6-MP and 6-thioguanine were developed in the 1950s by George Hitchings and Gertrude B. Elion (Giner-Sorolla, 1988). Hitchings had theorized that antagonists of DNA bases might be able to inhibit the growth of rapidly dividing cells, and Elion screened several purine derivatives for the inhibition of purine utilization (Elion, 1989). 6-MP turned out to be active against tumors in animal models and acute leukemia in children. Azathioprine is a 1-methyl-4-nitroimidazole-5-yl derivative of 6-MP (Fig. 6.1), which was designed as a prodrug intended to release 6-MP slowly and preferably by a tumor-specific enzyme. For childhood leukemia, the chemotherapeutic index of azathioprine and 6-MP was similar, while for a carcinoma model in mice, azathioprine was similarly active but less toxic.

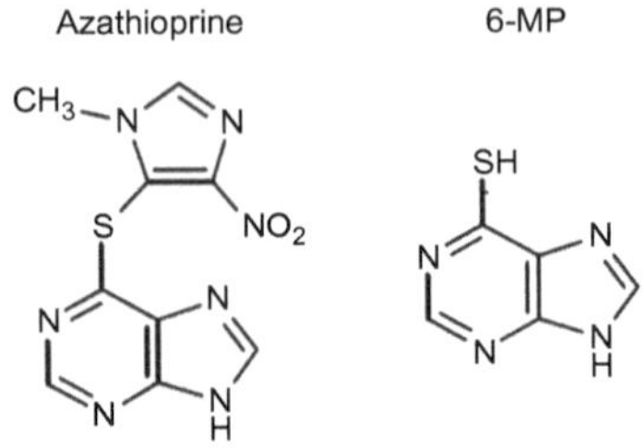

Figure 6.1 Structural formulas of azathioprine and 6-mercaptopurine (6-MP).

Robert Schwartz proposed that since the immunoblastic lymphocytes formed during an immune response were similar to leukemic lymphocytes, the purine analogs might inhibit the immune response (Schwartz & Dameshek, 1959). 6-MP given to rabbits simultaneously with antigen administration was indeed effective. Calne and coworkers (1962) proved that kidney transplantation in dogs was possible with immunosuppression by 6-MP, and later, studies showed that azathioprine was superior for preventing rejection. Since its introduction, the clinical use of azathioprine has been dominated by its ability to suppress the immune system. In the early 1960s, kidney transplantations in humans became successful by the administration of azathioprine and anti-inflammatory corticosteroids.

The pharmacological action of azathioprine is based on the release of 6-MP effected by the chemical elimination of the imidazole moiety. Glutathione is the most abundant low-molecular-mass thiol in the cell (Josephy & Mannervik, 2006), and the thiolysis of azathioprine by glutathione has been established as the pivotal biotransformation in crude liver preparations (Kaplowitz, 1976). Although the original work suggested the involvement of GST activity, it was not until studies involving purified enzymes were performed that the overwhelming contribution of the enzymatic reaction was established (Eklund et al., 2006).

2.2. Glutathione

The tripeptide glutathione (Fig. 6.2) is synthesized intracellularly, while the degradation of glutathione conjugates is initiated extracellularly by γ-glutamyl transferase on the cell membrane (Ballatori, Hammond, Cunningham, Krance, & Marchan, 2005). The concentration of reduced glutathione has been estimated at about 1–2 μM in human plasma and about 1 mM in whole blood (Andersson, Lindgren, Arnadottir,

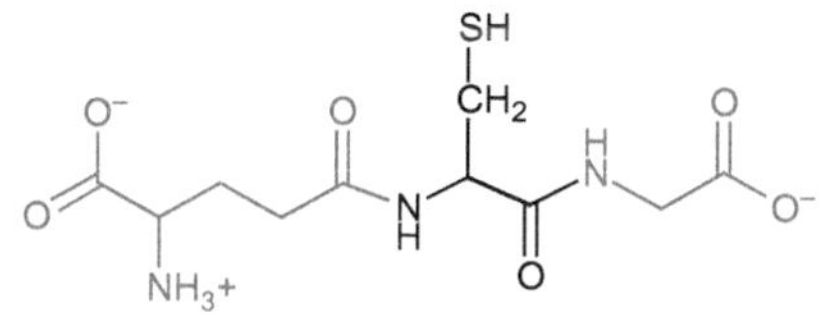

Figure 6.2 Glutathione (γ-L-glutamyl-L-cysteinylglycine or γ-L-Glu-L-Cys-Gly).

Prytz, & Hultberg, 1999) while estimated to be in the range 0.5–10 mM inside mammalian cells (Meister & Anderson, 1983). The thiol group of glutathione is a nucleophile and the ionized thiolate is an even more reactive chemical group that can attack electrophilic centers in other molecules. In addition to conjugation reactions, glutathione can serve as an antioxidant and reduce peroxides and other reactive oxidative species.

2.3. Glutathione transferases

2.3.1 Function

GSTs are enzymes that catalyze the reaction of glutathione with electrophiles of both endogenous and xenobiotic origins. The main biological roles of GSTs encompass detoxification and protection against oxidative stress. By conjugating glutathione with toxic electrophilic substrates, the resulting molecules generally become less reactive and more soluble, thus facilitating their excretion from cells and the organism. The GSTs are promiscuous in their substrate acceptance, and collectively, they can deal with various xenobiotics or metabolic by-products that otherwise could be harmful and damage DNA and other cell constituents (Josephy & Mannervik, 2006). It is noteworthy that certain GSTs are upregulated in long-lived strains of nematodes, fruit flies, and mice (McElwee et al., 2007). The protective effects of GSTs against xenobiotics are also exemplified by herbicide resistance in plants (Dixon, Lapthorn, & Edwards, 2002), insecticide resistance in flies (Low et al., 2007), and drug resistance in tumors (Morgan et al., 1998). There are four families of GSTs: cytosolic, mitochondrial, microsomal/membrane-associated, and fosfomycin/glyoxalase (Pearson, 2005). In humans, there are 17 different cytosolic GSTs, divided into 7 classes (alpha, mu, omega, pi, sigma, theta, and zeta) based on sequence similarity; the genes of the different classes are located on seven distinct chromosomes (Mannervik, Board, Hayes, Listowsky, & Pearson, 2005).

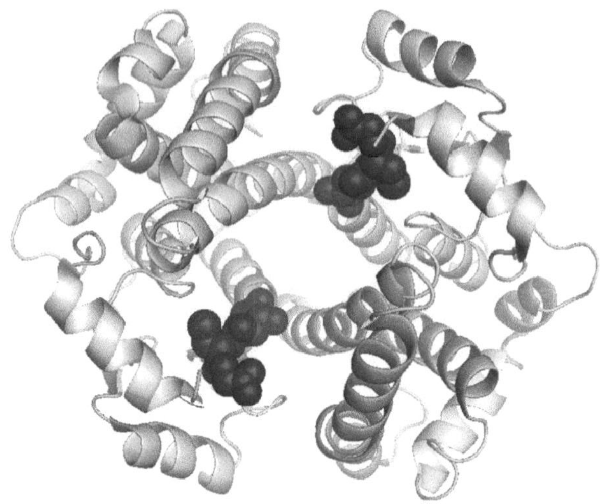

Figure 6.3 Structure of human GST A2-2 in complex with glutathione determined by X-ray crystallography (PDB entry 2WJU). View along the twofold symmetry axis with the two subunits to the left and to the right, each binding a glutathione molecule (rendered as CPK structure) in the G-site.

2.3.2 Structure

The cytosolic GSTs are dimeric proteins with one catalytic center in each subunit, located near the interface between the subunits (Fig. 6.3).

The G-site binds and activates glutathione, while the H-site accommodates the hydrophobic second substrate (Mannervik, Guthenberg, Jakobson, & Warholm, 1978). The amino acids forming the H-site are located in three separate regions of the primary structure. Each subunit consists of an N-terminal thioredoxin-like domain with secondary structure βαβαββα and an all-helical C-terminal domain. The alpha class has five homodimeric members in humans, GSTs A1-1 to A5-5, and the genes of the subunits are located in a cluster on chromosome 6. Most alpha class genes encode GSTs with 222 amino acids and each subunit has a molecular mass of around 26 kDa. In addition to homodimers, the subunits can also form heterodimers, for example, GST A1-2 (Mannervik et al., 2005). Among 14 human GSTs investigated, only three showed relatively high activity with azathioprine: GSTs A1-1, A2-2, and M1-1 (Eklund et al., 2006). GST M1-1 is polymorphic with a null allele and approximately half the human population is lacking this enzyme (Warholm, Guthenberg, Mannervik, & von Bahr, 1981). This chapter will focus primarily on the homodimeric GST A2-2, which has the highest catalytic efficiency of all GSTs.

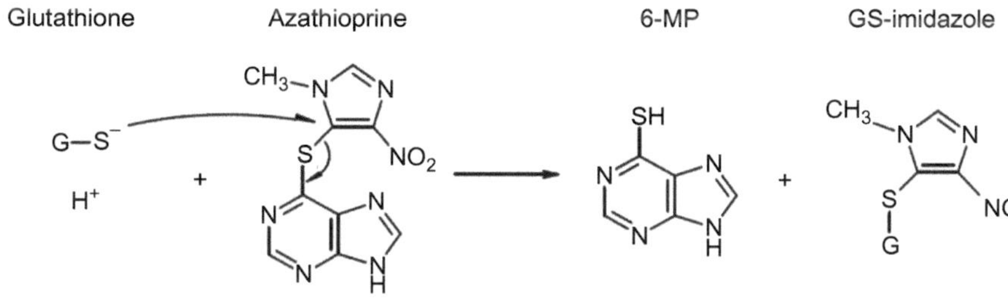

Figure 6.4 The reaction of glutathione with azathioprine. The electrophilic carbon of azathioprine is attacked by the sulfur of the glutathionylate, forming the glutathione–imidazole conjugate and 6-mercaptopurine (6-MP).

2.3.3 Mechanism of catalysis

Glutathione bound to an alpha class GST is first activated by deprotonation of its thiol group. The glutamyl α-carboxylate group of the glutathione receives the proton, while active-site water assists as a bridge (Dourado, Fernandes, Mannervik, & Ramos, 2008). The thiolate formed is stabilized by hydrogen bonding to a conserved tyrosine (Tyr9) and thus ready for nucleophilic attack on the electrophilic second substrate. The reactions catalyzed by GSTs can be substitutions, additions, or isomerizations, depending on the enzyme and the nature of the second substrate (Josephy & Mannervik, 2006). The reaction with azathioprine is an aromatic substitution reaction (Fig. 6.4).

2.3.4 Tissue distribution of human GSTs

Human GSTs A1-1 and A2-2 are expressed at high levels in the liver, small intestine, testis, kidney, adrenal gland, and pancreas and at lower levels in other tissues (Table 6.1).

GST A3-3 is expressed in steroidogenic tissues (Johansson & Mannervik, 2001; Larsson, Mannervik, & Raffalli-Mathieu, 2011), and GST A4-4 is expressed in a wide range of tissues, whereas no expression of A5-5 has been noted. Variable concentrations of GSTs have been reported for alpha class GSTs in several tissues (Table 6.2). The ratios between the minimum and the maximum expression levels were at the most 50-fold for GST A1-1 and 15-fold for GST A2-2 (Coles & Kadlubar, 2005).

2.3.5 Biomarker applications

The presence of alpha class GSTs in blood plasma is low in healthy individuals (median 2.6 $\mu g\, l^{-1}$, range 0.2–20.4 $\mu g\, l^{-1}$, and $N=350$; Mulder et al., 1999). Alpha class GSTs can therefore be used as a sensitive marker of liver disease

Table 6.1 Tissue expression and specific activities with azathioprine of human GSTs

GST	Small intestine	Erythrocytes	Liver	Specific activity (μmol mg^{-1} min^{-1})
A1-1	+	–	+	0.24
A2-2	+	–	+	0.53
A3-3	–	–	–	0.01
A4-4	+	–	+	0.01
M1-1	(+)[a]	–	(+)[a]	0.17
M2-2	–	–	–	0.05
M3-3	+	–	+	<0.001
M4-4	–	–	+	<0.001
M5-5	–	–	–	<0.003
K1-1	+	–	+	0.05
O1-1	+	–	+	0
P1-1	+	+	–	0
T1-1	(+)[a]	(+)[a]	(+)[a]	0
Z1-1	+	–	+	0

[a]If not a null individual.
–, not detected
Based on Eklund et al. (2006).

due to raised plasma levels during liver damage (Mulder, Peters, Court, & Jansen, 1996). The alpha and pi class GSTs are localized in different parts of the kidney (Rozell, Hansson, Guthenberg, Tahir, & Mannervik, 1993), and they can be used as differential biomarkers in the urine for various pathological kidney conditions (Sundberg, Nilsson, Appelkvist, & Dallner, 1995).

3. POLYMORPHISMS

3.1. Genetic polymorphism of human GST A2-2

The genes encoding the human alpha class GSTs are all localized in a cluster on chromosome 6. Four single-nucleotide polymorphisms (SNPs) in the gene of human GST A2-2 have been described, leading to five genotypes. Using genotype A2*A (P110/S112/K196/E210) as a reference, the other GST variants differ in positions E210A (A2*B),

Table 6.2 Variation of expression levels, as μg GST per mg protein, in various human tissue types

Liver	
GST alpha class	Mean 25.1 ± 9.4
	(*N* = 35; Mulder, Court, & Peters, 1999)
Liver	
GST A1-1	Mean 6, 10–90th percentiles, range 3–10
GST A2-2	Mean 3, 10–90th percentiles, range 1–7
	(*N* = 52; Coles & Kadlubar, 2005)
Pancreas	
GST A1-1	Mean 1.4, range 0.2–5.1
GST A2-2	Mean 4.3, range 0.7–10.8
	(*N* = 43; Coles et al., 2000)
Gastrointestinal tract (proximal section of the small intestine)	
GST A1-1	Mean 7.7 ± 3.4, range 0.3–12.6
GST A2-2	Mean 1.5 ± 0.6, range 0.3–2.7
	(*N* = 15; Coles, Chen, Kadlubar, & Radominska-Pandya, 2002)

S112T (A2*C), K196N/E210A (A2*D), and P110S (A2*E). Genotype A2*D has so far been found only in a database of expressed sequence tags and not in population studies.

The allele frequencies of the other genotypes reported for different populations are 30–86% of variant A2*A, 8–43% of A2*B, 1–57% of A2*C, and 1–13% of A2*E (Coles & Kadlubar, 2005; Ning et al., 2004; Silva et al., 2009; Tetlow & Board, 2004; Tetlow, Liu, & Board, 2001). Variant A2*C is the most common allele in Caucasians, while A2*A is the most common in Chinese and Australian populations. The allele frequency of A2*E is 1% in Bantu Africans, 4% in Portuguese Caucasians, 5% in Australians, 12% in Chinese, and 13% in Creole Africans.

3.2. Phenotypic differences in expression of different GST genotypes

The expression level in tissues is dependent on the genotype of the GST. For human GST A1-1, polymorphism in the 5′-noncoding region causes variable expression levels in the liver, with median 8 μg mg^{-1} (range 1–13 μg mg^{-1}) for A1*A versus median 4 μg mg^{-1} (range 2–7 μg mg^{-1}) for A1*B (Coles & Kadlubar, 2005). A trend of inverse relationships has been noted between the expressions of GSTs A1-1 and A2-2 linked to this polymorphism, with GST A2-2 expression having median 2 μg mg^{-1} (range 0–5 μg mg^{-1}, $N=20$) versus median 7 μg mg^{-1} (range 5–9 μg mg^{-1}, $N=10$) for the two different homozygous genotypes A1*A and A1*B, respectively. The differences in expression in relation to the genotypes were more pronounced for GST A2-2 than for GST A1-1, where the individual variation is larger.

For the hepatic expression levels, there was a similar relationship for the GST A2-2 polymorphism. The homozygous genotype A2*C (median 2 μg mg^{-1}, range 0–5 μg mg^{-1}, and $N=20$) had threefold lower expression than combinations of the genotypes A2*A/B/E (median 7 μg mg^{-1}, range 3–9 μg mg^{-1}, and $N=11$) (Ning et al., 2004). In the pancreas, however, there was no significant difference between the median expression levels of the different *GSTA2* genotypes. The variability within each genotype was larger than for the liver samples. Part of the difference in expression levels can be attributed to known polymorphisms, with threefold lower hepatic expression for A2*C (Ning et al., 2004) and differences in expression levels for GSTs A1-1 and A2-2 depending on a *GSTA1* polymorphism in the 5′-noncoding region (Coles & Kadlubar, 2005). Genotyping of patients might thus aside from identifying the specific activities of the GSTs with azathioprine also provide some hints about their expression levels. Based on the allele frequencies mentioned earlier, the occurrence of A2*E homozygosity could be up to 1.7% in some populations and the A2*E heterozygosity up to 23%. In the Portuguese Caucasian population, the occurrence of homozygous A2*E individuals with higher specific activity toward azathioprine would be 0.6%, while 32% would have A2*C homozygosity with lower hepatic expression.

3.3. Thermal inactivation of allelic GST 2-2 variants

Thermostability of the five GST A2-2 variants has been assessed at the physiological glutathione concentrations of 2 mM. At a temperature of 37 °C, the half-life of the thermal inactivation of variant A2*C was about 2 weeks, while the others were more stable. At 48 °C, it was shown that A2*E was more stable than the other four variants (Zhang, Modén, & Mannervik, 2010). However, the half-lives might be even shorter *in vivo*

(García-Mata, Conde, & Sanllorenti, 1998; Shen, Ranganathan, Kuzmich, & Tew, 1995) indicating that thermostability is not as such a major determinant of the intracellular GST concentrations.

4. AZATHIOPRINE AND IMMUNOSUPPRESSION

4.1. Azathioprine and inflammatory bowel disease

4.1.1 Thiopurine treatment

IBD is a class of chronic inflammatory disorders of the gastrointestinal tract including Crohn's disease and ulcerative colitis. Azathioprine has proven efficacy in the induction of remission, maintaining remission and reducing steroid dependence (Chouchana, Narjoz, Beaune, Loriot, & Roblin, 2012; Pearson, May, Fick, & Sutherland, 1995). Stronger anti-inflammatory corticosteroids are often used for the induction of remission followed by the administration of immunosuppressive thiopurines as steroid-sparing agents or when the steroids no longer work or produce adverse effects (Lakatos & Kiss, 2011; Prefontaine, Sutherland, Macdonald, & Cepoiu, 2009). According to a meta-analysis of relapse rates, it is significantly more advantageous to continue than to stop the treatment, at least up to 18 months after induction of remission (French, Dalzell, Srinivasan, & El-Matary, 2011). For patients continuing with azathioprine treatment after achieving remission, 62% were still in remission after five years and 19% had only had a short relapse (Fraser, Orchard, & Jewell, 2002).

4.1.2 Monoclonal antibodies

Monoclonal antibodies against tumor necrosis factor-α (TNF-α) represent another alternative for inducing or maintaining remission in IBD and are sometimes used in conjunction with azathioprine or as a third option in patients when steroids and azathioprine have failed (Lakatos & Kiss, 2011). For example, infliximab (Knight et al., 1993) binds all forms of TNF-α, even receptor-bound, leading to caspase-3 activation and apoptosis (Van den Brande et al., 2003).

4.1.3 Bioavailability and activation of azathioprine

The time to maximum concentrations of azathioprine and 6-MP after oral administration of azathioprine varies significantly among subjects (range 15–240 min and $N=8$), and the difference between the maximum times also varies up to 4 h (Odlind et al., 1986). The different rates of activation might be related to the distribution and toxicity of the metabolites. The estimated GST-catalyzed activation of azathioprine in the liver, pancreas, and small intestine is potentially very high with large variation among individuals and might be one factor involved in effectiveness of treatment as well as the commonness of adverse effects in these tissues.

4.2. Azathioprine metabolism

The metabolism of azathioprine is complex, involving multiple pathways of activation or inactivation, variation among patients, and uncertainty of which metabolites actually exert therapeutic or adverse effects and by what mechanisms. An overview is presented in Fig. 6.5.

4.2.1 Activation

The prodrug azathioprine is activated by conjugation with thiols, mainly glutathione, and has widely been considered to take place nonenzymatically,

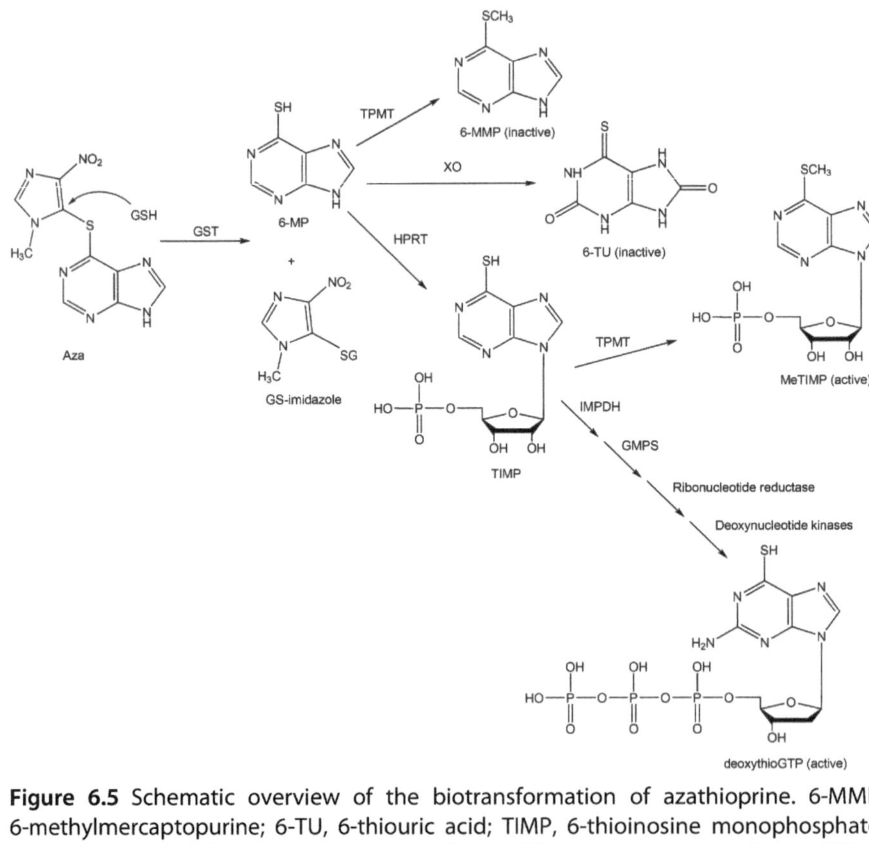

Figure 6.5 Schematic overview of the biotransformation of azathioprine. 6-MMP, 6-methylmercaptopurine; 6-TU, 6-thiouric acid; TIMP, 6-thioinosine monophosphate; MeTIMP, 6-methyl-thioinosine monophosphate; GST, glutathione transferase; TPMT, thiopurine methyltransferase; XO, xanthine oxidase; HPRT, hypoxanthine phosphoribosyltransferase; IMPDH, inosine monophosphate dehydrogenase; GMPS, guanine monophosphate synthetase. *From Zhang et al. (2010).*

even though it was known that GSTs could be involved. Recently, it was shown that the bioactivation in human tissues was performed mainly enzymatically by GSTs A2-2, A1-1, and M1-1, and the uncatalyzed reaction of azathioprine with glutathione was estimated to contribute with less than 1% of the total biotransformation (Eklund et al., 2006). Upon conjugation with glutathione, 6-MP is released (Fig. 6.4) and the metabolites are most likely responsible for the therapeutic effects.

Among the three human GSTs with the dominant azathioprine activities, GST A2-2 has the highest catalytic efficiency. There is polymorphism in the genes encoding these enzymes, and notably, there is one *GSTM1-null* genotype, which does not produce any enzyme. Its presence seems to be negatively correlated with adverse effects in patients receiving azathioprine (Stocco et al., 2007). Since GST A2-2 has an even higher azathioprine activity, it was of interest to examine the effects of the allelic variants of this enzyme (Zhang et al., 2010).

4.2.2 Cytotoxicity

After the initial GST-catalyzed activation, 6-MP is metabolized in several steps into 6-thioguanine nucleotides (6-TGNs). The incorporation of 6-TGNs into DNA can lead to methylation, DNA strand breaks, and erroneous base pairing and, with an active mismatch-repair system, can result in cell death (Karran, 2006; Swann et al., 1996).

4.2.3 Inactivation

There are two major metabolic enzymes involved in the inactivation of 6-MP (Roberts & Barclay, 2012): Xanthine oxidase (XO) catalyzes the formation of 6-thiouric acid that is excreted. TPMT methylates 6-MP and 6-thioinosine monophosphate (TIMP) into 6-methylmercaptopurine (6-MMP) and 6-methyl-thioinosine monophosphate (MeTIMP), respectively. Higher activities of these inactivating enzymes lead to lower concentrations of active 6-TGN. Aldehyde oxidase (AO) also acts on azathioprine, 6-MP, and other metabolites, but the functional significance of the products is poorly understood (Smith et al., 2009).

4.2.4 Purine starvation

An additional effect of the thiopurines is the fact that MeTIMP is a potent inhibitor of phosphoribosyl pyrophosphate amidotransferase, leading to blocking of the primary pathway of purine nucleotide biosynthesis (Tay, Lilley, Murray, & Atkinson, 1969). It is possible that the effect of the purine starvation has been exaggerated by *in vitro* studies involving cell lines with mismatch-repair defects, and it is not clear to which extent the purine starvation contributes to the cellular toxicity (Karran, 2006). On the other hand, patients taking 6-thioguanine require higher levels of 6-TGN in order to reach therapeutic efficiency compared to azathioprine and 6-MP, indicating either that the distribution and activation profiles are different or that there are other effects in addition to that of 6-TGN (Cara et al., 2004).

4.2.5 Other 6-MP metabolites

The metabolite TIMP can be phosphorylated twice leading to 6-thioinosine triphosphate (TITP). Inosine triphosphate pyrophosphatase (ITPA) then converts it back to TIMP. In the case of ITPA deficiency, the accumulation of TITP could potentially lead to toxicity, perhaps via the incorporation of 6-thioinosine phosphate into DNA (Chouchana et al., 2012). The methylated 6-MMP and 6-methylated mercaptopurine ribonucleotides, 6-MMPR, might also contribute to the toxicity of the thiopurines.

4.2.6 Changes in metabolite levels

When nonresponders with high 6-MMP levels were given diminished azathioprine doses and coadministration of the XO inhibitor allopurinol, the mean levels of 6-MMP decreased from 10,600 to 2000 pmol/8×10^8 red blood cells, while the mean levels of 6-TGN increased from 190 to 400 pmol/8×10^8 red blood cells (Sparrow, Hande, Friedman, Cao, & Hanauer, 2007). For patients given a 50% dose increase, the mean 6-TGN levels increased almost proportionally, from 150 to 260 pmol/8×10^8 red blood cells, while the mean 6-MMP levels increased from 280 to 2260 pmol/8×10^8 red blood cells (Kennedy et al., 2013). When using split dosing for a group of mostly 6-MP users, half the dose was administered twice a day. The mean levels of 6-MMP decreased with 55% from 11,800 to 5300 pmol/8×10^8 red blood cells, while the mean levels of 6-TGN did not change significantly, only from 239 to 216 pmol/8×10^8 red blood cells (Shih et al., 2012). The changes in metabolite levels were accompanied by a reduction in adverse effects. The notion that the dose or the rate of accumulation of early metabolites can influence the distribution of the metabolites downstream along the pathways suggests that differences in activation in the first step of azathioprine metabolism by GSTs could influence metabolite concentrations in later steps.

4.2.7 Metabolite concentration thresholds

A 6-TGN concentration threshold of 235 pmol/8 $\times 10^8$ red blood cells has been suggested, with a higher proportion of response for patients with higher 6-TGN concentration (Dubinsky et al., 2000). It was also reported that the mean level of 6-MMP in patients with hepatotoxicity was elevated and that 5700 pmol/8 $\times 10^8$ red blood cells could be a suitable threshold for 6-MMP-related liver damage. However, others have failed to find a difference in 6-TGN concentrations between subjects with active IBD and subjects in remission (Lowry et al., 2001). An analysis of different possible thresholds noticed no correlations between levels of 6-TGN and 6-MMP or the ratio 6-MMP/6-TGN on the one hand and efficacy of treatment or adverse effects on the other (González-Lama et al., 2011). In a large study, TPMT activity correlated with 6-MMP levels and the 6-MMP/6-TGN ratio, but not with 6-TGN levels (van Egmond, Chin, Zhang, Sies, & Barclay, 2012). However, the distributions of TPMT activities for the group with high or low 6-MMP/6-TGN ratio were almost overlapping with a small but significant difference in mean. Even though there was a small difference in mean, there could be no clinically relevant threshold level that could separate the metabolite level groups based on TPMT activity. Most of the variations in these metabolite concentrations are thus not explained by the TPMT status.

4.2.8 Cellular signaling

Upon CD28 costimulation of T lymphocytes, the small GTPase Rac1 mediates antiapoptotic signals by activating bcl-x_L. 6-Thio-GTP binds to Rac1 but blocks the corresponding Vav guanosine exchange activity leading to the accumulation of inactive 6-thio-GDP–Rac1 complexes (Poppe et al., 2006). The molecular mechanism of azathioprine and 6-TGN in immunosuppression is thus blocking of the Vav guanosine exchange activity on Rac1, thereby inhibiting antigen presentation and antigen-specific immune responses, as well as inducing apoptosis via the mitochondrial pathway and caspase-9 activation (Tiede et al., 2003).

4.2.9 Glutathione conjugation

Aside from the 6-MP metabolites, it is possible that the glutathione conjugate of the reaction with azathioprine exerts effects during therapy. When the 6-MP component of azathioprine was replaced by different nontoxic thiols, several of the prodrugs were more potent than azathioprine in the human mixed lymphocyte reaction, which tests inhibitory potency

(Crawford, Maddocks, Jones, & Szawlowski, 1996). Two of the variants performed better than azathioprine in prolonging graft survival in mice, without showing similar toxicity as azathioprine. Immunosuppressive effects were later confirmed in rabbits by determining the antibody titer response (Al-Safi, Tashtoush, & Hassan, 2003).

Glutathione depletion is also a possible cause of toxicity during therapy with azathioprine or the aforementioned prodrugs. Cell studies often show changes in viability after a few days of incubation with azathioprine, while therapeutic results usually appear on a larger timescale (Chouchana et al., 2012). It is possible that the glutathione depletion can be the cause of some of these effects, especially in the *in vitro* experiments (DeLeve, Wang, Kuhlenkamp, & Kaplowitz, 1996; Lee & Farrell, 2001).

5. ADVERSE EFFECTS OF AZATHIOPRINE

5.1. Dose adjustment

Dose is often adjusted during therapy in order to reach therapeutic effect or to decrease adverse effects and can be adjusted before administration on the basis of the thiopurine methyltransferase (TPMT) status (Stocco et al., 2007). Among IBD patients who had started azathioprine therapy, about 9% had discontinued because of lack of effect and 11% because surgery was needed (Barabino et al., 2002; Fraser et al., 2002; Zabala-Fernández et al., 2011).

5.2. Various adverse effects

About 28% of patients with IBD receiving azathioprine experienced adverse effects, and the severity led to withdrawal in about 20% of the patients under treatment (Barabino et al., 2002; Connell, Kamm, Ritchie, & Lennard-Jones, 1993; Gardiner, Gearry, Begg, Zhang, & Barclay, 2008; Schwab et al., 2002; Stocco et al., 2007; van Dieren et al., 2005; Weersma et al., 2004; Wroblova et al., 2012; Xiong et al., 2010; Zabala-Fernández et al., 2011). Reported adverse effects were bone marrow suppression (7%), gastrointestinal toxicity (6%), hepatotoxicity (4%), pancreatitis (4%), flu-like symptoms (3%), rash (2%), and others (2%).

5.3. Cancer

The use of DNA base analogs was intended to inhibit the growth of fast-dividing cells, but the mutations introduced could potentially also lead to cancer. The risk of developing lymphoproliferative disorder was estimated for IBD patients receiving thiopurines to be 0.1% per year, which was five times higher than for IBD patients not receiving thiopurine treatment (Beaugerie et al., 2009). Since thiopurines absorb ultraviolet A radiation, leading to oxidative stress and DNA damage, the combination of sunlight and thiopurine treatment might synergistically increase the risk of skin cancer (Maddox & Soltani, 2008; O'Donovan et al., 2005).

5.4. Selectivity

About 60% of azathioprine-intolerant patients could tolerate 6-MP instead (Lees, Maan, Hansoti, Satsangi, & Arnott, 2008). Several of those that did not stand the switch to 6-MP experienced different types of adverse effects as compared to the effects during the stage of azathioprine intolerance. The selectivity of the toxicity depends on the levels of individual metabolic enzymes, mitotic rate, drug transport, and metabolite pool sizes (Elion, 1989). Increased knowledge about the enzymes involved in the metabolism could provide insights relevant to dosing adjustment, efficacy, and adverse effects during azathioprine therapy.

6. POLYMORPHISMS IN THE METABOLIC PATHWAYS OF AZATHIOPRINE

6.1. Clinical observations

The metabolism of azathioprine involves several steps of activation and multiple inactivation routes (Fig. 6.5). Genetic variation in TPMT can have drastic effects on the outcome of clinical treatment of patients with the drug. For double null mutants, the dose administered must be radically lowered to avoid acute bone marrow degeneration (Roberts & Barclay, 2012). However, the TPMT status can explain only a small part of the adverse effects (Dong, Zheng, Zhu, Tong, & Ran, 2010), and the results of polymorphism in other steps have not been thoroughly examined. The specific activity of azathioprine biotransformation was estimated in the liver samples from eight donors to be $4.3 \pm 2.1\ \text{min}^{-1}\ \text{mg}^{-1}$, with a range of 1.6–7.6 nmol $\text{min}^{-1}\ \text{mg}^{-1}$ (von Bahr, Glaumann, Gudas, & Kaplowitz, 1980). The urinary excretion of thiouric acid reached a maximum within the first 4 h after oral administration of azathioprine (1 mg kg^{-1} body mass) in three individuals (Chalmers, Knight, & Atkinson, 1967). The time to maximum plasma concentrations of azathioprine and 6-MP was determined in eight subjects receiving 100 mg azathioprine orally to be 75 ± 77 min and 114 ± 83 min, respectively, with a range of 15–240 min for both azathioprine and 6-MP (Odlind et al.,

1986). The time to azathioprine and 6-MP maximum coincided for some subjects and were separated up to 4 h in other subjects.

With such rate differences in the first step of the activation of azathioprine, it is reasonable to believe that the activity and expression of the relevant GSTs might affect the distributions and toxicities of the prodrug metabolites.

6.2. GST polymorphism

The first metabolic reaction involves glutathione conjugation by GSTs, in human mainly by GSTs A2-2, A1-1, and M1-1 (Eklund et al., 2006). Different rates of activation could be related to adverse effects as seen in a study monitoring the presence of *GSTM1-null* subjects (Stocco et al., 2007). However, no studies have been published investigating the effects of differences in phenotype for GSTs A2-2 and A1-1, despite their high activity with azathioprine and high expression levels in several tissue types. In view of their significance, the polymorphic variants of GST A2-2 were characterized with respect to catalytic efficiency with azathioprine and alternative substrates as well as thermostability (Zhang et al., 2010).

6.2.1 Kinetic characterization of allelic GST A2-2 variants

Five genotypes of human GST A2-2 have been found (Table 6.2), and all but A2*D have been confirmed in population studies. The coding regions of the five published GST A2-2 genotypes were cloned, expressed, and characterized. The steady-state kinetic parameters were determined and were similar for all except A2*E, which had three- to fourfold higher k_{cat} and k_{cat}/K_M values (Zhang et al., 2010).

Assuming *in vivo* azathioprine concentrations $<10\ \mu M$ (Odlind et al., 1986) and K_M values >1 mM, the data indicated that the activation rate under physiological conditions is proportional to k_{cat}/K_M. Variant A2*E turned out to catalyze this reaction three- to fourfold faster than the other variants of A2-2 and about sixfold faster than GSTs A1-1 and M1-1.

Structurally, A2*E differs from the other variants by the P110S substitution. Prolines can influence secondary structure elements, and substitutions in this position could lead to reorientation of the adjacent H-site residues. The proline residue in position 110 is otherwise conserved in all human alpha class GSTs. Serine in position 110 is present in another alpha class GST from *Monodelphis domestica*. Positions 107, 108, 110, and 111 are all potential H-site residues, located near the end of helix α4.

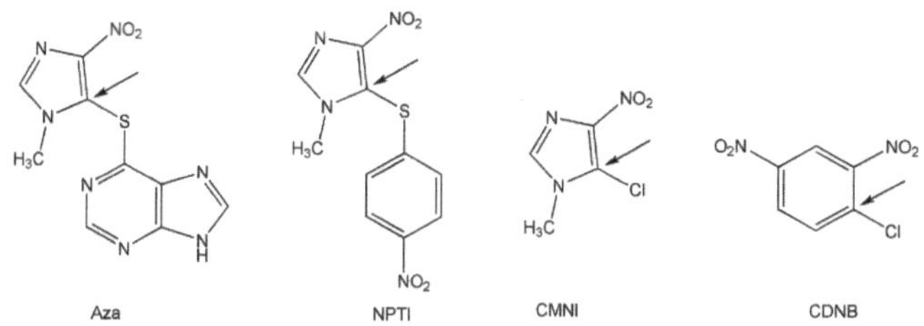

Figure 6.6 Structures of azathioprine (Aza), NPTI, CMNI, and CDNB. Arrows indicate the positions of nucleophilic substitution by GSH. All compounds except CDNB give the same imidazolyl–GSH conjugate.

6.2.2 Characterization of GST A2-2 using alternative substrates

Previously, the azathioprine analog 1-methyl-4-nitro-5-(4-nitrophenylthio)-1H-imidazole (NPTI) was developed as a surrogate for screening of azathioprine activity (Kurtovic, Grehn, Karlsson, Hellman, & Mannervik, 2008). NPTI is an azathioprine analog in which the leaving group is 4-nitrothiophenol instead of 6-MP, facilitating colorimetric screening (Fig. 6.6). This substrate and 5-chloro-1-methyl-4-nitroimidazole (CMNI) and 1-chloro-2,4-dinitrobenzene (CDNB) were used as alternative substrates for comparison with azathioprine, in order to obtain further information about structure–function relationships in GST A2-2 variants (Modén, Zhang, & Mannervik, 2012; Zhang et al., 2010). CMNI has the same substituted imidazole ring as azathioprine and NPTI, but only a chloride ion as leaving group. CMNI has been tested as a radiosensitizer (Watts, Hodgkiss, Sehmi, & Woodcock, 1980) with the primary purpose to deplete cellular glutathione before radiation therapy (Sauer, Hantke, & Wilmanns, 1988). CDNB has a size similar to CMNI and is the standard substrate for measurement of GST activity.

The catalytic efficiency of A2*E with NPTI was about threefold higher than those of the other variants, while it was lower with CMNI and CDNB (Zhang et al., 2010). Azathioprine, NPTI, and CMNI all form the same glutathione conjugate, and since k_{cat} was different for the three substrates, the results indicate that the reaction rates were not limited by the release of this conjugate. Principal component analysis of the normalized catalytic efficiencies with the four substrates was performed (Zhang et al., 2010). The analysis confirmed that variant A2*E was functionally different from the other four variants, mainly due to its higher catalytic efficiency with azathioprine and NPTI and lower catalytic efficiency with CDNB and CMNI.

6.2.3 Structural comparison of A2*C and A2*E

The naturally occurring allelic variants of GST A2-2, A2*C, and A2*E, differ in two positions: P110S/T112S. By comparing the three variants A2*A, A2*C, and A2*E (Tables 6.3 and 6.4), it can be concluded that the T112S mutation influenced the activity with azathioprine only marginally, while P110S enhanced the catalytic efficiency three- to fourfold (Zhang et al., 2010; Table 6.4). The crystal structure of A2*C has been determined (Tars, Olin, & Mannervik, 2010), but the structure of A2*E is available only in the mutant form GDH (Zhang, Modén, Tars, & Mannervik, 2012). Even in the absence of a structure of wild-type A2*E, it is clear that the side chains of residues 110 and 112 protrude from helix α4 in the same general direction. In A2*C and the GDH mutant, also the side chains of E104 and F111 in helix α4 are in almost the same locations. In the case of the GDH mutant, it appears clear that the enhanced azathioprine activity is associated with an enlarged H-site (Zhang et al., 2010), and it is possible that replacing P110

Table 6.3 Amino acid variations in allelic variants of human GST A2-2

Residue	110	112	196	210
GST A2*A	Pro	Ser	Lys	Glu
GST A2*B	Pro	Ser	Lys	Ala
GST A2*C	Pro	Thr	Lys	Glu
GST A2*D	Pro	Ser	Asn	Ala
GST A2*E	Ser	Ser	Lys	Glu

Residues 110, 112, and 210 contribute to the active site.

Table 6.4 Specific activities of allelic GST A2-2 variants

GST	Azathioprine	NPTI	CDNB	CMNI
A2*A	0.45±0.01	12.8±2.6	25.9±2.3	12.9±1.3
A2*B	0.46±0.02	16.5±1.1	32.3±2.7	14.2±1.6
A2*C	0.60±0.03	16.8±2.5	25.7±0.6	11.2±0.2
A2*D	0.54±0.02	18.1±2.2	34.8±0.7	12.9±0.2
A2*E	1.79±0.03	48.5±2.9	16.0±0.7	8.4±0.1

Values are means ± SD (μmol mg^{-1} min^{-1}).
From Zhang et al. (2010).

and T112 in A2*C by S110 and S112 in wild-type A2*E widens the binding site enough to cause the threefold higher activity of A2*E.

6.2.4 Combining activity and expression in different tissues

The rate of activation of azathioprine in human liver cytoplasm has been estimated (Eklund et al., 2006). The enzymatic activation dominates over the uncatalyzed reaction, and the enzymes contributing the most are GSTs A2-2, A1-1, and M1-1. A similar estimation was performed using the GST concentrations from Rowe, Nieves, and Listowsky (1997) combined with the catalytic efficiencies of the allelic variants determined (Zhang et al., 2010). The sum of the apparent rate constants (min^{-1}) was estimated to be 2.8–4.5 (liver), 2.4–3.6 (testis), 0.6–1.8 (pancreas), 0.6–1.1 (kidney), 0.4–0.7 (adrenal glands), 0.2–0.3 (brain), 0.05–0.07 (lung), 0.04–0.07 (pituitary gland), and 0.03 (heart). The range of the rates in different tissues represented the variation among individuals, assuming the same concentration of GSTs but with either A2*E (higher value) or another A2-2 variant (lower value). The GST concentrations were determined as the mean of two samples from different individuals.

Due to the higher catalytic efficiency of A2*E, there is a 1.5- to 2-fold difference in the sum of the apparent rate constants in most tissue types, except the pancreas with a threefold difference and the heart with no GST A2-2 expression. The uncatalyzed reaction contributes only about 1% or less in the five tissue types with the highest apparent rate constant. A *GSTM1-null* homozygous mutant would only influence the apparent rate constant by a 5–20% decrease in these tissue types. In the four other tissue types, the apparent rate constant of the GSTs and the uncatalyzed reaction would be 1.5- to 3-fold lower in subjects with the *GSTM1-null* genotype.

6.2.5 Interindividual variations of activity and expression

As judged from the studies of the expression of GSTs in specific tissues using a larger sample of individuals, it is clear that the interindividual differences in expression levels are also of notable magnitude. The ratios between the minimum and maximum expression levels are at the most 50-fold for GST A1-1 and 15-fold for GST A2-2. The sum of the apparent rate constants could vary by that order of magnitude between individuals with very high and very low concentrations of azathioprine-active GSTs, even if the GSTs were of the same genotype. von Bahr et al. (1980) reported measurement with azathioprine using human liver cytosol with a specific activity of 4.3 ± 2.1 nmol min^{-1} mg^{-1} ($N=8$, range 1.6–7.6 nmol min^{-1} mg^{-1}).

6.3. TPMT polymorphism

Polymorphism in the gene for TPMT results in 89% of subjects having high TPMT activity, 11% being heterozygous with intermediate activity, and 1 of 300 being TPMT-deficient (Weinshilboum & Sladek, 1980). Several variant genes have been described leading to the absence of TPMT activity (Yates et al., 1997). Determination of TPMT genotype was one of the first pharmacogenetic tests to be incorporated into routine clinical practice, in order to avoid potentially fatal myelotoxicity in phenotypically null patients (Gearry & Barclay, 2005). These patients generally receive a 10-fold dose reduction or no treatment with thiopurines (Roberts & Barclay, 2012). Of several examined TPMT variants with nonsynonymous SNPs, only one was devoid of enzymatic activity, one had no expression at all, and the rest had about the same kinetic properties but lower expression levels than the wild-type enzyme (Wang & Weinshilboum, 2006). The difference in expression levels was probably caused by faster degradation involving ubiquitination and proteasome degradation or misfolding and aggregation.

A meta-analysis showed that bone marrow toxicity was six times more common in patients with at least one TPMT mutation compared to patients homozygous for functional TPMT (Dong et al., 2010). Still only 30% of the cases of bone marrow toxicity could be explained by the TPMT mutation and only 21% of patients with the TPMT mutation had bone marrow toxicity compared to 5% of the control population, indicating that other factors than TMPT status are influencing the outcome.

S-Adenosyl-L-methionine is a cofactor in the reaction catalyzed by TPMT, and polymorphism in enzymes affecting *S*-adenosyl-L-methionine concentrations might decrease the TPMT activity of subjects with wild-type TPMT genotype (Arenas et al., 2005; Roberts & Barclay, 2012). On the other hand, polymorphism in inosine monophosphate dehydrogenase (IMPDH) can in individuals with wild-type TPMT direct the metabolism preferentially into 6-MMPR, which might lead to toxicity (Roberts & Barclay, 2012).

6.4. ITPA polymorphism

One polymorphic variant of ITPA, C94A, results in loss of ITPA activity. Some studies have indicated a correlation between the occurrence of this mutation and adverse effects or lack of response (Marinaki et al., 2004; von Ahsen et al., 2005), while others have not been able to confirm these findings (Allorge, Hamdan, Broly, Libersa, & Colombel, 2005; Breen,

Marinaki, Arenas, & Hayes, 2005; Gearry, Roberts, Barclay, & Kennedy, 2004; van Dieren et al., 2005). When examining ITPA activity instead of genotype, it seems that lower activity could be correlated with adverse effects (Shipkova et al., 2011; Xiong et al., 2010). Few homozygous C94A individuals were included in the mentioned studies and many were suffering from adverse effects (Marinaki et al., 2004; Shipkova et al., 2011; Xiong et al., 2010) suggesting that homozygosity with the deficient allele is a risk factor, while heterozygous or wild-type individuals have adverse effects not directly related to the ITPA status.

6.5. Polymorphism in transporters

6-MP metabolites can be transported by multidrug resistance protein 4 (MRP4) and multidrug resistance protein 5 (Wielinga et al., 2002). One known MRP4 mutant exhibits decreased cell surface expression leading to enhanced sensitivity to 6-MP (Roberts & Barclay, 2012). Its allele frequency is 15–22% in Japanese, 8% in Han Chinese, and less than 1% in Caucasians and Africans. The concentrations of 6-TGN were higher and adverse effects in the form of leucopenia tended to be more common in mutant carriers (Ban et al., 2010).

6.6. XO and AO polymorphisms

One case report indicated that high XO activity could prevent response to azathioprine therapy, including no detectable levels of 6-TGN or 6-MMP, while administration of allopurinol and 6-MP instead could result in high metabolite levels and toxicity (Wong, Derijks, den Dulk, Gemmeke, & Hooymans, 2007). Different genetic variants of XO have been described and deficiency in XO activity is very rare (Kudo et al., 2008). One SNP of AO with 11–16% allele frequency has been reported to be involved in improved response to azathioprine therapy (Smith et al., 2009). They also reported that polymorphisms in XO and the enzyme inserting the molybdenum cofactor of AO and XO might contribute to fewer adverse effects.

6.7. Rac1 polymorphism

Polymorphism in the promoter region of Rac1 can influence the expression of Rac1, but no significant correlation between the occurrence of polymorphism and therapy response or leucopenia has been noted (Bourgine et al., 2011).

6.8. Polymorphism summary

Several factors have been examined that could influence metabolite levels, therapeutic response, and adverse effects during azathioprine treatment, especially in the case of enzyme deficiency and homozygosity of mutant alleles. Still only a minority of adverse effects can be explained or predicted using existing knowledge from genotyping and metabolite monitoring, and some of the indications need to be replicated and confirmed by additional studies. The adverse effects of azathioprine therapy are multifactorial in nature with more than just differences in TPMT and ITPA activities contributing (Zabala-Fernández et al., 2011).

7. STRUCTURAL REQUIREMENTS FOR HIGH GST ACTIVITY WITH AZATHIOPRINE

7.1. Chimeric GST variants obtained by DNA shuffling of homologous sequences

Structure–activity relationships were sought in order to understand the higher catalytic efficiency of human GST A2-2 compared to the other GSTs examined. DNA shuffling (Stemmer, 1994) of the coding sequence of GST A2-2 with homologous alpha class GSTs generated a library of related GST chimeras that could be heterologously expressed and characterized (Kurtovic, Modén, Shokeer, & Mannervik, 2008). Three human GSTs, A1-1, A2-2 (allelic variant A*C), and A3-3, and bovine GST A1-1 and rat GSTs A2-2 and A3-3 were used for the construction of the library. The latter three mammalian sequences added increased structural diversity to the chimeras. Human GST A2-2 shares 95% amino acid sequence identity with human GST A1-1, 89% with human GST A3-3, 81% with bovine GST A1-1, 73% with rat GST A2-2, and rat GST A3-3. Between themselves, these two rat GST sequences are not more than 68% identical. By including enzymes with larger sequence divergence and with both high and low activity toward azathioprine, the regions of importance for the function could be identified. Only human GSTs A1-1 and A2-2 have relatively high azathioprine activity.

To introduce additional variability, a previously created mutant library of human GST A1-1 was included in the shuffling. This library had 10 potential H-site residues mutated with the degenerate codon NNN to encode all 20 amino acids (Hansson, Widersten, & Mannervik, 1997; Widersten & Mannervik, 1995). Point mutations in the active site have probably a higher chance of drastically altering the function of the enzyme compared to mutations further away. Shuffling of homologous sequences introduces mutations in block-wise segments and affects both the active site and more distant parts of the enzymes. Human GST A4-4 had too low sequence identity with human GST A2-2 (54%) to be included in the shuffling process (Hubatsch, Ridderström, & Mannervik, 1998). The fifth human alpha class gene encoding GST A5-5 can be processed to a mature mRNA, which is translation-competent, producing a catalytically active enzyme (Singh, Zimniak, & Zimniak, 2010), but the protein has not so far been identified *in vivo*. The H-site residues of GST A5-5 are highly similar to those of GST A2-2, but the G-site substitution R15S in human GST A5-5 might be detrimental to the catalytic activity, considering the conservation of a basic residue in this position and its importance for catalysis in alpha class sequences (Björnestedt et al., 1995; Dourado, Fernandes, Mannervik, & Ramos, 2010). For these reasons, GSTs A4-4 and A5-5 were not included in the chimeragenesis.

7.2. Stochastic chimeragenesis

The construction of the shuffled library started by cleaving the DNA with DNase I into segments of 50–100 bp (Kurtovic, Modén, et al., 2008). During cycles of primerless PCR, pieces of homologous single-stranded DNA were stochastically annealed followed by elongation and denaturation. In the end, full-length sequences were amplified using end primers. The resulting chimeric sequences had mosaic patterns of segments with different origins (Fig. 6.7). The chimeras also had point mutations based on the error-prone *Taq* polymerase used in the amplification of the DNA.

7.3. Sequence analysis

In total, 1570 bacterial clones expressing GST chimeras were analyzed, and the majority of the bacterial lysates showed no measurable activity with azathioprine. A total of 62 clones were subjected to DNA sequencing, with representatives of high, medium, and low activities with azathioprine in lysates. Thirteen sequences were identical to that of human GST A2-2. Since GST A2-2 was the most active parental enzyme, all of the colonies expressing human GST A2-2 were most likely found as hits in the screening. By assuming that wild-type occurrence in the final library was roughly the same for the different wild-type templates, the total occurrence of wild-type sequences in the library was estimated to be $5 \times 13/1570 \approx 4\%$. (A frame

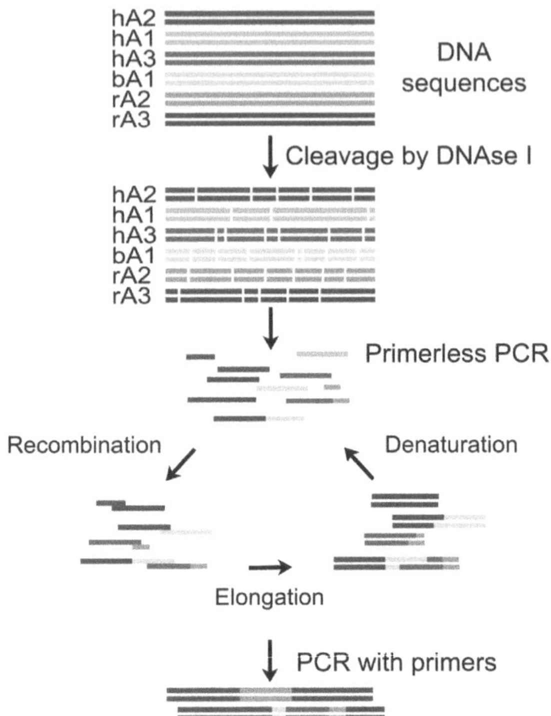

Figure 6.7 Scheme representing shuffling of DNA encoding several parental enzymes. cDNAs encoding six homologous alpha class GSTs were cleaved randomly by DNase I to form small segments of 50–100 bp. The DNA pieces were annealed to each other by primerless PCR, which generates recombinant DNA molecules with mosaic structures, in which different parental DNA segments of diverse lengths are building blocks. This stochastic recombination generates a library of new enzyme variants, which can be screened for targeted activities. *From Kurtovic, Modén, et al. (2008).* (See the color plate.)

shift in the 5′-region of the bovine GST A1-1 template sequence prevented 1/6 of the library from encoding full-length GSTs.)

Aside from the wild-type sequences found, the number of crossover points at the DNA level ranged from 1 to 13, with a mean of 5.9 ± 2.6 per chimera. The chimeric enzymes had mosaic sequences originating from on average 4.1 different templates (Kurtovic, Modén, et al., 2008). The library additionally had about 0.6 nonsilent point mutations per sequence, but none of them occurred in the active site.

7.4. Structure–activity relationships

The active site of the GSTs is composed of two binding cavities, the G-site and the H-site (Mannervik et al., 1978). The G-site accommodates

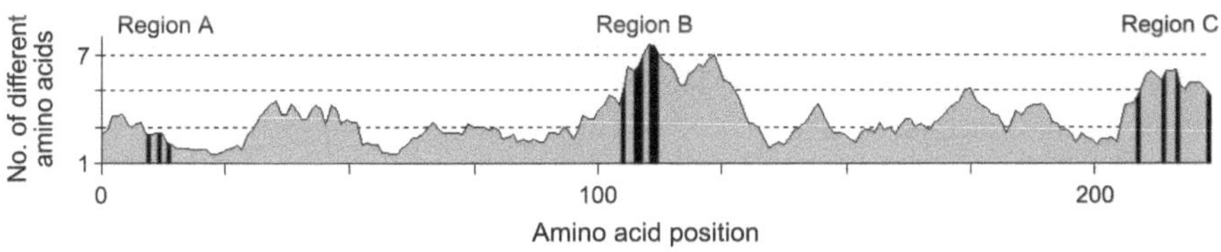

Figure 6.8 Conserved and variable amino acid residues in the sequences of naturally occurring alpha class GSTs. The number of different residues present at each position in a multiple alignment of 30 known and predicted alpha class sequences is shown as a moving average with a step size of seven. H-site positions are colored black and are located mainly in the parts with more variation. *From Kurtovic, Modén, et al. (2008).*

glutathione and resides primarily in the N-terminal thioredoxin-like domain, and the H-site binds the largely hydrophobic second substrate in the complementary C-terminal domain (Sinning et al., 1993). In the primary structure of approximately 220 amino acid residues, the H-site is based on three distinct regions A, B, and C, which in GST A2-2 encompass residues 10–12, 107–111, and 208–222, respectively (Fig. 6.8).

7.5. Some H-site residues conserved among the active chimeric mutants

The analyzed amino acid sequences were divided into 23 segments representing primary structure regions recombined during the shuffling process. Segments originating from all wild-type GSTs were present among the four most active chimeric GSTs (Kurtovic, Modén, et al., 2008). Segments 1–2 containing H-site region A and segments 20–22 overlapping region C were on the other hand conserved among the four most active chimeras as in the most active wild-type enzyme, human GST A2-2. Segments 1–2 were located in the N-terminus, and segment 2 contained two H-site residues (S10/I12). Segments 20–22 comprised the C-terminus except for the two last residues 221 and 222 and contained one H-site residue each in region C (M208/L213/S216). Thus, these conserved segments contained H-site residues and regions A and C were thus considered the primary determinants of azathioprine activity in the extant human alpha class GSTs.

7.6. Other segments of interest

A basic assumption in the analysis is that variable segments in the most active mutants are not important, whereas conserved regions are relevant regions to the function of interest. However, depending on the recombination resolution and the number of chimeras sequenced, segments of importance might in the analysis be linked to other segments that do not necessarily contribute to the function. For example, segment 1, lacking H-site residues, might just be linked to the H-site residues of segment 2 due to the recombination process.

The finding of H-site residues in almost all conserved segments among the most active enzymes demonstrated the power of family shuffling in identifying important functional elements. However, segment 8, having 5 H-site residues of region C, did not seem to be correlated with azathioprine activity in this alpha class context. It was later found that mutations of residues in region C could greatly enhance azathioprine activity (Zhang et al., 2012), but none of the parental GSTs used in the DNA shuffling happened to contain the necessary amino acid residues.

7.7. Structural interpretation

Segments 2, 8, and 20–23 occur in the three parts of the H-site cavity called regions A, B, and C (Kurtovic, Modén, et al., 2008) (Fig. 6.9). Segments 21–22 are located in the C-terminal helix α9 with segment 20 in the preceding hinge region and segment 23 at the end of the helix. Human GST A1-1 crystal structures have shown that the C-terminal helix is disordered without a ligand bound in the active site (Grahn et al., 2006). The flexible C-terminus forms a lid closing on the H-site, coordinating the binding of both glutathione and the second substrate (Nilsson, Edalat, Pettersson, & Mannervik, 2002).

Several of the chimeras with some activity found in the screening, but showing low specific activity with azathioprine when measured with purified enzymes, had segments from human GST A2-2 in the N-terminal region but not in the C-terminal region. One mutant had the sequence originating from human GST A2-2 in the C-terminal region but not in the N-terminal and did not have any detectable activity in lysate. The findings suggest that the residues in both N-terminal and C-terminal regions should preferably replicate those of the GST A2-2 structure.

7.8. An example of an active GST lacking C-terminal amino acid 222

With more chimeric enzymes included in the sequence analysis, the resolution of a crossover pattern becomes more detailed, possibly facilitating a more detailed understanding of the sequence–function relationships. One

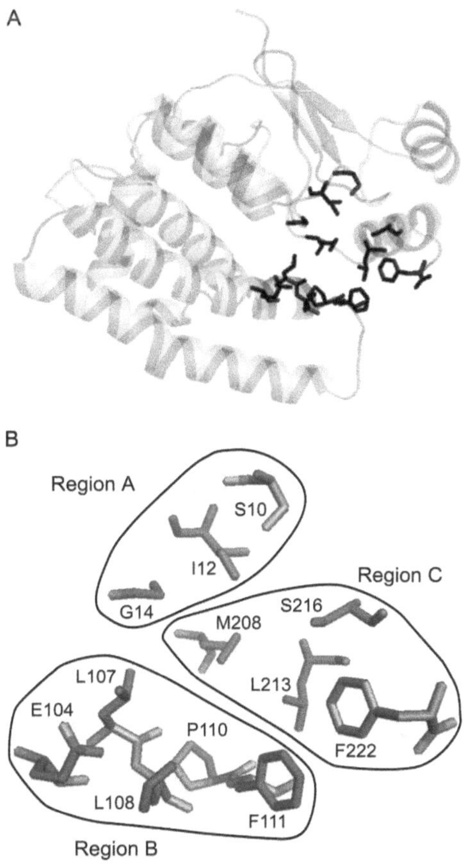

Figure 6.9 Structural model of human GST A2-2. (A) Subunit of human GST A2-2, modeled from human GST A1-1 (PDB entry 1PKW), with the H-site residues in black. (B) H-site residues in the model of human GST A2-2, as visualized from the direction of the bound glutathione. *From Kurtovic, Modén, et al. (2008).* (See the color plate.)

example was a mutant differing from human GST A2-2 only in the last two amino acids (S221 plus stop codon, compared to wild-type R221 plus F222). Position 222 is in the H-site. The mutant still had 42% of the human GST A2-2 specific activity with azathioprine indicating that those two amino acids were not essential for the activity. The division into smaller segments on the other hand also makes it harder to draw conclusions about the contribution of individual segments to the function, compared to methods using site-directed recombination.

8. RATIONAL DESIGN OF CHIMERAS

8.1. Strategy

Comparison of the sequences and their corresponding enzymatic activities with azathioprine indicated that segments 1–2 corresponding to region A in the N-terminus and segments 20–22 corresponding to region C in the C-terminus were important for the activity with this substrate. These segments contained several H-site residues that could be expected to make physical contact the substrate azathioprine. Choosing to focus on only the segments that were conserved among the active mutants in the library of GST chimeras has provided further information about features of the primary structure of importance for the azathioprine activity of human GST A2-2. The human GST A2-2 variant named A2*C was used in this work.

8.2. Design of chimeric GSTs with flanking N- and C-terminal sequences from GST A2-2

In order to find out if the N- and C-terminal portions of GST A2-2 carried the information necessary to boost azathioprine activity in other GSTs with low intrinsic activity, chimeric mutants were rationally engineered from human GST A3-3 and rat GST A3-3, respectively. These enzymes both have low azathioprine activity in their wild-type form. Two chimeric GSTs were rationally designed by the replacement of the N- and C-terminal sequences of human GST A3-3 and rat GST A3-3 by the homologous sequences of human GST A2-2 (Fig. 6.10). These mutants were produced by PCR using long primers covering the inserted human GST A2-2 fragments, some bases complementary to the templates, and the restriction sites outside of the coding sequence. The amino acid sequences of human GST

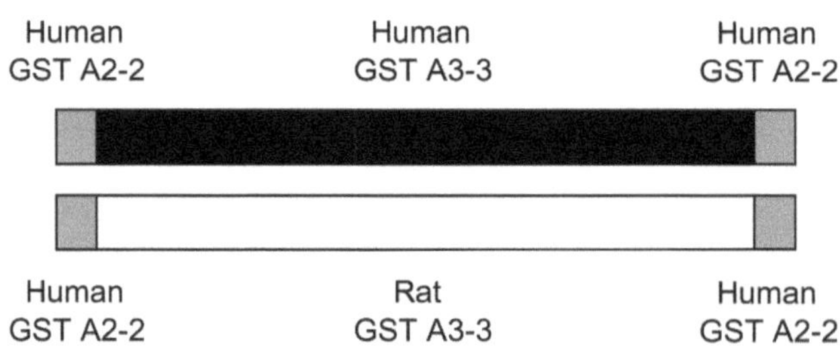

Figure 6.10 Two chimeric mutants constructed by insertion of the N- and C-terminal sequences from human GST A2-2 in place of the homologous sequences of human GST A3-3 and rat GST A3-3.

A3-3 and rat GST A3-3 differ in 25 and 60 positions, respectively, from human GST A2-2. The inserted C- and N-terminal sequences introduced 7 and 15 mutations, respectively, into the chimeric mutants based on human A3-3 and rat GST A3-3 (Modén et al., 2012).

8.3. Characterization of the designed chimeras

The chimeric mutants were folded, and stable GSTs and their specific activities toward azathioprine were determined to be 0.19 and 0.44 μmol min^{-1} mg^{-1}, about 10-fold higher than that of their respective templates human GST A3-3 and rat GST A3-3 and almost up to the level of human GST A2-2, 0.67 μmol min^{-1} mg^{-1} (Modén et al., 2012).

The chimeric mutants were also examined using alternative substrates, CDNB, CMNI, and NPTI, mimicking azathioprine. The least difference among the five enzymes was found with CDNB, the general GST substrate, with only small changes upon insertion of the flanking sequences from human GST A2-2. Human GST A2-2 has more than 10-fold higher specific activity with CMNI than human GST A3-3 and rat GST A3-3, similarly to the case of azathioprine, but the chimeric mutants only displayed up to 14% of the specific activity of human GST A2-2 with CMNI. With NPTI, the chimeric mutants had 100- and 15-fold higher specific activities than the templates human GST A3-3 and rat GST A3-3. It is noteworthy that the chimeric mutant with human GST A3-3 in the central portion even displayed 10-fold higher specific activity than human GST A2-2, with a value of 660 μmol min^{-1} mg^{-1}.

9. SATURATION MUTAGENESIS OF TWO H-SITE RESIDUES IN THE C-TERMINAL REGION

9.1. Synthesis of mutant library

The importance of the conservation of the C-terminal region of human GST A2-2 was further examined by site-directed saturation mutagenesis of residues 208 and 213 in variant A2*C (Modén et al., 2012). These two H-site residues, M208/L213, were conserved among the active mutants and in human GST A1-1, which has 60% of the human GST A2-2 activity with azathioprine. L213 is located in the α9 helix of the C-terminus with M208 in the hinge segments preceding the helix. M208 in GST A1-1 was previously selected for saturation mutagenesis owing to its proximity to the bound active-site ligand (Widersten, Björnestedt, & Mannervik, 1994). In the alpha class GSTs, the α9 helix acts as a lid of the H-site and forms a wall of one side of the binding cavity. Depending on the dynamics of the enzyme, the effects of residues 208 and 213 might display synergistic contributions to catalysis. These two conserved positions were targeted by saturation mutagenesis with NNS codons (Modén et al., 2012).

The M208X/L213X mutant library based on human GST A2-2 was prepared using an inverted polymerase chain reaction. Full-length linear plasmids were generated using adjacent 5′-phosphorylated primers pointing in opposite directions. The PCR product was purified after agarose gel electrophoresis and its blunt ends ligated before transformation into *Escherichia coli*. The use of degenerate primers with NNS codons in two positions led to a library of 400 possible mutants with amino acids in both positions and some mutants with early stop codons.

9.2. Screening and mutant characterization

Lysates from 1880 bacterial colonies were screened with azathioprine and the library coverage was estimated by Monte Carlo simulation to be $93 \pm 1\%$ or 374 ± 5 unique mutants without stop codons in the mutated positions (Modén et al., 2012). 1504 of the lysates were also screened with CDNB, and 752 of them with NPTI and CMNI as well. During screening, only three lysates had higher activity with any of the substrates, with values of 120–150% of the human GST A2-2 CDNB activity.

Sequencing and purification of selected variants confirmed that the wild-type human GST A2-2 had the highest activity with azathioprine. In assays with CDNB, CMNI, and NPTI, there were mutants with about the same specific activities as wild-type GST A2-2, even though fewer lysates were screened with these substrates. The general trend confirmed by principal component analysis was that the mutations led to decreased catalytic activities with all four tested substrates, even though some mutants had distinctly different substrate profiles compared to the parental GST A2-2 (Modén et al., 2012).

Analysis of the mutants with higher azathioprine activity indicated that hydrophobic amino acid residues were preferred, not only in position 208 but also in position 213 (Fig. 6.11).

One exception was the M208D/L213R mutant, with 40% of the wild-type specific activity toward azathioprine, and possibly, the opposite charges of the mutated residues were stabilizing the active site. A truncated variant with valine in position 208 and a stop codon in 213 still had 31% of the specific activity of human GST A2-2 with CDNB, indicating that the α9 helix

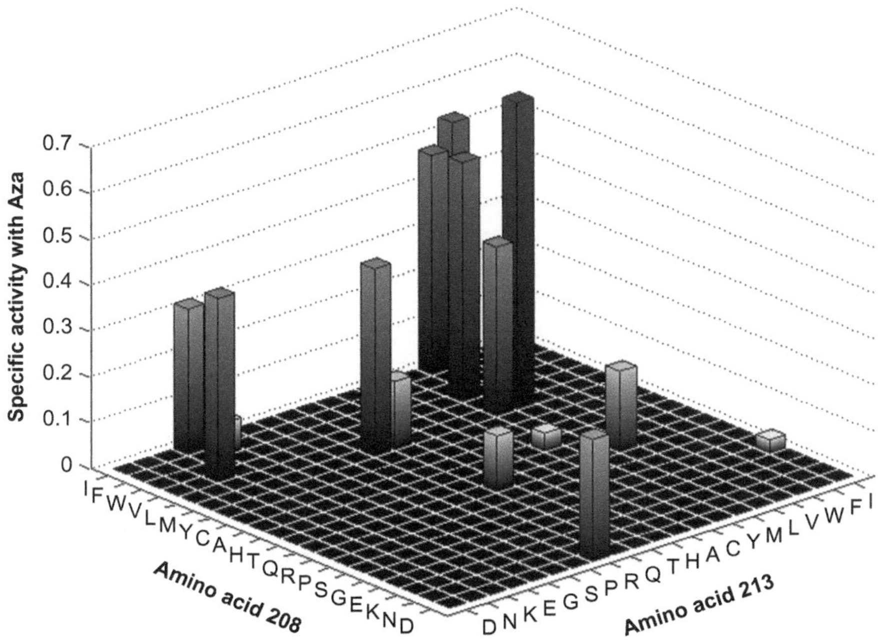

Figure 6.11 Specific activity with azathioprine for mutants purified from the M208X/L213X library. The *x*- and *y*-axes represent the amino acids in positions 208 and 213, respectively, ordered by hydrophobicity. Cells with no bars represent mutants that were not emerging as active with azathioprine in the screening of the library. *From Modén et al. (2012).* (See the color plate.)

is not necessary for general catalytic competence, even though truncation of the helix is more detrimental for the catalysis with other substrates.

Notably, the H-site residues M208/L213 were conserved among the wild-types and GST variants that in the chimeric mutants displayed high azathioprine activity (Kurtovic, Modén, et al., 2008).

10. REDESIGN OF GST A2-2 FOR ENHANCED AZATHIOPRINE ACTIVITY

10.1. Targeted H-site residues

The high GST activity with substrates other than azathioprine also led to the question whether GST mutants could obtain a similarly high activity with azathioprine, and this was approached by further mutational strategies. The structure–activity relationships showed that regions A and C in human GST A2-2, which were conserved among the chimeric mutants with high activity, could sustain high azathioprine activity within the context of the examined alpha class sequences, while the conserved H-site residues within these segments did not tolerate mutations with retention of the activity toward azathioprine. Mutational strategies attempting to improve the catalytic efficiency with this substrate could preferably instead focus on the variable segments among the mutants with high activity.

The rationale for improving azathioprine activity was based on targeting H-site amino acid residues that were likely to influence catalysis. Variant A2*E was used as template due to its higher catalytic efficiency with azathioprine compared to the other GST A2-2 variants (Zhang et al., 2010). Residues in probable proximity to the substrate bound in the active site were indicated by molecular docking (Zhang et al., 2012). Five residues were identified and two of them, in positions 111 and 213, were individually randomized, but no mutant with enhanced azathioprine activity was identified. On the other hand, the three positions 107, 108, and 222 were chosen for a combined mutant library created by using reduced sets of codons. The alternative of using the degenerate codon NNK (or NNS) simultaneously in all three positions would theoretically generate a mutant library of all combinations of the 20 amino acids from a total of 32,768 codon variants, which would be challenging to screen.

10.2. Mutant library based on reduced codon sets

The three-position library was produced using reduced codon sets in the targeted positions 107, 108, and 222 in order to limit the size of the library to facilitate screening of a larger proportion of the theoretically possible amino acid combinations. In total, the library contained 864 potential amino acid combinations. Furthermore, the library design provided high quality in terms of an even frequency in the occurrence of different amino acids. NDT (coding for amino acids Arg, Asn, Asp, Cys, Gly, His, Ile, Leu, Phe, Ser, Tyr, and Val) was used for positions 107 and 108, giving a range of different types of amino acids, all with only one codon per amino acid. The degenerate codon set YHC (coding for His, Leu, Phe, Pro, Ser, and Tyr) was used for position 222, since an F222A mutation in a human GST A1-1 mutant had detrimental effects on catalysis indicating that large amino acids might be required in this position (Nilsson et al., 2002).

The library was prepared using the dual-tube megaprimer approach (Tseng, Lin, Wei, & Fang, 2008; Zhang & Mannervik, 2013). By first amplifying the region between the mutated sites (corresponding to amino acids

Table 6.5 Molecular diversity and specific activities of 12 selected GST variants

Residues at the mutagenized positions 107 108 222	Specific activity ($\mu mol^{-1} mg\ min^{-1}$)
L L F (human A2*E)	2.3 ± 0.1
V G S	59 ± 3
C S H	16 ± 1
S D P	49 ± 1
G D P	86 ± 1
G N P	48 ± 2
H S H	49 ± 2
C G H	19 ± 1
G S H	34 ± 1
V G P	50 ± 1
G D H	128 ± 4
S G P	64 ± 5
H G P	65 ± 5

From Zhang et al. (2012).

107/108 and 222), the purified PCR product was then used as primer during the amplification of the whole plasmid. After DpnI digestion of the template plasmids and transformation into *E. coli*, sequencing of randomly sampled colonies confirmed adequate codon variability and no overrepresentation of template sequences.

Lysates of 1901 bacterial colonies were assayed corresponding to an estimated 89 ± 1% coverage of the library. The distribution of activities was skewed toward higher values and 44% of the lysates had azathioprine activity values significantly higher than the parental GST A2-2*E. In general, the mutants with the highest activities had small or polar residues in the mutated positions (Table 6.5).

The most active mutant, L107G/L108D/F222H, had a 70-fold improved catalytic efficiency with the targeted substrate. Determination of the structure of the mutant by X-ray crystallography showed a wider H-site, suggesting that that the transition state could be accommodated in a mode better suited for catalysis.

10.3. Kinetic characterization of mutants isolated

Mutants with high activity toward azathioprine were sequenced and purified. Kinetic parameters were determined and the specific activities with azathioprine are given in Table 6.5. Mutant G107/D108/H222 (named "GDH") had the highest specific activity toward azathioprine, 128 μmol mg^{-1} min^{-1}. The catalytic efficiency k_{cat}/K_M was 292 mM^{-1} s^{-1}, 70 times larger than of A2*E, with 11-fold higher k_{cat} and sixfold lower K_M (Kurtovic, Modén, et al., 2008). The pH profiles of the examined GSTs show an activity maximum around 7.

10.4. Comparison with other alpha class sequences

The GSTs with highest activity in the screening had mainly hydrophilic or small amino acids in the three positions mutated (Table 6.6). These positions are primarily occupied by large hydrophobic amino acids in the wild-type alpha class GSTs. P222 is present in four of 30 alpha class GSTs, while G107 and D108 are rare with only one occurrence of G107 in the sequences examined (Kurtovic, Modén, et al., 2008).

10.5. Structural comparisons

10.5.1 Comparison of GST A1-1, A2*C, and GDH

In human A2*C and the A2*E-derived GDH mutant, F111 is pointing in to the middle of the H-site, making the cavity less deep than that of the H-site of human GST A1-1. On the other hand, the GDH mutant lacks the large hydrophobic residues L107/L108 that in GST A2-2*C and GST A1-1 form one wall of the H-site. Instead G107 is small and D108 is pointing away from the H-site, thus making it larger in the mutant GDH than in A2*C (Kurtovic, Modén, et al., 2008).

Table 6.6 Amino acids represented in the mutated positions of the GSTs with highest activity in the screening with azathioprine

Amino acid position	107	108	222
H-site region according to Figure 6.9	B	B	C
Present in the most active	GHSVC	DGSN	HPS
Absent in the most active	DFILNRY	CFHILRVY	FLY
Present frequently in alpha class	LIMAG	LMQHYVIGS	F—PIML

Note that mutants with C107 had higher expression and lower specific activity relative to the other mutants examined. Some alpha class GSTs end before amino acid 222 and the absence is marked as "—."

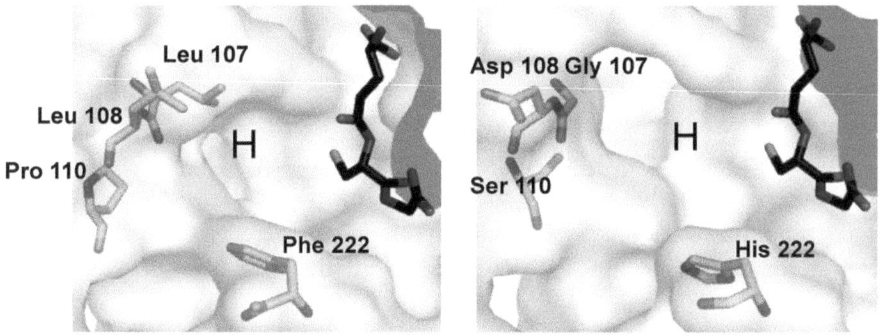

Figure 6.12 Comparison of the active-site topographies of allelic variant C of human GST A2-2 (PDB entry 2WJU) and mutant GDH of allelic variant E (PDB entry 4ACS). Glutathione (rendered as a stick model with black carbons) points its nucleophilic sulfur toward the H-site (marked "H"), which binds the electrophilic second substrate. Four residues (107, 108, 110, and 222) differing between the two structures (stick models with green carbons) contribute to the H-site cavity, in which azathioprine is supposed to bind with the imidazole leaving group close to the attacking sulfur of glutathione. The large purine moiety of azathioprine has more space in the H-site of the GDH mutant since the bulky Leu side chains of residues 107 and 108 have been removed. *From Zhang et al. (2012).* (See the color plate.)

Besides, the side chain of H222 in mutant GDH is a bit shorter and a bit further away from the H-site compared to F222 in A2*C and GST A1-1, making the opposite side of the active-site pocket slightly larger as well (Fig. 6.12). Probably azathioprine and in particular the transition state can fit more efficiently into the wider pocket of the GDH mutant. This is in agreement with the substitutions in the other mutants that displayed high activity with azathioprine, which mainly had smaller amino acid residues in positions 107, 108, and 222 compared to the residues available in the mutagenesis of those positions (Table 6.3).

10.5.2 Three mutated positions

In the design of the mutagenesis, it was speculated that introduction of novel functional groups might enhance the catalytic activity of A2*E. However, in the most active mutant GDH, the side chain of D108 is not directed toward the active site, and the introduction of H222 does not significantly influence the pH dependence of the activity, which could be expected if H222 were contributing as a novel acid/base in catalysis (Zhang et al., 2012). Supporting this notion, the related mutant GDP lacking the possible acid/base functionality was almost as active as GDH (Table 6.5). Furthermore, one of the purified mutants with very high specific activity with azathioprine, V107/G108/P222, lacked hydrophilic groups in the three mutated positions, indicating that it was not the insertion of new polar or charged catalytic groups that caused the rate enhancement. Instead, the structurally visible effect of the GDH mutation on the A2*E enzyme was the widening of the H-site, while the mutated amino acids did not necessarily interact directly with the substrate (Zhang et al., 2012). In the structure determination by X-ray crystallography, the ligand electron density was not well defined, neither for azathioprine nor for the imidazolyl group of the corresponding glutathione conjugate. This suggests that there is not a fixed binding conformation of the substrate in line with the recently proposed "local molten globule behavior" that optimizes binding of the transition state (Honaker, Acchione, Zhang, Mannervik, & Atkins, 2013).

10.5.3 Other positions involved

Since the mutated residues 107, 108, and 222 appeared not to provide new catalytic groups, they possibly complemented already existing residues by optimizing the shape and electrostatics of the active site for transition-state stabilization (Warshel, 2003). Residue Y9 is already known to stabilize the deprotonated sulfur on the activated glutathione (Dourado et al., 2008). The neighboring R15 is also a hydrogen bond donor, which is a conserved basic residue in alpha class GSTs (Björnestedt et al., 1995), and its positive charge is of importance for the activation of glutathione (Dourado et al., 2010). F111 could possibly stabilize the purine ring of azathioprine depending on the conformation of the substrate and the transition state. Perhaps replacements to larger or polar amino acids in other H-site positions could also lead to new interactions with the substrate while removal of bulky residues might position the substrate or interacting amino acids even more favorably. Mutants with very high azathioprine activity in the three-position library had small or polar residues that were rare in wild-type alpha class sequences in these positions. This could be a reason why the shuffling of alpha class sequences did not identify mutants with higher specific activity than human A2-2 toward azathioprine.

11. CONCLUDING REMARKS

The prodrug azathioprine is metabolized along several pathways and the first step is activated in humans by GSTs A2-2, A1-1, and M1-1. Approximately 30% of the patients undergoing azathioprine therapy experience adverse side effects, and genotyping can identify patients with TPMT gene deficiency with a high risk for severe myelosuppression. However, only a minor proportion of the cases of adverse effects can be predicted. The differences in the rate of azathioprine activation might affect metabolite concentrations and distributions. A null genotype of the azathioprine-activating GST M1-1 has been indicated to be related to differences in the occurrence of adverse effects.

The activation of azathioprine is catalyzed most efficiently by GST A2-2 compared to other human GSTs, and differences in the GST expression levels are known. Five allelic variants of human GST A2-2 were characterized and variant A2*E turned out to have a three- to fourfold higher catalytic efficiency with azathioprine than the other ones. DNA shuffling and directed mutagenesis identified N- and C-terminal regions essential to support the high activity of GST A2-2.

Redesign of variant A2*E by protein engineering produced mutant GSTs with up to 70-fold enhanced azathioprine activity approaching the highest activities known with any GST substrate. The crystal structure of one of the mutants suggests that widening of the substrate binding site, rather than introduction of novel catalytic functionalities, explains the increased catalytic efficiency. The design of GSTs with enhanced catalytic functions opens possibilities for the use of the enzymes as biotherapeutics.

ACKNOWLEDGMENTS

This chapter is largely based on the PhD thesis of Olof Modén (2013). Work from the Mannervik laboratory was supported by the Swedish Cancer Society and the Swedish Research Council.

REFERENCES

Allorge, D., Hamdan, R., Broly, F., Libersa, C., & Colombel, J. F. (2005). ITPA genotyping test does not improve detection of Crohn's disease patients at risk of azathioprine/6-mercaptopurine induced myelosuppression. *Gut*, *54*(4), 565.

Al-Safi, S., Tashtoush, B., & Hassan, M. (2003). Comparison of the effects of azathioprine and its novel non-mercaptopurine analog on antibody response in rabbits. *Polish Journal of Pharmacology*, *55*(2), 239–243.

Andersson, A., Lindgren, A., Arnadottir, M., Prytz, H., & Hultberg, B. (1999). Thiols as a measure of plasma redox status in healthy subjects and in patients with renal or liver failure. *Clinical Chemistry*, *45*(7), 1084–1086.

Arenas, M., Simpson, G., Lewis, C. M., Shobowale-Bakre, el-M, Escuredo, E., & Fairbanks, L. D. (2005). Genetic variation in the MTHFR gene influences thiopurine methyltransferase activity. *Clinical Chemistry*, *51*(12), 2371–2374.

Ballatori, N., Hammond, C. L., Cunningham, J. B., Krance, S. M., & Marchan, R. (2005). Molecular mechanisms of reduced glutathione transport: Role of the MRP/CFTR/ABCC and OATP/SLC21A families of membrane proteins. *Toxicology and Applied Pharmacology*, *204*(3), 238–255.

Ban, H., Andoh, A., Imaeda, H., Kobori, A., Bamba, S., Tsujikawa, T., et al. (2010). The multidrug-resistance protein 4 polymorphism is a new factor accounting for thiopurine sensitivity in Japanese patients with inflammatory bowel disease. *Journal of Gastroenterology, 45*(10), 1014–1021.

Barabino, A., Torrente, F., Ventura, A., Cucchiara, S., Castro, M., & Barbera, C. (2002). Azathioprine in paediatric inflammatory bowel disease: An Italian multicentre survey. *Alimentary Pharmacology & Therapeutics, 16*(6), 1125–1130.

Barba, J., Rosell, D., Rincon, A., Robles, J. E., Zudaire, J., Berian, J. M., et al. (2012). New immunosuppressive therapies and surgical complications after renal transplantation. *Transplantation Proceedings, 44*(5), 1275–1280.

Beaugerie, L., Brousse, N., Bouvier, A. M., Colombel, J. F., Lémann, M., Cosnes, J., et al. (2009). Lymphoproliferative disorders in patients receiving thiopurines for inflammatory bowel disease: A prospective observational cohort study. *Lancet, 374*(9701), 1617–1625.

Björnestedt, R., Stenberg, G., Widersten, M., Board, P. G., Sinning, I., Jones, T. A., et al. (1995). Functional significance of arginine 15 in the active site of human class alpha glutathione transferase A1-1. *Journal of Molecular Biology, 247*(4), 765–773.

Bourgine, J., Garat, A., Allorge, D., Crunelle-Thibaut, A., Lo-Guidice, J. M., Colombel, J. F., et al. (2011). Evidence for a functional genetic polymorphism of the Rho-GTPase Rac1. Implication in azathioprine response? *Pharmacogenetics and Genomics, 21*(6), 313–324.

Breen, D. P., Marinaki, A. M., Arenas, M., & Hayes, P. C. (2005). Pharmacogenetic association with adverse drug reactions to azathioprine immunosuppressive therapy following liver transplantation. *Liver Transplantation, 11*(7), 826–833.

Calne, R. Y., Alexandre, G. P., & Murray, J. E. (1962). A study of the effects of drugs in prolonging survival of homologous renal transplants in dogs. *Annals of the New York Academy of Sciences, 99*, 743–761.

Cara, C. J., Pena, A. S., Sans, M., Rodrigo, L., Guerrero-Esteo, M., Hinojosa, J., et al. (2004). Reviewing the mechanism of action of thiopurine drugs: Towards a new paradigm in clinical practice. *Medical Science Monitor, 10*(11), RA247–RA254.

Chalmers, A. H., Knight, P. R., & Atkinson, M. R. (1967). Conversion of azathioprine into mercaptopurine and mercaptoimidazole derivatives in vitro and during immunosuppressive therapy. *The Australian Journal of Experimental Biology and Medical Science, 45*(6), 681–691.

Chouchana, L., Narjoz, C., Beaune, P., Loriot, M. A., & Roblin, X. (2012). Review article: The benefits of pharmacogenetics for improving thiopurine therapy in inflammatory bowel disease. *Alimentary Pharmacology & Therapeutics, 35*(1), 15–36.

Clayton, P. A., McDonald, S. P., Chapman, J. R., & Chadban, S. J. (2012). Mycophenolate versus azathioprine for kidney transplantation: A 15-year follow-up of a randomized trial. *Transplantation, 94*(2), 152–158.

Coles, B. F., Anderson, K. E., Doerge, D. R., Churchwell, M. I., Lang, N. P., & Kadlubar, F. F. (2000). Quantitative analysis of interindividual variation of glutathione S-transferase expression in human pancreas and the ambiguity of correlating genotype with phenotype. *Cancer Research, 60*(3), 573–579.

Coles, B. F., Chen, G., Kadlubar, F. F., & Radominska-Pandya, A. (2002). Interindividual variation and organ-specific patterns of glutathione S-transferase alpha, mu, and pi expression in gastrointestinal tract mucosa of normal individuals. *Archives of Biochemistry and Biophysics, 403*(2), 270–276.

Coles, B. F., & Kadlubar, F. F. (2005). Human alpha class glutathione S-transferases: Genetic polymorphism, expression, and susceptibility to disease. *Methods in Enzymology, 401*, 9–42.

Connell, W. R., Kamm, M. A., Ritchie, J. K., & Lennard-Jones, J. E. (1993). Bone marrow toxicity caused by azathioprine in inflammatory bowel disease: 27 years of experience. *Gut, 34*(8), 1081–1085.

Crawford, D. J., Maddocks, J. L., Jones, D. N., & Szawlowski, P. (1996). Rational design of novel immunosuppressive drugs: Analogues of azathioprine lacking the 6-mercaptopurine substituent retain or have enhanced immunosuppressive effects. *Journal of Medicinal Chemistry*, *39*(14), 2690–2695.

DeLeve, L. D., Wang, X., Kuhlenkamp, J. F., & Kaplowitz, N. (1996). Toxicity of azathioprine and monocrotaline in murine sinusoidal endothelial cells and hepatocytes: The role of glutathione and relevance to hepatic venoocclusive disease. *Hepatology*, *23*(3), 589–599.

Dixon, D. P., Lapthorn, A., & Edwards, R. (2002). Plant glutathione transferases. *Genome Biology*, *3*(3), reviews3004.1–3004.10.

Dong, X. W., Zheng, Q., Zhu, M. M., Tong, J. L., & Ran, Z. H. (2010). Thiopurine S-methyltransferase polymorphisms and thiopurine toxicity in treatment of inflammatory bowel disease. *World Journal of Gastroenterology*, *16*(25), 3187–3195.

Dourado, D. F., Fernandes, P. A., Mannervik, B., & Ramos, M. J. (2008). Glutathione transferase: New model for glutathione activation. *Chemistry*, *14*(31), 9591–9598.

Dourado, D. F., Fernandes, P. A., Mannervik, B., & Ramos, M. J. (2010). Glutathione transferase A1-1: Catalytic importance of arginine 15. *The Journal of Physical Chemistry B*, *114*(4), 1690–1697.

Dubinsky, M. C., Lamothe, S., Yang, H. Y., Targan, S. R., Sinnett, D., Théorêt, Y., et al. (2000). Pharmacogenomics and metabolite measurement for 6-mercaptopurine therapy in inflammatory bowel disease. *Gastroenterology*, *118*(4), 705–713.

Eklund, B. I., Moberg, M., Bergquist, J., & Mannervik, B. (2006). Divergent activities of human glutathione transferases in the bioactivation of azathioprine. *Molecular Pharmacology*, *70*(2), 747–754.

Elion, G. B. (1989). The purine path to chemotherapy. *Science*, *244*(4900), 41–47.

Fraser, A. G., Orchard, T. R., & Jewell, D. P. (2002). The efficacy of azathioprine for the treatment of inflammatory bowel disease: A 30 year review. *Gut*, *50*(4), 485–489.

French, H., Dalzell, A., Srinivasan, R., & El-Matary, W. (2011). Relapse rate following azathioprine withdrawal in maintaining remission for Crohn's disease: A meta-analysis. *Digestive Diseases and Sciences*, *56*(7), 1929–1936.

García-Mata, R., Conde, R. D., & Sanllorenti, P. M. (1998). Contribution of breakdown in the regulation of mouse liver glutathione S-transferase subunits during protein depletion and re-feeding. *Biochimica et Biophysica Acta*, *1448*(1), 46–50.

Gardiner, S. J., Gearry, R. B., Begg, E. J., Zhang, M., & Barclay, M. L. (2008). Thiopurine dose in intermediate and normal metabolizers of thiopurine methyltransferase may differ three-fold. *Clinical Gastroenterology and Hepatology*, *6*(6), 654–660, quiz 604.

Gearry, R. B., & Barclay, M. L. (2005). Azathioprine and 6-mercaptopurine pharmacogenetics and metabolite monitoring in inflammatory bowel disease. *Journal of Gastroenterology and Hepatology*, *20*(8), 1149–1157.

Gearry, R. B., Roberts, R. L., Barclay, M. L., & Kennedy, M. A. (2004). Lack of association between the ITPA 94C>A polymorphism and adverse effects from azathioprine. *Pharmacogenetics*, *14*(11), 779–781.

Giner-Sorolla, A. (1988). The excitement of a suspense story, the beauty of a poem: The work of Hitchings and Elion. *Trends in Pharmacological Sciences*, *9*, 437–438.

González-Lama, Y., Bermejo, F., López-Sanromán, A., García-Sánchez, V., Esteve, M., Cabriada, J. L., et al. (2011). Thiopurine methyl-transferase activity and azathioprine metabolite concentrations do not predict clinical outcome in thiopurine-treated inflammatory bowel disease patients. *Alimentary Pharmacology & Therapeutics*, *34*(5), 544–554.

Grahn, E., Novotny, M., Jakobsson, E., Gustafsson, A., Grehn, L., Olin, B., et al. (2006). New crystal structures of human glutathione transferase A1-1 shed light on glutathione binding and the conformation of the C-terminal helix. *Acta Crystallographica Section D: Biological Crystallography*, *62*(Pt 2), 197–207.

Hansson, L. O., Widersten, M., & Mannervik, B. (1997). Mechanism-based phage display selection of active-site mutants of human glutathione transferase A1-1 catalyzing SNAr reactions. *Biochemistry, 36*(37), 11252–11260.

Honaker, M. T., Acchione, M., Zhang, W., Mannervik, B., & Atkins, W. M. (2013). Enzymatic detoxification, drug metabolism and the role of molten globule active sites. *The Journal of Biological Chemistry, 288*, 18599–18611.

Hubatsch, I., Ridderström, M., & Mannervik, B. (1998). Human glutathione transferase A4-4: An alpha class enzyme with high catalytic efficiency in the conjugation of 4-hydroxynonenal and other genotoxic products of lipid peroxidation. *The Biochemical Journal, 330*(Pt 1), 175–179.

Johansson, A. S., & Mannervik, B. (2001). Human glutathione transferase A3-3, a highly efficient catalyst of double-bond isomerization in the biosynthetic pathway of steroid hormones. *The Journal of Biological Chemistry, 276*(35), 33061–33065.

Josephy, P. D., & Mannervik, B. (2006). *Molecular toxicology* (2nd ed.). New York; Oxford: Oxford University Press.

Kaplowitz, N. (1976). Enzymatic thiolysis of azathioprine in vitro. *Biochemical Pharmacology, 25*, 2421–2426.

Karran, P. (2006). Thiopurines, DNA damage, DNA repair and therapy-related cancer. *British Medical Bulletin, 79–80*, 153–170.

Kennedy, N. A., Asser, T. L., Mountifield, R. E., Doogue, M. P., Andrews, J. M., & Bampton, P. A. (2013). Thiopurine metabolite measurement leads to changes in management of inflammatory bowel disease. *Internal Medicine Journal, 43*(3), 278–286.

Knight, D. M., Trinh, H., Le, J., Siegel, S., Shealy, D., McDonough, M., et al. (1993). Construction and initial characterization of a mouse-human chimeric anti-TNF antibody. *Molecular Immunology, 30*(16), 1443–1453.

Kudo, M., Moteki, T., Sasaki, T., Konno, Y., Ujiie, S., Onose, A., et al. (2008). Functional characterization of human xanthine oxidase allelic variants. *Pharmacogenetics and Genomics, 18*(3), 243–251.

Kurtovic, S., Grehn, L., Karlsson, A., Hellman, U., & Mannervik, B. (2008). Glutathione transferase activity with a novel substrate mimics the activation of the prodrug azathioprine. *Analytical Biochemistry, 375*(2), 339–344.

Kurtovic, S., Modén, O., Shokeer, A., & Mannervik, B. (2008). Structural determinants of glutathione transferases with azathioprine activity identified by DNA shuffling of alpha class members. *Journal of Molecular Biology, 375*, 1365–1379.

Lakatos, P. L., & Kiss, L. S. (2011). Current status of thiopurine analogues in the treatment in Crohn's disease. *World Journal of Gastroenterology, 17*(39), 4372–4381.

Larsson, E., Mannervik, B., & Raffalli-Mathieu, F. (2011). Quantitative and selective polymerase chain reaction analysis of highly similar human alpha-class glutathione transferases. *Analytical Biochemistry, 412*(1), 96–101.

Lee, A. U., & Farrell, G. C. (2001). Mechanism of azathioprine-induced injury to hepatocytes: Roles of glutathione depletion and mitochondrial injury. *Journal of Hepatology, 35*(6), 756–764.

Lees, C. W., Maan, A. K., Hansoti, B., Satsangi, J., & Arnott, I. D. (2008). Tolerability and safety of mercaptopurine in azathioprine-intolerant patients with inflammatory bowel disease. *Alimentary Pharmacology & Therapeutics, 27*(3), 220–227.

Low, W. Y., Ng, H. L., Morton, C. J., Parker, M. W., Batterham, P., & Robin, C. (2007). Molecular evolution of glutathione S-transferases in the genus Drosophila. *Genetics, 177*(3), 1363–1375.

Lowry, P. W., Franklin, C. L., Weaver, A. L., Pike, M. G., Mays, D. C., Tremaine, W. J., et al. (2001). Measurement of thiopurine methyltransferase activity and azathioprine metabolites in patients with inflammatory bowel disease. *Gut, 49*(5), 665–670.

Maddox, J. S., & Soltani, K. (2008). Risk of nonmelanoma skin cancer with azathioprine use. *Inflammatory Bowel Diseases*, *14*(10), 1425–1431.
Mannervik, B., Board, P. G., Hayes, J., Listowsky, I., & Pearson, W. R. (2005). Nomenclature for mammalian soluble glutathione transferases. *Methods in Enzymology*, *401*, 1–8.
Mannervik, B., Guthenberg, C., Jakobson, I., & Warholm, M. (1978). Glutathione conjugation: Reaction mechanism of glutathione S-transferase A. In A. Aitio (Ed.), *Conjugation reactions in drug biotransformation* (pp. 101–110). Amsterdam: Elsevier/North-Holland.
Marinaki, A. M., Duley, J. A., Arenas, M., Ansari, A., Sumi, S., Lewis, C. M., et al. (2004). Mutation in the ITPA gene predicts intolerance to azathioprine. *Nucleosides, Nucleotides & Nucleic Acids*, *23*(8–9), 1393–1397.
McElwee, J. J., Schuster, E., Blanc, E., Piper, M. D., Thomas, J. H., Patel, D. S., et al. (2007). Evolutionary conservation of regulated longevity assurance mechanisms. *Genome Biology*, *8*(7), R132.
Meister, A., & Anderson, M. E. (1983). Glutathione. *Annual Review of Biochemistry*, *52*, 711–760.
Modén, O. (2013). *Mutational analysis and redesign of alpha-class glutathione transferases for enhanced azathioprine activity*. Uppsala: Acta Universitatis Upsaliensis.
Modén, O., Zhang, W., & Mannervik, B. (2012). Mutational analysis of human glutathione transferase A2-2 identifies structural elements supporting high activity with the prodrug azathioprine. *Protein Engineering, Design & Selection*, *25*, 189–197.
Morgan, A. S., Sanderson, P. E., Borch, R. F., Tew, K. D., Niitsu, Y., Takayama, T., et al. (1998). Tumor efficacy and bone marrow-sparing properties of TER286, a cytotoxin activated by glutathione S-transferase. *Cancer Research*, *58*(12), 2568–2575.
Mulder, T. P., Court, D. A., & Peters, W. H. (1999). Variability of glutathione S-transferase alpha in human liver and plasma. *Clinical Chemistry*, *45*(3), 355–359.
Mulder, T. P., Peters, W. H., Court, D. A., & Jansen, J. B. (1996). Sandwich ELISA for glutathione S-transferase Alpha 1–1: Plasma concentrations in controls and in patients with gastrointestinal disorders. *Clinical Chemistry*, *42*(3), 416–419.
Nilsson, L. O., Edalat, M., Pettersson, P. L., & Mannervik, B. (2002). Aromatic residues in the C-terminal region of glutathione transferase A1-1 influence rate-determining steps in the catalytic mechanism. *Biochimica et Biophysica Acta*, *1598*(1–2), 199–205.
Ning, B., Wang, C., Morel, F., Nowell, S., Ratnasinghe, D. L., Carter, W., et al. (2004). Human glutathione S-transferase A2 polymorphisms: Variant expression, distribution in prostate cancer cases/controls and a novel form. *Pharmacogenetics*, *14*(1), 35–44.
Odlind, B., Hartvig, P., Lindström, B., Lönnerholm, G., Tufveson, G., & Grefberg, N. (1986). Serum azathioprine and 6-mercaptopurine levels and immunosuppressive activity after azathioprine in uremic patients. *International Journal of Immunopharmacology*, *8*(1), 1–11.
O'Donovan, P., Perrett, C. M., Zhang, X., Montaner, B., Xu, Y. Z., Harwood, C. A., et al. (2005). Azathioprine and UVA light generate mutagenic oxidative DNA damage. *Science*, *309*(5742), 1871–1874.
Pearson, W. R. (2005). Phylogenies of glutathione transferase families. *Methods in Enzymology*, *401*, 186–204.
Pearson, D. C., May, G. R., Fick, G. H., & Sutherland, L. R. (1995). Azathioprine and 6-mercaptopurine in Crohn disease. A meta-analysis. *Annals of Internal Medicine*, *123*(2), 132–142.
Poppe, D., Tiede, I., Fritz, G., Becker, C., Bartsch, B., Wirtz, S., et al. (2006). Azathioprine suppresses ezrin-radixin-moesin-dependent T cell-APC conjugation through inhibition of Vav guanosine exchange activity on Rac proteins. *Journal of Immunology*, *176*(1), 640–651.

Prefontaine, E., Sutherland, L. R., Macdonald, J. K., & Cepoiu, M. (2009). Azathioprine or 6-mercaptopurine for maintenance of remission in Crohn's disease. *Cochrane Database of Systematic Reviews*, (1), Art. No. CD000067, pp. 1–20.

Roberts, R. L., & Barclay, M. L. (2012). Current relevance of pharmacogenetics in immunomodulation treatment for Crohn's disease. *Journal of Gastroenterology and Hepatology*, *27*(10), 1546–1554.

Rowe, J. D., Nieves, E., & Listowsky, I. (1997). Subunit diversity and tissue distribution of human glutathione S-transferases: Interpretations based on electrospray ionization-MS and peptide sequence-specific antisera. *The Biochemical Journal*, *325*(Pt 2), 481–486.

Rozell, B., Hansson, H.-A., Guthenberg, C., Tahir, M. K., & Mannervik, B. (1993). Glutathione transferases of classes Alpha, Mu and Pi show selective expression in different regions of the rat kidney. *Xenobiotica*, *23*, 835–849.

Sauer, H., Hantke, U., & Wilmanns, W. (1988). Azathioprine lymphocytotoxicity. Potentially lethal damage by its imidazole derivatives. *Arzneimittel-Forschung*, *38*(6), 820–824.

Schwab, M., Schäffeler, E., Marx, C., Fischer, C., Lang, T., Behrens, C., et al. (2002). Azathioprine therapy and adverse drug reactions in patients with inflammatory bowel disease: Impact of thiopurine S-methyltransferase polymorphism. *Pharmacogenetics*, *12*(6), 429–436.

Schwartz, R., & Dameshek, W. (1959). Drug-induced immunological tolerance. *Nature*, *183*, 1682–1683.

Shen, H., Ranganathan, S., Kuzmich, S., & Tew, K. D. (1995). Influence of ethacrynic acid on glutathione S-transferase pi transcript and protein half-lives in human colon cancer cells. *Biochemical Pharmacology*, *50*(8), 1233–1238.

Shih, D. Q., Nguyen, M., Zheng, L., Ibanez, P., Mei, L., Kwan, L. Y., et al. (2012). Split-dose administration of thiopurine drugs: A novel and effective strategy for managing preferential 6-MMP metabolism. *Alimentary Pharmacology & Therapeutics*, *36*(5), 449–458.

Shipkova, M., Franz, J., Abe, M., Klett, C., Wieland, E., & Andus, T. (2011). Association between adverse effects under azathioprine therapy and inosine triphosphate pyrophosphatase activity in patients with chronic inflammatory bowel disease. *Therapeutic Drug Monitoring*, *33*(3), 321–328.

Silva, S. N., Azevedo, A. P., Teixeira, V., Pina, J. E., Rueff, J., & Gaspar, J. F. (2009). The role of GSTA2 polymorphisms and haplotypes in breast cancer susceptibility: A case–control study in the Portuguese population. *Oncology Reports*, *22*(3), 593–598.

Singh, S. P., Zimniak, L., & Zimniak, P. (2010). The human hGSTA5 gene encodes an enzymatically active protein. *Biochimica et Biophysica Acta*, *1800*, 16–22.

Sinning, I., Kleywegt, G. J., Cowan, S. W., Reinemer, P., Dirr, H. W., Huber, R., et al. (1993). Structure determination and refinement of human alpha class glutathione transferase A1-1, and a comparison with the mu and pi class enzymes. *Journal of Molecular Biology*, *232*, 192–212.

Smith, M. A., Marinaki, A. M., Arenas, M., Shobowale-Bakre, M., Lewis, C. M., Ansari, A., et al. (2009). Novel pharmacogenetic markers for treatment outcome in azathioprine-treated inflammatory bowel disease. *Alimentary Pharmacology & Therapeutics*, *30*(4), 375–384.

Sparrow, M. P., Hande, S. A., Friedman, S., Cao, D., & Hanauer, S. B. (2007). Effect of allopurinol on clinical outcomes in inflammatory bowel disease nonresponders to azathioprine or 6-mercaptopurine. *Clinical Gastroenterology and Hepatology*, *5*(2), 209–214.

Stemmer, W. P. (1994). Rapid evolution of a protein in vitro by DNA shuffling. *Nature*, *370*(6488), 389–391.

Stocco, G., Martelossi, S., Barabino, A., Decorti, G., Bartoli, F., Montico, M., et al. (2007). Glutathione-S-transferase genotypes and the adverse effects of azathioprine in young patients with inflammatory bowel disease. *Inflammatory Bowel Diseases*, *13*(1), 57–64.

Sundberg, A. G., Nilsson, R., Appelkvist, E. L., & Dallner, G. (1995). ELISA procedures for the quantitation of glutathione transferases in the urine. *Kidney International, 48*(2), 570–575.

Swann, P. F., Waters, T. R., Moulton, D. C., Xu, Y. Z., Zheng, Q., Edwards, M., et al. (1996). Role of postreplicative DNA mismatch repair in the cytotoxic action of thioguanine. *Science, 273*(5278), 1109–1111.

Tars, K., Olin, B., & Mannervik, B. (2010). Structural basis for featuring of steroid isomerase activity in Alpha class glutathione transferases. *Journal of Molecular Biology, 397*, 332–340.

Tay, B. S., Lilley, R. M., Murray, A. W., & Atkinson, M. R. (1969). Inhibition of phosphoribosyl pyrophosphate amidotransferase from Ehrlich ascites-tumour cells by thiopurine nucleotides. *Biochemical Pharmacology, 18*(4), 936–938.

Tetlow, N., & Board, P. G. (2004). Functional polymorphism of human glutathione transferase A2. *Pharmacogenetics, 14*(2), 111–116.

Tetlow, N., Liu, D., & Board, P. (2001). Polymorphism of human Alpha class glutathione transferases. *Pharmacogenetics, 11*(7), 609–617.

Tiede, I., Fritz, G., Strand, S., Poppe, D., Dvorsky, R., Strand, D., et al. (2003). CD28-dependent Rac1 activation is the molecular target of azathioprine in primary human CD4+ T lymphocytes. *The Journal of Clinical Investigation, 111*(8), 1133–1145.

Tseng, W. C., Lin, J. W., Wei, T. Y., & Fang, T. Y. (2008). A novel megaprimed and ligase-free, PCR-based, site-directed mutagenesis method. *Analytical Biochemistry, 375*(2), 376–378.

Van den Brande, J. M., Braat, H., van den Brink, G. R., Versteeg, H. H., Bauer, C. A., Hoedemaeker, I., et al. (2003). Infliximab but not etanercept induces apoptosis in lamina propria T-lymphocytes from patients with Crohn's disease. *Gastroenterology, 124*(7), 1774–1785.

van Dieren, J. M., van Vuuren, A. J., Kusters, J. G., Nieuwenhuis, E. E., Kuipers, E. J., & van der Woude, C. J. (2005). ITPA genotyping is not predictive for the development of side effects in AZA treated inflammatory bowel disease patients. *Gut, 54*(11), 1664.

van Egmond, R., Chin, P., Zhang, M., Sies, C. W., & Barclay, M. L. (2012). High TPMT enzyme activity does not explain drug resistance due to preferential 6-methylmercaptopurine production in patients on thiopurine treatment. *Alimentary Pharmacology & Therapeutics, 35*(10), 1181–1189.

von Ahsen, N., Armstrong, V. W., Behrens, C., von Tirpitz, C., Stallmach, A., Herfarth, H., et al. (2005). Association of inosine triphosphatase 94C>A and thiopurine S-methyltransferase deficiency with adverse events and study drop-outs under azathioprine therapy in a prospective Crohn disease study. *Clinical Chemistry, 51*(12), 2282–2288.

von Bahr, C., Glaumann, H., Gudas, J., & Kaplowitz, N. (1980). Inhibition of hepatic metabolism of azathioprine by furosemide in human liver in vitro. *Biochemical Pharmacology, 29*(10), 1439–1441.

Wang, L., & Weinshilboum, R. (2006). Thiopurine S-methyltransferase pharmacogenetics: Insights, challenges and future directions. *Oncogene, 25*(11), 1629–1638.

Warholm, M., Guthenberg, C., Mannervik, B., & von Bahr, C. (1981). Purification of a new glutathione *S*-transferase (transferase μ) from human liver having high activity with benzo(a)pyrene-4,5-oxide. *Biochemical and Biophysical Research Communications, 98*, 512–519.

Warshel, A. (2003). Computer simulations of enzyme catalysis: Methods, progress, and insights. *Annual Review of Biophysics and Biomolecular Structure, 32*, 425–443.

Watts, M. E., Hodgkiss, R. J., Sehmi, D. S., & Woodcock, M. (1980). Rapid-mix studies on the anomalous radiosensitization of mammalian cells by 5-chloro-1-methyl-4-nitromidazole. *International Journal of Radiation Biology and Related Studies in Physics, Chemistry, and Medicine, 38*(6), 673–675.

Weersma, R. K., Peters, F. T., Oostenbrug, L. E., van den Berg, A. P., van Haastert, M., Ploeg, R. J., et al. (2004). Increased incidence of azathioprine-induced pancreatitis in Crohn's disease compared with other diseases. *Alimentary Pharmacology & Therapeutics*, *20*(8), 843–850.

Weinshilboum, R. M., & Sladek, S. L. (1980). Mercaptopurine pharmacogenetics: Monogenic inheritance of erythrocyte thiopurine methyltransferase activity. *The American Journal of Human Genetics*, *32*(5), 651–662.

Widersten, M., Björnestedt, R., & Mannervik, B. (1994). Contribution of amino acid residue 208 in the hydrophobic binding site to the catalytic mechanism of human glutathione transferase A1-1. *Biochemistry*, *33*(39), 11717–11723.

Widersten, M., & Mannervik, B. (1995). Glutathione transferases with novel active sites isolated by phage display from a library of random mutants. *Journal of Molecular Biology*, *250*, 115–122.

Wielinga, P. R., Reid, G., Challa, E. E., van der Heijden, I., van Deemter, L., de Haas, M., et al. (2002). Thiopurine metabolism and identification of the thiopurine metabolites transported by MRP4 and MRP5 overexpressed in human embryonic kidney cells. *Molecular Pharmacology*, *62*(6), 1321–1331.

Wong, D. R., Derijks, L. J., den Dulk, M. O., Gemmeke, E. H., & Hooymans, P. M. (2007). The role of xanthine oxidase in thiopurine metabolism: A case report. *Therapeutic Drug Monitoring*, *29*(6), 845–848.

Wroblova, K., Kolorz, M., Batovsky, M., Zboril, V., Suchankova, J., Bartos, M., et al. (2012). Gene polymorphisms involved in manifestation of leucopenia, digestive intolerance, and pancreatitis in azathioprine-treated patients. *Digestive Diseases and Sciences*, *57*(9), 2394–2401.

Xiong, H., Xin, H., Wu, X., Li, Q., Xiong, L., & Yu, A. (2010). Association between inosine triphosphate pyrophosphohydrolase deficiency and azathioprine-related adverse drug reactions in the Chinese kidney transplant recipients. *Fundamental & Clinical Pharmacology*, *24*(3), 393–400.

Yates, C. R., Krynetski, E. Y., Loennechen, T., Fessing, M. Y., Tai, H. L., Pui, C. H., et al. (1997). Molecular diagnosis of thiopurine S-methyltransferase deficiency: Genetic basis for azathioprine and mercaptopurine intolerance. *Annals of Internal Medicine*, *126*(8), 608–614.

Zabala-Fernández, W., Barreiro-de Acosta, M., Echarri, A., Carpio, D., Lorenzo, A., Castro, J., et al. (2011). A pharmacogenetics study of TPMT and ITPA genes detects a relationship with side effects and clinical response in patients with inflammatory bowel disease receiving Azathioprine. *Journal of Gastrointestinal and Liver Diseases*, *20*(3), 247–253.

Zhang, W., & Mannervik, B. (2013). An improved dual-tube megaprimer approach for multi-site saturation mutagenesis. *World Journal of Microbiology and Biotechnology*, *29*, 667–672.

Zhang, W., Modén, O., & Mannervik, B. (2010). Differences among allelic variants of human glutathione transferase A2-2 in the activation of azathioprine. *Chemico-Biological Interactions*, *186*, 110–117.

Zhang, W., Modén, O., Tars, K., & Mannervik, B. (2012). Structure-based redesign of GST A2-2 for enhanced catalytic efficiency with azathioprine. *Chemistry & Biology*, *19*, 414–421.

CHAPTER SEVEN

Thioredoxin and Hematologic Malignancies

Ningfei An, Yubin Kang[1,2]
Division of Hematology and Oncology, Department of Medicine, Medical University of South Carolina, Charleston, South Carolina, USA
[1]Current address: Division of Hematologic Malignancy and Cellular Therapy/Adult BMT, Department of Medicine, Duke University Medical Center, North Carolina, USA
[2]Corresponding author: e-mail address: kangy@musc.edu

Contents

Abstract

Thioredoxin (Trx) is an inflammation-inducible small oxidoreductase protein ubiquitously expressed in all organisms. Trx acts both intracellularly and extracellularly and is involved in a wide range of physiological cellular responses. Inside the cell, Trx alleviates oxidative stress by scavenging reactive oxygen species (ROS), regulates a variety of redox-sensitive signaling pathways as well as ROS-independent genes, and exerts cytoprotective effects. Outside the cell, Trx acts as growth factors or cytokines and promotes cell growth and many other cellular responses. Trx is also implicated in tumorigenesis. Trx is a proto-oncogene and is overexpressed in many cancers and correlates with poor prognosis. Trx stimulates cancer cell survival, promotes tumor angiogenesis, and inhibits both spontaneous apoptosis and drug-induced apoptosis. Inhibitors

Advances in Cancer Research, Volume 122
ISSN 0065-230X
http://dx.doi.org/10.1016/B978-0-12-420117-0.00007-4

targeting Trx pathway provide a promising therapeutic strategy for cancer prevention and intervention. More recently, data from our laboratory demonstrate an important role of Trx in expanding long-term repopulating hematopoietic stem cells. In this chapter, we first provide an overview of Trx including its isoforms, compartmentation, and functions. We then discuss the roles of Trx in hematologic malignancies. Finally, we summarize the most recent findings from our lab on the use of Trx to enhance hematopoietic reconstitution following hematopoietic stem cell transplantation.

1. OVERVIEW OF THIOREDOXIN

Various chemical, physical, or biological stimuli generate reactive oxygen species (ROS) and cause oxidative stress in cells. Cells contain several antioxidant systems and respond against those oxidative stresses to maintain cellular redox homeostasis. Stable redox homeostasis is critical for many physiobiological processes like cell viability, activation, and proliferation (Handy & Loscalzo, 2012; Trachootham, Lu, Ogasawara, Nilsa, & Huang, 2008; Wang et al., 2013). The thioredoxin (Trx) system is one of the major disulfide reductase antioxidant systems important in maintaining the redox environment in the cell (Collet & Messens, 2010). Trx system consists of nicotinamide adenine dinucleotide phosphate (NADPH), Trx, and selenoprotein thioredoxin reductase (TrxR). Trx has a molecular weight of 12 kDa and is a ubiquitous enzyme that catalyzes thiol–disulfide exchange reactions via two Cys residues in the conserved active-site sequence (-Cys32-Gly-Pro-Cys35-). In this Trx system, Trx receives electrons from NADPH and transfers them to the active site of Trx. The active-site disulfide in the oxidized Trx is reduced to a dithiol by NADPH and TrxR (Fig. 7.1). Activated (i.e., reduced) Trx uses these electrons to reduce its target proteins at disulfides (Fig. 7.1; Collet & Messens, 2010). Therefore, the primary function of Trx is considered to be an oxidoreductase to maintain redox homeostasis and to protect proteins from oxidative damage or inactivation (Collet & Messens, 2010).

1.1. Historic overview

In 1964, Laurent et al. first identified Trx from extracts of *Escherichia coli* B as a hydrogen donor from NADPH to ribonucleotide reductase (RNR), an essential enzyme for DNA synthesis (Holmgren, 1985; Laurent, Moore, & Reichard, 1964). In 1968, *E. coli* Trx was sequenced, revealing the highly conserved prototypical dithiol Cys-Gly-Pro-Cys active-site motif (Holmgren, 1968). The folding structure of Trx, defined as a distinct structural motif consisting of a four-stranded β-sheet and three flanking α-helices,

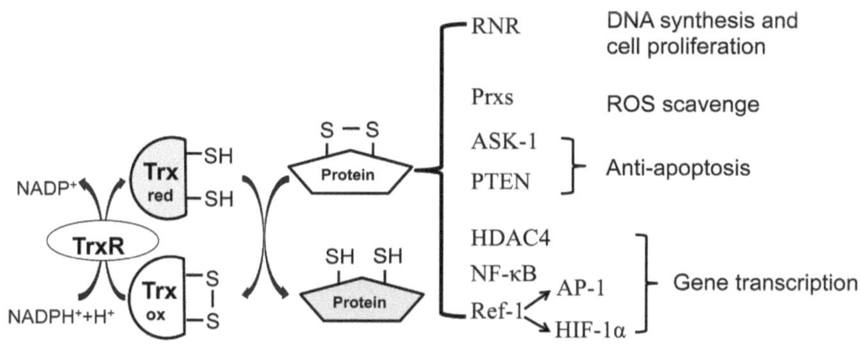

Figure 7.1 Schematic representation of Trx system. Disulfide form of Trx1 (-S-S-, oxidized) is reduced to dithiol form (-SH_2, reduced) by Trx reductase (TrxR) and NADPH. Reduced Trx catalyzes the reduction of disulfides (-S-S-) within multiple oxidized cellular proteins. Proteins that are known to be directly reduced by Trx include RNR (Holmgren, 1985), Prxs (Rhee, Kang, Chang, Jeong, & Kim, 2001), ASK1 (Nadeau, Charette, Toledano, & Landry, 2007), PTEN (Lee et al., 2002), HDAC4 (Ago et al., 2008), NF-κB (Matthews, Wakasugi, Virelizier, Yodoi, & Hay, 1992), and Ref-1 (Silber et al., 2002). Transcription factors AP-1 and HIF-1α are indirectly activated by Trx through intermediate Ref-1.

was first described in 1975 (Holmgren, Soderberg, Eklund, & Branden, 1975). In 1974, Yodoi et al. purified an interleukin-2 (IL-2) receptor-inducing factor from human T-lymphotropic virus type I (HTLV-I)-infected leukemic T-cell line (ATL-2) and designated this product as adult T-cell leukemia-derived factor (ADF; Yodoi, Takatsuki, & Masuda, 1974), which was later proved to be identical to human thiol-related oxidoreductase Trx (Tagaya et al., 1989; Teshigawara et al., 1985; Wakasugi et al., 1990). Human Trx was also cloned independently by Wakasugi et al. in 1990 as an autocrine IL-1 like growth factor produced by Epstein–Barr virus-transformed cells (Wakasugi et al., 1990).

1.2. Members in Trx system

There are two main isoforms of Trx in mammalian cells: Trx1 and Trx2. Trx1 is the prototypical member of this family and is mainly localized in the cytosol but can be translocated into the nucleus upon stress conditions or secreted out of the cell under certain circumstances (Fig. 7.2). Trx2 is the mitochondrial-restricted isoform of Trx. In addition to the Cys-Gly-Pro-Cys active site, Trx2 has an extra N-terminal mitochondrial translocation signal peptide that has been implicated to play an important role in mitochondria-mediated apoptosis (Conrad et al., 2004; Tanaka et al., 2002). Besides the two conserved cysteine residues in the active site, human

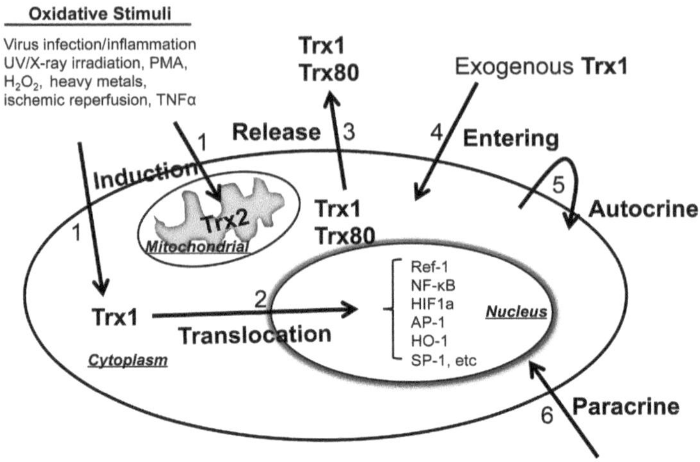

Figure 7.2 Induction, compartmentation, and secretion of Trx in mammalian cells. Three Trx isoforms are present in the mammalian cells: mitochondrial form Trx2, cytosolic form Trx1, and Trx80. Trx1 and Trx2 are induced by a variety of oxidative stimuli (1). These stimuli also induce the translocation of Trx1 from the cytosol to the nucleus (2) where Trx1 regulates multiple transcription factors. Trx1 and Trx80 can be released from cell to extracellular environment (3). Exogenous Trx1 can enter into the cells (4). Secreted Trx1 can act as cytokine or growth factor in an autocrine (5) or paracrine fashion (6).

Trx1 has three additional cysteine residues, Cys62, Cys69 and Cys73, which are absent in Trx2. These three cysteines are involved in protein transnitrosylation and denitrosylation modulation (Benhar, Forrester, Hess, & Stamler, 2008; Mitchell, Morton, Fernhoff, & Marletta, 2007). Cys69 S-nitrosylation is involved in redox regulation and antiapoptotic functions in the endothelial cells (Haendeler et al., 2002). Trx2 is also distinct from Trx1 by the fact that Trx2 participates in ROS detoxification through mitochondrial-specific peroxiredoxins and is reduced by its own reductase, TrxR2, thereby constituting a functionally separate Trx system restricted to the mitochondria (Zhang, Go, & Jones, 2007). Another member of Trx, named spermatocyte/spermatid-specific thioredoxin-3 (SPTRX-3), is found predominately in the Golgi apparatus in sperms (Jimenez et al., 2004; Miranda-Vizuete et al., 2004). The function of SPTRX-3 is related to the posttranslational modification of proteins required for germ cell-specific functions, such as acrosomal biogenesis. Posttranslational modification of Trx also results in a 10-kDa C-terminal truncated form of Trx (called Trx80), which contains the first 80–84 N-terminal amino acids of the initial protein (Rosen et al., 1995). Trx80 is a potent cytokine for monocytes and promotes a Th1 response through IL-12 production (Pekkari et al., 2001). Trx80 cannot be reduced by TrxR due to the fact that Trx80 lacks the C-terminal portion of Trx, which is crucial for its interaction with Trx reductase (Pekkari & Holmgren, 2004).

TrxRs, including cytosolic-localized TrxR1 and mitochondrial-located TrxR2, are important members of the Trx antioxidant system. TrxRs contain a conserved -Cys-Val-Asn-Val-Gly-Cys- redox catalytic site that is involved in catalyzing the NADPH-dependent reduction of the Trx and many other proteins (Lee et al., 1999; Mustacich & Powis, 2000; Nakamura, 2004). More recently, thioredoxin glutathione reductase (TGR) is grouped in the Trx system as an extended variant of TrxR (Sun, Kirnarsky, Sherman, & Gladyshev, 2001). As its name suggests, TGR reduces glutathione disulfide in addition to Trx (Sun et al., 2001).

1.3. Animal models

Various animal models (summarized in Table 7.1) have been generated to study the biological significance of the Trx system *in vivo*. The deletion of *TRX1* gene is lethal *in utero*, indicating its essential role in embryogenesis (Matsui et al., 1996). Knockout mice lacking Trx2 (Nonn et al., 2003), TrxR1 (Conrad et al., 2004), or TrxR2 (Jakupoglu et al., 2005) also died at early embryogenesis stage, suggesting the non-overlapping and indispensable role of Trx system in early embryo development. Transgenic mice overexpressing Trx have been generated by several groups but yield different results. For instance, Trx transgenic mice overexpressing human Trx driven by β-actin promoter (Tg(act-TRX1)) showed an extended life span in comparison to control wild-type mice under a variety of oxidative stress-associated disorders (Mitsui et al., 2002; Nakamura, Mitsui, & Yodoi, 2002; Shioji et al., 2002). On the other hand, Ikeno et al. generated Trx1 transgenic mice (Trx1 Tg) using a fragment of the human *TRX1* gene but found that at normal conditions, Trx1 Tg mice did not show much beneficial effects on life span, even among aged mice (Lisa et al., 2012). Widder and colleagues generated Trx2 transgenic mice (Tg hTrx2 mice) and showed that those mice had hypertension and vascular dysfunction in response to angiotensin II (Widder et al., 2009). Endothelial cell-specific Trx2 overexpressing animals showed beneficial effects of Trx2 on the prevention of atherosclerosis (Zhang, Luo, et al., 2007). Go et al. tagged the human *TRX1* with a nuclear localization signal (NLS) and generated an

Table 7.1 Trx family members and genetically engineered animal models

Trx antioxidant system		**Localization**	**Phenotypes of genetically engineered animal models**
Trx1	Full length Trx1	Cytosol, nucleus, extracellular	• Knockout mice: embryonic lethal (Matsui et al., 1996); • Tg (act-TRX1) mice: better survival under stress (Mitsui et al., 2002; Perez et al., 2011; Shioji et al., 2002; Zhang, Luo, et al., 2007); • Trx1 Tg: no survival benefits (Lisa et al., 2012); • NLS-hTrx1Tg: worse disease presentation (Go, Kang, Roede, Orr, & Jones, 2011)
	Trx80	Plasma membrane (Sahaf et al., 1997), extracellular	Knockout or Tg mice not reported
Trx2		Mitochondrial	• Knockout mice: embryonic lethal (Nonn, Williams, Erickson, & Powis, 2003); • Tg hTrx2 mice: vascular dysfunction and hypertension (Zhang, Luo, et al., 2007); • Endothelial cell-specific Trx2 Tg mice: protected against oxidant induced cell damage and atherosclerosis (He et al., 2008; Zhang, Luo, et al., 2007)
Trx3 (Sptrx3 or p32TrxL)		Golgi apparatus of Sperm	Knockout or Tg mice not reported
TrxR1		Cytosol	Knockout mice: embryonic lethal (Conrad et al., 2004)
TrxR2		Mitochondrial	Knockout mice: embryonic lethal (Jakupoglu et al., 2005)
TGR		Mainly male germ cells	Knockout or Tg mice not reported

hTrx1-overexpressing animal model (NLShTrx1Tg) specifically in the nucleus (Go et al., 2011). They demonstrated that NLS-hTrx1Tg mice exhibited an increased severity of disease manifestation likely due to the augmentation of redox-sensitive transcription factor activation upon influenza H1N1 infection. These results were somewhat contradictory to the observations seen with Trx1 (cytosolic) or Trx2 transgenic mice, both of which showed protection against various stresses (Nakamura, Tamura, Watanabe, Iwasaki, & Yodoi, 2002; Zhang, Luo, et al., 2007).

1.4. Induction, translocation, and secretion of Trx1

Trx1 is known as a stress-inducible protein and has been considered as a marker for oxidative stress. Trx1 can be induced by a broad range of stresses and pathophysiological stimuli including hypoxia, lipopolysaccharide, O_2, hydrogen peroxide, phorbol ester, viral infection, X-ray radiation, and UV irradiation (reviewed in Hirota, Nakamura, Masutani, & Yodoi, 2002; Powis & Montfort, 2001; Fig. 7.2). In the erythroleukemic cell line K562, Trx1 can be induced by hemin mainly through the binding of nuclear factor erythroid 2-related factor (Nrf2) to the antioxidant-responsive element (ARE) present in the Trx promoter (Kim et al., 2001). Trx1 level is upregulated by other inducers like DJ-1 (also called Parkinson protein 7 or PARK7) or sulforaphane through the same Nrf2–ARE mechanism (Im, Lee, Woo, Junn, & Mouradian, 2012; Tanito et al., 2005). Both Trx1 and Trx80 are induced by IL-1β and tumor necrosis factor-alpha (TNF-α) or by hydrogen peroxide (H_2O_2) in synoviocytes isolated from rheumatoid arthritis (RA) patients (Lemarechal et al., 2007). Most of these aforementioned stimuli also have been reported to induce Trx1 translocation from the cytoplasm (at normal physiological conditions) into the nucleus to facilitate DNA binding of transcription factors like NF-κB and glucocorticoid receptor (GR) and to potentiate signaling in immune cells (reviewed in Powis & Montfort, 2001; Fig. 7.2). The mechanism of Trx1 nuclear translocation is still not well defined, but may be related to the interaction of Trx1 with nuclear import sequence-containing proteins.

Although lacking a typical secretory signal sequence (Barash, Wang, & Shi, 2002), Trx1 is released from various types of mammalian cells including fibroblasts, airway epithelial cells, activated B and T cells, chronic lymphocytic leukemia cell lines, hepatocytes, monocytes, and many other cell types (Angelini et al., 2002; Ericson, Horling, Wendel-Hansen, Holmgren, & Rosen, 1992; Kondo et al., 2004; Rosen et al., 1995; Rubartelli, Bajetto, Allavena, Wollman, & Sitia, 1992; Rubartelli & Sitia, 1991; Sahaf & Rosen, 2000). Those observations suggest that Trx1 has an active secretion mechanism. A recent study demonstrated that human regulatory T cells (Tregs) express and secrete higher levels of Trx1 than conventional CD4$^+$ T cells (Mougiakakos, Johansson, Jitschin, Bottcher, & Kiessling, 2011). In addition, T-cell receptor (TCR)/CD28 stimulation led to a stronger Trx1 upregulation in Tregs compared to that in CD4$^+$ T cells (Mougiakakos et al., 2011).

The exact mechanisms for Trx1 secretion are currently unknown. Rubartelli et al. showed that blocking transport through the exocytotic pathway by either brefeldin A or dinitrophenol does not inhibit the secretion of Trx1 (Rubartelli et al., 1992). The data suggest that Trx1 is not secreted through the conventional endoplasmic reticulum/Golgi route (Nickel, 2003), but shares several features with the alternative, leaderless secretion pathway described for IL-1β (Rubartelli & Sitia, 1991). Considering the facts that the leaderless secretion is usually quite slow and inefficient and that the local concentration of Trx1 is likely low, it has been suggested that the secreted Trx1 largely acts in an autocrine or paracrine fashion and may not exert biological effects far from the site of production. Indeed, it has been shown that the released Trx1 can act as a paracrine or autocrine growth factor or as a chemoattractant for various tumor cells (Becker, Gromer, Schirmer, & Muller, 2000).

Trx1 also serves as the reservoir or precursor for Trx80. Trx80 is generated by the cleavage of full-length Trx1. It was found that both Trx1 and Trx80 are released from the synoviocytes of RA patients upon IL-1β and/or TNF-α stimulation (Lemarechal et al., 2007). Alpha-secretase was recently identified as the enzyme responsible for cleaving Trx1 into Trx80 (Gil-Bea et al., 2012). Trx80 is highly expressed on the cell surface of the monocyte/macrophage cell lines THP-1 and U937 (Sahaf et al., 1997) and is secreted in multiple cell culture systems (Pekkari et al., 2001; Powis & Montfort, 2001).

Elevated Trx1 secretion has been reported to be closely related to conditions associated with oxidative stress and inflammatory reactions and can be used as a marker for inflammatory responses. For example, the prognosis is much poorer in human immunodeficiency virus (HIV)-infected patients with higher plasma levels of Trx1 compared to those with normal levels of Trx1 (Nakamura et al., 1996). In hepatitis C virus infection, serum levels of Trx1 predict the efficiency of interferon therapy (Sumida et al., 2000); and patients with RA showed increased Trx1 in the serum/plasma samples (Maurice et al., 1997; Yoshida et al., 1999). In addition, increased serum/plasma Trx1 level was related to many cardiovascular diseases like coronary artery diseases (Miwa et al., 2003), acute myocardial infarction (Soejima et al., 2003), and myocardial damage prior to percutaneous coronary intervention (Shim et al., 2012).

1.5. Functions of Trx

Trx has a broad range of biological functions including antioxidative, anti-apoptotic, anti-inflammatory, and mitogenic activities. Trx also controls the expression or activity of other proteins through regulating transcription factor activities. Additionally, Trx can act as a growth factor, cytokine, or chemokine once secreted from host cells or administered exogenously to the biological system.

1.5.1 Functions of Trx inside the cell

Trx serves as a dithiol hydrogen donor for many target proteins. Trx was originally reported as an electron donor for RNR, which is important for DNA synthesis and cell proliferation. Trx is known for its function as a key protein in multiple tissues in defense against oxidative stress. Trx maintains a reduced environment inside the cells through thiol–disulfide exchange reactions and protects cells and tissues from oxidative stress (Koharyova & Kolarova, 2008; Nishinaka, Masutani, Nakamura, & Yodoi, 2001; Powis & Montfort, 2001). The key functions of Trx are summarized in the succeeding text.

1.5.1.1 Anti-inflammation

It appears to be a natural physiological response that Trx1 is induced upon oxidative and inflammatory stimuli, the main function of which is then to defend against those stresses and protect against a wide range of inflammatory disorders. Trx1 is upregulated in many inflammation-related diseases (for review, please see Matsuo & Yodoi, 2013). Trx1 has been demonstrated to have beneficial, protective effects on ischemic reperfusion injury animal models. Intravenous administration of Trx significantly suppressed lipopolysaccharide (LPS)-induced neutrophil extravasation in the air pouch model (Nakamura et al., 2001); Trx transgenic mice show a decreased ischemic neuronal injury in focal ischemic brain damage model (Takagi et al., 1999). Additionally, Trx transgenic mice were protected from a wide variety of inflammatory disorders (Nakamura, Hoshino, Okuyama, Matsuo, & Yodoi, 2009), which contributes to the longer life span seen in these mice (Mitsui et al., 2002; Perez et al., 2011). Similarly, intravenous administration of Trx protects the animal from brain damage following transient focal cerebral ischemia (Hattori et al., 2004).

One well-established pathway by which Trx1 attenuates the inflammatory response is through the regulation of redox-sensitive transcription factors such as NF-κB and Nrf2 (Billiet et al., 2005). It was found that Trx1 downregulated NF-κB activation by suppressing both p50 and p56 expressions on one hand and inducing I-κBα synthesis on the other hand (Billiet et al., 2005). Furthermore, Hadri et al. recently found that Trx1 is able to shift the balance between M1 and M2 macrophages and promote the differentiation of macrophages into anti-inflammatory phenotype (El Hadri et al., 2012). Moreover, recombinant human Trx1 suppresses neutrophil p38 mitogen-activated protein kinase (Nakamura et al., 2001), reduces leukocyte recruitment and activation (Inomata et al., 2008), and inhibits neutrophil cell adhesion on endothelial cells by removing L-selectin from the neutrophil surface (Nakamura et al., 2001). Additionally, circulating plasma Trx1 inhibits neutrophil extravasation into the inflammatory sites and suppresses the expression and release of proinflammatory macrophage migration inhibitory factor (MIF; Tamaki et al., 2006). All these mechanisms contribute to the anti-inflammatory activity seen with Trx. Recombinant Trx1, therefore, may provide a novel therapeutic approach for the treatment of acute inflammatory disorders.

It is noteworthy that unlike Trx1, Trx80 shows proinflammatory effects. Trx80 promotes the differentiation of LPS-challenged macrophages into inflammatory M1 macrophages. ApoE2-Ki mice receiving daily LPS and Trx80 developed severe atherosclerotic lesions that were enriched with macrophages expressing predominantly M1 over M2 markers (Mahmood et al., 2013).

1.5.1.2 Antiapoptosis

The reduced form of Trx binds to a variety of cellular proteins (Fig. 7.1) including apoptosis signal-regulating kinase 1 (ASK1; Nadeau et al., 2007). ASK1 is one of several mitogen-activated protein kinase kinase kinases (MAP3Ks) that are activated in response to proinflammatory stimuli, ROS, and other cellular stresses. ASK1 then activates the c-Jun N-terminal kinase (JNK) and p38 MAPK pathway and is required for TNF-α-induced apoptosis (Imoto et al., 2006; Matsukawa, Matsuzawa, Takeda, & Ichijo, 2004). This ASK1–JNK/p38 pathway has been shown to induce apoptosis mainly through mitochondrion-dependent caspase-9 and caspase-3 activation (Hatai et al., 2000). Reduced Trx binds to the N-terminal noncatalytic region of ASK1 through Cys32 and Cys35, suppresses ASK1–JNK/p38 signaling, and thus inhibits apoptosis in human embryonic kidney (HEK) cells (Saitoh et al., 1998). A later study by Liu et al. showed that the overexpression of wild-type Trx in endothelial cells induced ASK1 ubiquitination and degradation. It was further demonstrated that both Cys32 and Cys35 are required for ASK1 binding and degradation. Cys32S and Cys35S double mutant Trx cannot bind to ASK1 and therefore is unable to induce ASK1 ubiquitination and degradation (Liu & Min, 2002). Consistent with the antiapoptotic function of Trx, a recent study (Huh et al., 2013) found that in primary rat islet β-cells, Trx1 protein expression was downregulated when the cells underwent apoptosis induced by the immunosuppressive reagent mycophenolic acid. Furthermore, the knockdown of Trx1 increased cell death, whereas the overexpression of Trx1 blocked ROS generation, decreased the activations of JNK and caspase-3, and increased cell survival. Mitochondrial Trx2 also binds to ASK1 but at a site different from Trx1. Trx1 and Trx2 bind to Cys250 and Cys30 in the N-terminal domain of ASK1, respectively. Interestingly, the mutation of ASK1 at Cys250 enhanced ASK1-induced JNK activation and apoptosis, whereas the mutation of ASK1 at Cys30 increased ASK1-induced apoptosis without effects on JNK activation (Zhang et al., 2004). These data suggest that cytosolic Trx1 and mitochondrial Trx2 regulate ASK1 activity and apoptosis differently but cooperate with each other to achieve synergistic effects.

The antiapoptotic effect of Trx1 also involves the interaction with the tumor suppressor gene phosphatase and tensin homologue deleted on chromosome ten (PTEN; Lee et al., 2002; Meuillet, Mahadevan, Berggren, Coon, & Powis, 2004). PTEN is a major tumor suppressor of human cancer that prevents the activation of the survival signaling kinase Akt. Trx1 binds to PTEN through a disulfide bond between the active-site Cys32 of Trx1 and Cys212 of the C2 domain of PTEN. The binding of Trx1 to PTEN blocks PTEN activity and subsequently activates Akt pathway, leading to increased cell survival (Meuillet et al., 2004).

1.5.1.3 Transcription factor regulation

Trx exerts its antioxidant effect through its ability to reduce Trx peroxidases (i.e., peroxiredoxins) that scavenge H_2O_2 (Rhee et al., 2001) and to regulate the activities of various enzymes including those that function to counteract oxidative stress (Gromer, Urig, & Becker, 2004). In addition, Trx1 regulates several redox-sensitive transcription factors by modulating their binding to DNA, thus affecting various aspects of cellular responses including redox homeostasis, cell growth, and survival. Trx has been shown to be able to either directly or indirectly modulate the DNA binding ability of redox protein redox factor 1 (Ref-1), activator protein-1 (AP-1), and NF-κB.

Trx is the hydrogen donor for Ref-1, which possesses a reducing activity and a purine/pyrimidine endonuclease repair activity (Silber et al., 2002). Reduced Ref-1 then proceeds to reduce transcription factor AP-1 to enhance its DNA binding (Hirota et al., 1997). Trx1 is therefore involved in many cellular functions like cell differentiation, proliferation, and apoptosis because AP-1 is responsible for many of those processes (Ameyar, Wisniewska, & Weitzman, 2003; Shaulian & Karin, 2001). Under stimulation by inflammatory mediators, Trx upregulates the expression of heme oxygenase-1 (HO-1; Wiesel et al., 2000). The upregulation of HO-1 protects islet cells from apoptosis and improves their function after transplantation (Pileggi et al., 2001), indicating that Trx and HO-1 may coordinate to protect cells from inflammatory stress.

Hypoxia-inducible factor (HIF-1α) is another transcription factor that is regulated by Trx1 through Ref-1. HIFs are primary transcriptional factors that control the expression of hypoxic stress-responsive genes under hypoxic stress in mammalian system (Chen, Nelin, Locy, Jin, & Tipple, 2013; Majmundar, Wong, & Simon, 2010). Trx1 promotes HIF signaling to reduce hypoxia-related cell stresses (Arner & Holmgren, 2006; Welsh et al., 2003).

Trx also directly activates several transcription factors independent of Ref-1. For example, Trx1 directly binds to NF-κB and reduces Cys62 in the p50 subunit of NF-κB, thus activating NF-κB-dependent gene expression (Hirota et al., 1999; Matthews et al., 1992). Activated NF-κB is responsible for the expression of many genes that protect cells from apoptotic cell death (Stehlik et al., 1998; Wang, Mayo, Korneluk, Goeddel, & Baldwin, 1998). Additionally, Trx1 enhances the DNA binding ability of the transcription factor specificity protein-1 (Sp-1), which in turn can promote Trx gene expression by activating Trx gene promoter (Bloomfield, Osborne, Kennedy, Clarke, & Tonissen, 2003). The regulation of Trx on Sp-1 and HIF-1α plays an important role in tumor angiogenesis because vascular endothelial growth factor (VEGF) is induced by Sp-1 under normoxic condition and by HIF-1α under hypoxic condition (Welsh, Bellamy, Briehl, & Powis, 2002). Furthermore, Trx regulates p53 signaling by both directly acting on p53 and indirectly through Ref-1-mediated pathway (Ueno et al., 1999), thereby controlling the multitude of downstream events mediated by p53.

1.5.2 *Functions of Trx outside the cell (cytokine or chemokine-like effects)*

Trx1 was first found to be secreted from HTLV-I-infected leukemic T-cell line (Yodoi et al., 1974). Subsequent studies showed that Trx1 can also be released from chronic lymphocytic leukemia (B-CLL) cell line or human $CD4^+$ T-cell hybridoma MP6 cells and can significantly stimulate B-cell growth (Ericson et al., 1992; Rosen et al., 1995). These autocrine effects are considered cytokine or chemokine-like effects of Trx1. It remains to be determined how extracellular Trx1 enters the cell or whether Trx1 binds to specific cell surface receptors. Several investigations have indicated that Trx1 can easily cross cell membranes and effectively enter living cells within 24 h (Kondo et al., 2004; Spector et al., 1988). Studies also showed that Trx1 strongly enhances the expression of various other cytokines including IL-4, interferon-γ, and TNF-α (Schenk, Vogt, Droge, & Schulze-Osthoff, 1996) and those cytokines are known to be able to promote B-CLL cell survival (Tangye & Raison, 1997). Therefore, the cytokine-like effects of Trx1 are a combination of direct activity of Trx1 and the effects of other cytokines that are induced by Trx1.

A truncated form of Trx (i.e., Trx80) is produced by monocytes and $CD4^+$ T cells and also has mitogenic, cytokine-like effects on resting human peripheral blood mononuclear cells (Pekkari, Gurunath, Arner, & Holmgren, 2000). Trx80 has been shown to effectively stimulate the activation, proliferation, differentiation, and cytokine production of monocytes (Pekkari et al., 2000, 2003).

2. THIOREDOXIN IN HEMATOLOGIC MALIGNANCIES AND OTHER CANCERS

Trx functions as an important antioxidant, scavenging ROS and defending the organisms against oxidative stress-induced inflammation and cell death. On the other hand, elevated Trx levels inside the cell contribute to many of the hallmarks of cancer phenotypes such as increased proliferation, resistance to cell death, increased angiogenesis, and chemotherapeutic drug resistance.

2.1. Trx expression is upregulated in solid tumors and hematologic malignancies

Trx levels are elevated in a variety of benign diseases including asthma, hepatitis, HIV infection, RA, and bronchial metaplasia. Trx levels are also increased in many malignant diseases like pancreatic and liver cancers (Miyazaki et al., 1998; Nakamura et al., 2000) and other solid cancers (for review, please see Powis & Montfort, 2001).

Increased tumor Trx1 levels have been linked to aggressive tumor growth, poorer prognosis, and decreased patient survival in various cancer types. Raffel et al. examined Trx1 expression in paraffin-embedded human normal colonic mucosa, adenomatous polyps, and primary and metastatic colorectal cancer. They found that Trx1 expression was significantly increased in primary colorectal cancer compared with normal colonic mucosa. Additionally, elevated Trx1 expression was associated with poorer patient survival in colorectal cancer (Raffel et al., 2003). In oligodendroglial brain tumors, increased plasma Trx1 level correlated closely with poorer outcome (Jarvela et al., 2006). In hepatocellular carcinoma, Trx1 levels were higher in patients with more advanced stage, and serum Trx1 levels decreased after surgical removal of the tumor (Miyazaki et al., 1998). Increased Trx1 gene expression is associated with increases in both HIF-1α level and HIF-1α transactivation in cancer cells (Baker et al., 2008), which could lead to increased VEGF production and enhanced tumor angiogenesis (Welsh et al., 2002). Moreover, HIF-1α upregulation further activates the gene expression of COX-2, which confers a poor prognosis in lung cancer (Csiki et al., 2006).

In hematologic malignancies, Trx1 level is also elevated. Shao et al. analyzed Trx1 level in the mononucleated cell fraction of leukemic samples from 28 children with newly diagnosed T-cell acute lymphoblastic leukemia (T-ALL; Shao et al., 2001). They found that Trx1 expression was highly variable in these patient samples as well as among normal healthy donors. Interestingly, the leukemia clonogenic cells showed different sensitivity toward 1-methylpropyl 2-imidazolyl disulfide (PX-12), an inhibitor of Trx1, as compared with normal hematopoietic progenitors. It was suggested that the inhibition of Trx1 expression may sensitize leukemia cells to chemotherapeutic agents and overcome drug resistance of leukemia cells. In acute myeloid leukemia (AML) patients, increased Trx1 level strongly correlated with indoleamine 2,3-dioxygenase (IDO) level. Because IDO is an enzyme from tumors not being affected by the host immune response, the positive correlation of Trx1 with IDO again suggests that increased Trx1 expression is associated with poor prognosis in AML patients. Therefore, Trx1 may provide a useful biomarker to predict relapse in acute leukemia patients (Zhou et al., 2010).

In diffuse large B-cell lymphoma (DLBCL) patients, Tome et al. reported that decreased expression of Trx inhibitory protein, that is, vitamin D3-upregulated protein (VDUP1, also called Trx-interacting protein), is associated with a worse outcome (Tome et al., 2005). A more recent study by Li et al. showed that both DLBCL cell lines and primary DLBCL cells had significantly elevated Trx1 expression (Li et al., 2012). Increased Trx1 level closely correlated to advanced DLBCL disease stage and inferior patient survival (Li et al., 2012). Another independent study from Evens et al. showed that Trx1 and TrxR were elevated in lymphoma cell lines (Evens et al., 2008). Those observations suggested an important role of Trx1 in the pathophysiology of hematologic malignancies. Of note, continuous exposure of cancer cells to increasing doses of doxorubicin results in increased Trx1 expression and doxorubicin resistance, suggesting that Trx1 upregulation in cancer cells may reflect the adaptation of cancer cells to better tolerate oxidative stress.

2.2. Trx stimulates cancer cell growth and protects cancer cells from apoptosis

Trx1 is released from cells and functions as a growth factor in paracrine or autocrine fashion for normal and tumor cells (Oblong, Berggren, Gasdaska, & Powis, 1994; Powis et al., 1994; Wakasugi et al., 1990; Yodoi & Tursz, 1991). Elevated Trx promotes cancer cell growth through the regulation of DNA synthesis and transcription factor activity.

Early studies using human breast cancer cell line MCF-7 showed that Trx1-transfected cells had significantly increased ability to form colonies compared to control cells. This effect was completely blocked by the expression of a dominant negative, redox-inactive Trx1 (Gallegos et al., 1996). When Trx1-transfected MCF-7 cells were inoculated into immunodeficient mice, tumor formed. In contrast, tumor formation by the dominant negative, Cys32S/Cys35S Trx1-transfected MCF-7 cells was almost completely inhibited (Gallegos et al., 1996). Another study also showed that mutant redox-inactive form of Trx1 lacking the active-site Cys32 and Cys35 residues was devoid of growth-stimulating activity to murine fibroblasts (Oblong et al., 1994). Furthermore, neither dithiothreitol nor glutathione could duplicate the ability of Trx to stimulate cell proliferation, suggesting that redox active form of Trx1 is essential for its growth-stimulating property.

Bazzichi et al. investigated the effects of *E. coli* recombinant Trx1 (r-Trx1) on the proliferation of various human lymphoid cell lines (Bazzichi, Incaprera, & Garzelli, 1994). It was found that under serum-free culture conditions, r-Trx1 promoted DNA synthesis and cell growth, and this effect was more dramatic when cells were treated with r-Trx1 together with growth factors like IL-1 or IL-6. Exogenously added Trx1 also stimulated proliferation of T-ALL clonogenic cells expressing relatively lower levels of intracellular Trx1 (Shao et al., 2001).

Exogenous Trx1 or overexpressed Trx1 protects cancer cells from apoptosis. Baker et al. investigated the effects of Trx on spontaneous apoptosis and drug-induced apoptosis in mouse WEHI7.2 thymoma cells. WEHI7.2 cells transfected with human Trx1 were protected from apoptosis induced by dexamethasone, staurosporine, etoposide, and thapsigargin (Baker, Payne, Briehl, & Powis, 1997). Additionally, when the cells were inoculated into SCID mice, the Trx1-transfected cancer cells showed an increased tumor growth, decreased spontaneous apoptosis, and decreased sensitivity to apoptosis induced by dexamethasone (Baker et al., 1997). Trx1 protects cancer cells from apoptosis through redox-dependent and redox-independent mechanisms. Cancer cells have a higher metabolic rate and generate more ROS. With its antioxidant function, Trx1 helps cancer cell scavenge ROS, reduce oxidative stress, and survive from apoptosis (Valko, Rhodes, Moncol, Izakovic, & Mazur, 2006). Additionally, Trx1 modulates ASK1 and attenuates ASK1-mediated apoptotic signaling pathway. Those observations suggest that Trx1 represents a prosurvival gene required for cancer cell survival and may provide a potential target for the treatment of cancer.

It is noteworthy that Trx1 transgenic mice were found to prevent the development of benzene-induced, ROS-mediated hematopoietic diseases such as thymic lymphoma. Li et al. exposed C57Bl/6 mice and human Trx1 overexpressing transgenic mice to benzene for 26 weeks and monitored the lifetime incidence of hematopoietic diseases. Thirty percent of wild-type C57Bl/6 mice developed thymic lymphoma. In contrast, none of Trx1 transgenic mice developed thymic lymphoma (Li et al., 2006). It was postulated that the thymic lymphoma development was mediated by ROS production following benzene exposure and Trx1 attenuates the oxidative stress, thus preventing the development of hematopoietic disorders in this mouse model (Li et al., 2006).

2.3. Trx confers cancer cell drug resistance

Drug resistance remains a major challenge in cancer therapy. Trx1 plays an important role not only in maintaining the transformed phenotype of several human cancers but also in their resistance to chemotherapeutic drugs. *In vitro* and *in vivo* studies have demonstrated that the resistance to cancer drug treatment is closely correlated with Trx1 level. For instance, gastric cancer cell line St-4 and colon cancer cell line HT-29 that are resistant to cis-diamminedichloroplatinum (cDDP) showed a higher level of Trx1 expression (Yamada, Tomida, Yoshikawa, Taketani, & Tsuruo, 1996). The authors further examined Trx1 expression and drug resistance to cDDP in 11 human ovarian cell lines and found a positive correlation between Trx1 gene expression and cDDP resistance in these cell lines. Studies in doxorubicin-sensitive SKOV-3 and doxorubicin-resistant SKVLB human ovarian carcinoma cells also revealed that the development of doxorubicin resistance in ovarian carcinoma cells was associated with a dramatically increased expression of Trx1 and a moderate increase of Trx2 gene expression (Kalinina et al., 2007). Of note, transfection of HT-29 cells or human ovarian cancer A2780 with Trx did not confer the transfectants resistant to adriamycin, mitomycin C, or cDDP (Yamada, Tomida, Yoshikawa, Taketani, & Tsuruo, 1997), suggesting that drug resistance to chemotherapeutic agents in colon cancer and ovarian cancer is at least in part related to, but not completely dependent on, Trx1.

Kim et al. examined the correlation between Trx1 expression and resistance to docetaxel in 63 primary breast cancer patients. It was found that higher level of Trx1 expression in prechemotherapy breast cancer samples predicted drug resistance to docetaxel in these patients (Kim et al., 2005). Trx1 was also found to contribute to doxorubicin resistance to human bladder and prostatic cancer cells (Yokomizo et al., 1995) and adriamycin resistance to adult T-cell leukemia cells (Wang et al., 1997).

A gain of function in Trx1 locus has been observed in drug-resistant cells identified by comparative genomic hybridization (CGH) analysis (Efferth et al., 2002), suggesting that this locus is prone to genomic imbalances imposed by chemotherapy. Li et al. found that doxorubicin-resistant DLBCL cell line (McA-DR cells) expressed higher levels of Trx1 gene and that the downregulation of Trx1 in these cells reversed the chemoresistant phenotype (Li et al., 2012), suggesting that the higher expression of Trx1 in DLBCL may have been acquired during chemotherapy.

2.4. Trx constitutes an important component of tumor cell microenvironment

It is well recognized that the initiation, growth, invasion, and metastasis of tumor involve a complex interplay between the host environment and the cancer cell. Alterations in tissue microenvironment are clearly implicated in both the formation and the progression of human tumors (Balkwill, Charles, & Mantovani, 2005; Vakkila & Lotze, 2004). Trx1 is secreted from cancer cells and other cell types like B cells (Ericson et al., 1992), monocytes (Sahaf & Rosen, 2000), and regulatory T cells (Mougiakakos et al., 2011). Many of these cell types contribute to the formation of the microenvironment that surrounds the cancer cells. Extracellular Trx1 constitutes an important component in the cancer cell microenvironment and functions as cytokines and growth factors to promote the growth and survival of cancer cells. The role of Trx1 in regulating microenvironment can be demonstrated in the dendritic cell (DC)–T-cell interaction. DCs secrete Trx1 that contributes to the generation of a reducing microenvironment for T-cell proliferation (Angelini et al., 2002). When Trx1 neutralizing antibody was added to the culture, the reduced extracellular microenvironment was altered and T-cell proliferation and activation were inhibited. These data demonstrated that Trx1 contributes to the formation of microenvironment that is important for T-cell activation.

Backman et al. also examined the role of Trx1 in cancer microenvironment. They found that the primary leukemia cells in B-cell CLL patients expressed minimal amounts of Trx1. In contrast, the accessory cells at the lymph nodes of these patients such as stromal cells including fibroblastic reticular cells and follicular dendritic cells expressed high level of Trx1. Moreover, the expression of Trx1 was mainly localized to the proliferating center of CLL lymph nodes and surrounded the Ki-67^{+} proliferating leukemic cells (Backman et al., 2007). The authors further showed that the stromal cells protected CLL cells from apoptosis, and this protective effect of stromal cells was mediated by Trx because Trx neutralizing antibody abrogated this effect (Backman et al., 2007). Similarly, Li et al. demonstrated that Trx was highly expressed not only in DLBCL cells but also in histiocytes with macrophage-like or fibroblast-/dendritic-like morphology residing in the surrounding tumor microenvironment (Li et al., 2012). Taken together, these data support the notion that Trx1 produced by cancer-surrounding cells is a key component of tumor microenvironment that contributes to cancer cell growth and survival.

2.5. Development of Trx1 inhibitors for the treatment of cancer

Given the facts that Trx1 expression is upregulated in many cancers and the overexpression of Trx1 has been correlated with aggressive tumor growth, resistance to chemotherapeutic drugs, poorer prognosis, and decreased survival in patients (Fig. 7.3), a large number of reagents that target the Trx

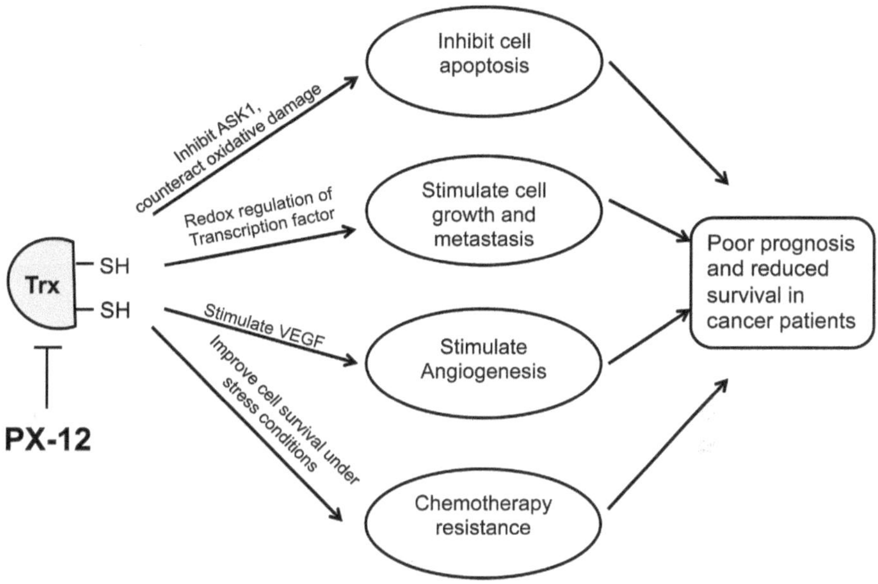

Figure 7.3 Trx in cancer. A wide variety of cancers have significantly increased levels of Trx. Increased Trx promotes cancer cell growth by multiple mechanisms: (1) Trx inhibits ASK1-induced apoptosis and counteracts oxidative damage, (2) Trx promotes cancer cell growth and activates directly or indirectly a number of transcription factors that are responsible for tumor growth and metastasis, (3) Trx stimulates VEGF and promotes angiogenesis, and (4) Trx improves cell survival under stress conditions and confers drug resistance to chemotherapy. These effects lead to poor prognosis and reduced survival of cancer patients.

system are being developed for the treatment of cancer either alone or in combination with other existing drugs.

To date, many different chemical inhibitors of Trx1 are being assessed for their antitumor activity. The most promising one is PX-12, which is currently under phase I/II clinical trials (phase II for pancreatic cancer (Ramanathan et al., 2007, 2011) and phase IB for advanced gastrointestinal cancers (Baker et al., 2013)).

PX-12 acts by binding to the Cys73 residue of Trx1, causing it to become irreversibly thioalkylated. Thioalkylated Trx1 is biologically inactive and is no longer able to act as a substrate for TrxR1 (Kirkpatrick et al., 1998). Furthermore, PX-12 decreases plasma VEGF levels, which may contribute to its antitumor activity. PX-12 inhibits the expression of VEGF both *in vitro* in cell culture and *in vivo* in human tumor xenograft models. The inhibition of VEGF by PX-12 was thought to be mediated by suppressing HIF-1α transcription factor (Welsh et al., 2003). PX-12 has also been shown to inhibit tubulin polymerization through cysteine oxidation (Huber et al., 2008). Tubulin polymerization is important for cancer cell migration, invasion, and metastasis.

Preclinical studies using HT-29 human colon cancer cell line have shown that PX-12 inhibited cancer cell growth (Welsh et al., 2003). *In vivo* model using HT-29 xenografts also demonstrated that PX-12 treatment caused a rapid (within 2 h postadministration) decrease in the average tumor microvascular permeability, and this effect lasted for at least 24 h (Jordan et al., 2005). PX-12 also inhibited the growth of A549 lung cancer cells and led to G2/M phase arrest and ROS-dependent apoptosis (You, Shin, & Park, 2014). Additional studies showed that PX-12 inhibited the growth of human MCF-7 breast cancer and HL-60 promyelocytic leukemia SCID murine xenografts in a dose-dependent manner (Kirkpatrick, Ehrmantraut, Stettner, Kunkel, & Powis, 1997). In clinical trials, PX-12 was found to decrease both Trx1 and VEGF levels in cancer patient plasmas (Baker et al., 2006).

Cancer cells and their normal counterparts exhibit differences in their antioxidant ability and their sensitivity to oxidative stress. These differences may offer a novel approach to selectively kill cancer cells without causing significant toxicity to normal cells (Irwin, Rivera-Del Valle, & Chandra, 2013; Schumacker, 2006). The increase of ROS in cancer cells plays an important role in the initiation and progression of cancer (Behrend, Henderson, & Zwacka, 2003; Wu, 2006). On the other hand, excessive levels of ROS can be toxic to the cells. Cancer cells with increased oxidative stress are likely to be more vulnerable to damage by additional ROS insults from exogenous agents (Pelicano, Carney, & Huang, 2004). Therefore, ROS-inducing agents could offer additional killing effects on cancer cells. Several ROS-inducing compounds like antimicrobial agents, pesticides, or natural products of plants have demonstrated anticancer activities in a number of cancer models including B-cell lymphoma (Armstrong, Hornung, Lecane, Jones, & Knox, 2001) and promyelocytic leukemia (Tada-Oikawa, Hiraku, Kawanishi, & Kawanishi, 2003). Many antitumor agents, such as vinblastine, cisplatin, mitomycin C, doxorubicin, camptothecin, inostamycin, and neocarzinostatin, induce cancer cell apoptosis in an ROS-dependent manner. These data suggest the potential use of ROS-inducing agents for the treatment of cancer. PX-12, by its anti-Trx activity, could be classified as an ROS-inducing agent.

2.6. Trx2 and cancer treatment

Relatively little is known about the function of Trx2 in cancer. One study showed that Trx2 overexpression inhibited TNF-α-induced apoptosis in cervical tumor cancer cell line HeLa cells (Hansen, Zhang, & Jones, 2006). The treatment of HeLa cells with cationic triphenylmethanes such as brilliant green downregulated Trx2 and induced cell death at nanomolar concentrations. Trx2-specific siRNA resulted in an increased sensitivity of HeLa cells to cationic triphenylmethanes (Zhang et al., 2011).

3. THIOREDOXIN IN HEMATOPOIESIS

Hematopoietic stem cells (HSCs) are a rare population of cells residing in the bone marrow (BM) and continuously replenish all mature blood cells throughout the life span. Alternations in the functions and regulations of HSCs result in a variety of hematologic disorders including bone marrow failure, anemia, myeloid leukemia, and myelodysplastic syndrome, among others. An accumulation of evidence demonstrates an important role of ROS in the regulation of HSC self-renewal and long-term repopulation (Owusu-Ansah & Banerjee, 2009).

ROS are unavoidable by-products of HSC metabolism. At physiological levels, ROS are important mediators for diverse biological functions. However, at above or below physiological levels, ROS are detrimental to cellular functions. Above physiological levels of ROS can lead to cell death and accelerate the aging process or activate redox-sensitive pathways. Below physiological levels of ROS may interrupt cell proliferation and lead to cell arrest. There is a fine-tuned balance between ROS production and antioxidant defense in HSCs to maintain certain physiological levels of ROS.

Studies have established a key role of ROS in HSCs' quiescence, self-renewal, and long-term repopulating capacity (Fig. 7.4). HSCs reside in specialized microenvironments, termed "niches" (Schofield, 1978), that provide cellular and molecular support for HSPCs to reside, proliferate, and differentiate (Li & Xie, 2005). Although still controversial (Eliasson & Jonsson, 2010), at least two HSC niches have been identified in the BM: osteoblastic niche and vascular niche. The osteoblastic niche is located next to the endosteal surface and the vascular niche is located in the medullary space next to sinusoid vessels. The roles and function of these two niches are not completely known. One of the theories considers the osteoblastic niche as the primary site for maintaining HSCs in a primitive,

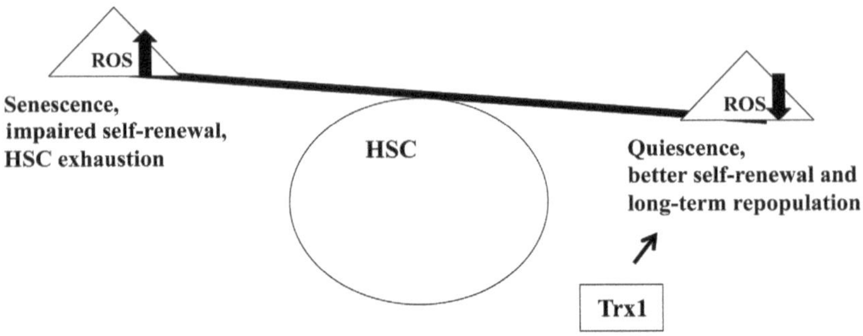

Figure 7.4 ROS regulates the functions of hematopoietic stem cells. Intracellular ROS level affects HSC function. Increased ROS result in HSC senescence, loss of self-renewal and long-term repopulation, and HSC exhaustion, whereas low ROS maintain HSC quiescence and the long-term repopulating capacity. Trx reduces ROS and enhances HSC long-term repopulation.

quiescent state and the vascular niche as the predominant site for HSCs that are activated for mobilization and differentiation. Consistent with the role of ROS in maintaining HSC quiescence and long-term repopulation, osteoblastic niche is at the lowest end of an oxygen gradient with the BM and protects HSCs from oxidative stress (Parmar, Mauch, Vergilio, Sackstein, & Down, 2007; Suda, Arai, & Hirao, 2005; Yahata et al., 2008).

Jang and Sharkis (2007) sorted murine BM HSCs based on their intracellular ROS level and then performed serial bone marrow transplantation with ROS^{low} population and ROS^{high} cell population. They found that murine ROS^{low} HSC population had a higher long-term repopulating ability than ROS^{high} HSC population. The ROS^{low} population demonstrated phenotypic characteristics of HSCs located within the osteoblastic niche. In contrast, the ROS^{high} HSC population expressed high levels of activated p38 MAPK and mammalian target of rapamycin (mTOR). The treatment of ROS^{high} HSC population with the antioxidant *N*-acetyl-L-cysteine (NAC), p38 inhibitor SB203580, or an mTOR inhibitor restored the functional activity of HSCs. Similar to murine HSCs, human HSCs can be separated into ROS^{low} and ROS^{high} population based on their intracellular ROS level, and ROS^{high} human HSCs exhausted the repopulating capacity by the tertiary transplant, while ROS^{low} HSCs retained their self-renewing and repopulation potential (Yahata et al., 2011).

It was further demonstrated that increasing intracellular ROS led to DNA oxidative damage, HSC senescence, and the loss of self-renewal capacity of HSCs (Jang & Sharkis, 2007; Wang et al., 2010; Yahata et al.,

2011). Yahata et al. cultured human HSCs with buthionine sulfoximine (BSO), an inhibitor of glutathione synthetase, to increase the intracellular ROS level and found that the repopulating capacity of BSO-treated HSCs was significantly diminished. Treatment with NAC restored the long-term repopulating capacity of BSO-treated HSCs (Yahata et al., 2011). These studies provide strong evidence for the roles of ROS in HSC's self-renewal and long-term repopulating potential.

Because of the effects of intracellular ROS on long-term repopulating HSCs, efforts have been made to manipulate redox pathway for the clinical benefits of hematopoietic stem cell transplantation (HSCT). NAC and glutathione have been shown to be able to protect HSCs from oxidative stress and promote the engraftment of long-term repopulating HSCs (Ito et al., 2006; Tothova et al., 2007). However, these compounds need to be given frequently (at least daily) and in large quantity (at least 100 mg/Kg in mice). There is also a concern that prolonged administration of these compounds may actually provide substrates for increased ROS production, rather than reducing ROS.

During the proteomic analysis of the BM extracellular fluid harvested from transplant recipient mice, we found that Trx1 expression was upregulated in AMD3100 (a CXCR4 antagonist)-treated transplant recipient mice (An et al., 2013). AMD3100-treated transplant recipient mice had a faster hematopoietic recovery, and their BM supernatant promoted colony-forming units (CFUs) *in vitro*, compared to PBS-treated transplant recipient mice. Therefore, we chose to explore further the use of Trx1 for enhancing HSC reconstitution, based on several considerations: (1) Trx1 has a very stable structure and functions both intracellularly and extracellularly. (2) Trx1 can easily cross cell membranes and effectively enter living cells (Fernando, Nanri, Yoshitake, Nagata-Kuno, & Minakami, 1992; Spector et al., 1988). Extracellular Trx1 can readily enter cells within 24 h (Kondo et al., 2004). Therefore, Trx1 can be simply added into cell culture system or given systemically. (3) Trx1 is an endogenous protein ubiquitously expressed, thus eliminating the concern of hosts recognizing it as foreign protein and generating neutralizing antibodies or immune responses. (4) Human recombinant Trx1 is available, and its structure and chemical properties are well understood. (5) Trx1 has diverse biological activities on hematopoietic cells. Trx1 can (i) induce the migration of monocytes, T cells, and polymorphonuclear neutrophils (i.e., chemotactic activity; Bertini et al., 1999; Bizzarri et al., 2005); (ii) stimulate the growth of lymphocytes (i.e., cytokine and chemokine activity; Bertini et al., 1999;

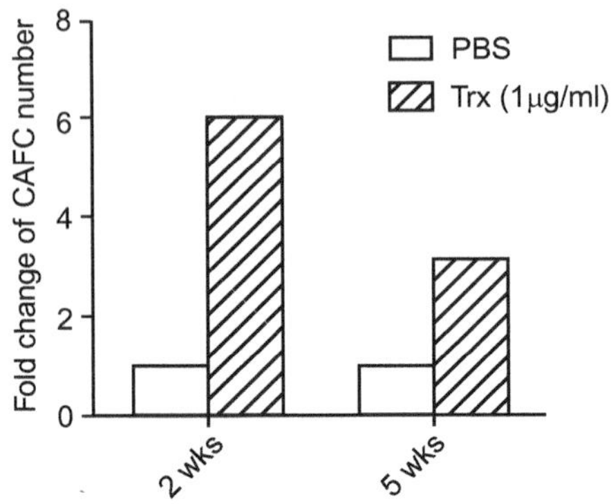

Figure 7.5 Recombinant Trx1 increases the cobblestone area-forming cells (CAFCs). The CAFC assay was performed as previously described (van Os, Dethmers-Ausema, & de Haan, 2008). Red blood cell-depleted bone marrow cells were obtained from C57Bl/6 mice, irradiated with 1.0 Gy, and then plated on feeder cells at 33 °C with 5% CO_2 in the absence or presence of 1 μg/ml human recombinant Trx1. Fresh media were added to the culture once weekly (half media change, no additional rhTrx1 was added to the culture after the first week). The frequencies of CAFCs were determined at week 2 and week 5 using L-Calc software (STEMCELL Technologies).

Nakamura, Nakamura, & Yodoi, 1997); and (iii) promote the growth of normal and tumor cells (i.e., autocrine growth factor activity; Gasdaska, Berggren, & Powis, 1995; Miyazaki et al., 1998; Wakasugi et al., 1990).

We subsequently found that recombinant Trx1 increased BFU-E and CFU-GEMM *in vitro* in a dose-dependent manner (An et al., 2013). Additionally, Trx1 enhanced cobblestone area-forming cell (CAFC) assay units, a miniature long-term BM culture assay to measure the function of hematopoietic stem/progenitor cells (Fig. 7.5). Moreover, when Trx1 was administered into 9.5 Gy-irradiated C57Bl/6 mice within 2 h of irradiation and then daily for 5 days, irradiation-induced death was reduced and the radiation-related hematologic injury was mitigated. Furthermore, *ex vivo* exposure of HSCs to Trx1 (10 μg/ml) for only 24 h enhanced HSC long-term repopulation as demonstrated in serial transplant recipients (An et al., 2013). Our studies provide justification for further exploring Trx1 for the use in HSCT settings to enhance hematopoietic reconstitution.

4. CLOSING REMARKS

It is merely the beginning to appreciate the roles of Trx in tumorigenesis and to utilize Trx for enhancing hematopoietic reconstitution in HSCT. Much more work needs to be done and future efforts should focus on (a) understanding how Trx enters cells, (b) dissecting Trx signaling pathways and the genes/molecules involved in the pathways, and (c) examining the mechanisms through which Trx enhances hematopoietic recovery. Extensive clinical studies should be performed to define the roles of Trx in predicting patient outcomes and the effectiveness of Trx inhibitors in the treatment of cancer. With collaborative efforts between clinicians and scientist, Trx will be proved to be an important therapeutic target in our fight to cure cancer and to improve HSCT outcome.

ACKNOWLEDGMENTS

The authors thank Dr. Woodrow J. Coker and Ms. Logan Roof for their critical reading of the manuscript. This work is supported by MUSC Hollings Cancer Center Startup Fund, Hollings Cancer Center ACS IRG, ASCO Conquer Cancer Foundation Career Development Award, NIH 1K08HL 103780-01A1, and NIH 3P30CA138313-01S3. The content is solely the responsibility of the authors and does not necessarily represent the official views of the National Institutes of Health or other funding agents.

Competing Interests: The authors declare no competing financial interests.

REFERENCES

Ago, T., Liu, T., Zhai, P., Chen, W., Li, H., Molkentin, J. D., et al. (2008). A redox-dependent pathway for regulating class II HDACs and cardiac hypertrophy. *Cell, 133*, 978–993.

Ameyar, M., Wisniewska, M., & Weitzman, J. B. (2003). A role for AP-1 in apoptosis: The case for and against. *Biochimie, 85*, 747–752.

An, N., Janech, M. G., Bland, A. M., Lazarchick, J., Arthur, J. M., & Kang, Y. (2013). Proteomic analysis of murine bone marrow niche microenvironment identifies thioredoxin as a novel agent for radioprotection and for enhancing donor cell reconstitution. *Experimental Hematology, 41*, 944–956.

Angelini, G., Gardella, S., Ardy, M., Ciriolo, M. R., Filomeni, G., Di Trapani, G., et al. (2002). Antigen-presenting dendritic cells provide the reducing extracellular microenvironment required for T lymphocyte activation. *Proceedings of the National Academy of Sciences of the United States of America, 99*, 1491–1496.

Armstrong, J. S., Hornung, B., Lecane, P., Jones, D. P., & Knox, S. J. (2001). Rotenone-induced G2/M cell cycle arrest and apoptosis in a human B lymphoma cell line PW. *Biochemical and Biophysical Research Communications, 289*, 973–978.

Arner, E. S., & Holmgren, A. (2006). The thioredoxin system in cancer. *Seminars in Cancer Biology, 16*, 420–426.

Backman, E., Bergh, A. C., Lagerdahl, I., Rydberg, B., Sundstrom, C., Tobin, G., et al. (2007). Thioredoxin, produced by stromal cells retrieved from the lymph node microenvironment, rescues chronic lymphocytic leukemia cells from apoptosis in vitro. *Haematologica, 92*, 1495–1504.

Baker, A. F., Adab, K. N., Raghunand, N., Chow, H. H., Stratton, S. P., Squire, S. W., et al. (2013). A phase IB trial of 24-hour intravenous PX-12, a thioredoxin-1 inhibitor, in patients with advanced gastrointestinal cancers. *Investigational New Drugs, 31*, 631–641.

Baker, A. F., Dragovich, T., Tate, W. R., Ramanathan, R. K., Roe, D., Hsu, C. H., et al. (2006). The antitumor thioredoxin-1 inhibitor PX-12 (1-methylpropyl 2-imidazolyl disulfide) decreases thioredoxin-1 and VEGF levels in cancer patient plasma. *The Journal of Laboratory and Clinical Medicine*, *147*, 83–90.

Baker, A. F., Koh, M. Y., Williams, R. R., James, B., Wang, H., Tate, W. R., et al. (2008). Identification of thioredoxin-interacting protein 1 as a hypoxia-inducible factor 1alpha-induced gene in pancreatic cancer. *Pancreas*, *36*, 178–186.

Baker, A., Payne, C. M., Briehl, M. M., & Powis, G. (1997). Thioredoxin, a gene found overexpressed in human cancer, inhibits apoptosis in vitro and in vivo. *Cancer Research*, *57*, 5162–5167.

Balkwill, F., Charles, K. A., & Mantovani, A. (2005). Smoldering and polarized inflammation in the initiation and promotion of malignant disease. *Cancer Cell*, 7, 211–217.

Barash, S., Wang, W., & Shi, Y. (2002). Human secretory signal peptide description by hidden Markov model and generation of a strong artificial signal peptide for secreted protein expression. *Biochemical and Biophysical Research Communications*, *294*, 835–842.

Bazzichi, A., Incaprera, M., & Garzelli, C. (1994). Effect of Escherichia-coli thioredoxin, a homolog to the adult T-cell leukemia-derived factor, on epstein-barr virus-transformed B-lymphocytes. *International Journal of Oncology*, *5*, 41–46.

Becker, K., Gromer, S., Schirmer, R. H., & Muller, S. (2000). Thioredoxin reductase as a pathophysiological factor and drug target. *European Journal of Biochemistry*, *267*, 6118–6125.

Behrend, L., Henderson, G., & Zwacka, R. M. (2003). Reactive oxygen species in oncogenic transformation. *Biochemical Society Transactions*, *31*, 1441–1444.

Benhar, M., Forrester, M. T., Hess, D. T., & Stamler, J. S. (2008). Regulated protein denitrosylation by cytosolic and mitochondrial thioredoxins. *Science*, *320*, 1050–1054.

Bertini, R., Howard, O. M., Dong, H. F., Oppenheim, J. J., Bizzarri, C., Sergi, R., et al. (1999). Thioredoxin, a redox enzyme released in infection and inflammation, is a unique chemoattractant for neutrophils, monocytes, and T cells. *The Journal of Experimental Medicine*, *189*, 1783–1789.

Billiet, L., Furman, C., Larigauderie, G., Copin, C., Brand, K., Fruchart, J. C., et al. (2005). Extracellular human thioredoxin-1 inhibits lipopolysaccharide-induced interleukin-1beta expression in human monocyte-derived macrophages. *The Journal of Biological Chemistry*, *280*, 40310–40318.

Bizzarri, C., Holmgren, A., Pekkari, K., Chang, G., Colotta, F., Ghezzi, P., et al. (2005). Requirements for the different cysteines in the chemotactic and desensitizing activity of human thioredoxin. *Antioxidants & Redox Signaling*, 7, 1189–1194.

Bloomfield, K. L., Osborne, S. A., Kennedy, D. D., Clarke, F. M., & Tonissen, K. F. (2003). Thioredoxin-mediated redox control of the transcription factor Sp1 and regulation of the thioredoxin gene promoter. *Gene*, *319*, 107–116.

Chen, B., Nelin, V. E., Locy, M. L., Jin, Y., & Tipple, T. E. (2013). Thioredoxin-1 mediates hypoxia-induced pulmonary artery smooth muscle cell proliferation. *American Journal of Physiology. Lung Cellular and Molecular Physiology*, *305*, L389–L395.

Collet, J. F., & Messens, J. (2010). Structure, function, and mechanism of thioredoxin proteins. *Antioxidants & Redox Signaling*, *13*, 1205–1216.

Conrad, M., Jakupoglu, C., Moreno, S. G., Lippl, S., Banjac, A., Schneider, M., et al. (2004). Essential role for mitochondrial thioredoxin reductase in hematopoiesis, heart development, and heart function. *Molecular and Cellular Biology*, *24*, 9414–9423.

Csiki, I., Yanagisawa, K., Haruki, N., Nadaf, S., Morrow, J. D., Johnson, D. H., et al. (2006). Thioredoxin-1 modulates transcription of cyclooxygenase-2 via hypoxia-inducible factor-1alpha in non-small cell lung cancer. *Cancer Research*, *66*, 143–150.

Efferth, T., Verdorfer, I., Miyachi, H., Sauerbrey, A., Drexler, H. G., Chitambar, C. R., et al. (2002). Genomic imbalances in drug-resistant T-cell acute lymphoblastic CEM leukemia cell lines. *Blood Cells, Molecules & Diseases*, *29*, 1–13.

El Hadri, K., Mahmood, D. F., Couchie, D., Jguirim-Souissi, I., Genze, F., Diderot, V., et al. (2012). Thioredoxin-1 promotes anti-inflammatory macrophages of the M2 phenotype and antagonizes atherosclerosis. *Arteriosclerosis, Thrombosis, and Vascular Biology, 32*, 1445–1452.
Eliasson, P., & Jonsson, J. I. (2010). The hematopoietic stem cell niche: Low in oxygen but a nice place to be. *Journal of Cellular Physiology, 222*, 17–22.
Ericson, M. L., Horling, J., Wendel-Hansen, V., Holmgren, A., & Rosen, A. (1992). Secretion of thioredoxin after in vitro activation of human B cells. *Lymphokine and Cytokine Research, 11*, 201–207.
Evens, A. M., Schumacker, P. T., Helenowski, I. B., Singh, A. T., Dokic, D., Keswani, A., et al. (2008). Hypoxia inducible factor-alpha activation in lymphoma and relationship to the thioredoxin family. *British Journal of Haematology, 141*, 676–680.
Fernando, M. R., Nanri, H., Yoshitake, S., Nagata-Kuno, K., & Minakami, S. (1992). Thioredoxin regenerates proteins inactivated by oxidative stress in endothelial cells. *European Journal of Biochemistry, 209*, 917–922.
Gallegos, A., Gasdaska, J. R., Taylor, C. W., Paine-Murrieta, G. D., Goodman, D., Gasdaska, P. Y., et al. (1996). Transfection with human thioredoxin increases cell proliferation and a dominant-negative mutant thioredoxin reverses the transformed phenotype of human breast cancer cells. *Cancer Research, 56*, 5765–5770.
Gasdaska, J. R., Berggren, M., & Powis, G. (1995). Cell growth stimulation by the redox protein thioredoxin occurs by a novel helper mechanism. *Cell Growth & Differentiation, 6*, 1643–1650.
Gil-Bea, F., Akterin, S., Persson, T., Mateos, L., Sandebring, A., Avila-Carino, J., et al. (2012). Thioredoxin-80 is a product of alpha-secretase cleavage that inhibits amyloid-beta aggregation and is decreased in Alzheimer's disease brain. *EMBO Molecular Medicine, 4*, 1097–1111.
Go, Y. M., Kang, S. M., Roede, J. R., Orr, M., & Jones, D. P. (2011). Increased inflammatory signaling and lethality of influenza H1N1 by nuclear thioredoxin-1. *PLoS One, 6*, e18918.
Gromer, S., Urig, S., & Becker, K. (2004). The thioredoxin system—From science to clinic. *Medicinal Research Reviews, 24*, 40–89.
Haendeler, J., Hoffmann, J., Tischler, V., Berk, B. C., Zeiher, A. M., & Dimmeler, S. (2002). Redox regulatory and anti-apoptotic functions of thioredoxin depend on S-nitrosylation at cysteine 69. *Nature Cell Biology, 4*, 743–749.
Handy, D. E., & Loscalzo, J. (2012). Redox regulation of mitochondrial function. *Antioxidants & Redox Signaling, 16*, 1323–1367.
Hansen, J. M., Zhang, H., & Jones, D. P. (2006). Mitochondrial thioredoxin-2 has a key role in determining tumor necrosis factor-alpha-induced reactive oxygen species generation, NF-kappaB activation, and apoptosis. *Toxicological Sciences, 91*, 643–650.
Hatai, T., Matsuzawa, A., Inoshita, S., Mochida, Y., Kuroda, T., Sakamaki, K., et al. (2000). Execution of apoptosis signal-regulating kinase 1 (ASK1)-induced apoptosis by the mitochondria-dependent caspase activation. *The Journal of Biological Chemistry, 275*, 26576–26581.
Hattori, I., Takagi, Y., Nakamura, H., Nozaki, K., Bai, J., Kondo, N., et al. (2004). Intravenous administration of thioredoxin decreases brain damage following transient focal cerebral ischemia in mice. *Antioxidants & Redox Signaling, 6*, 81–87.
Hirota, K., Matsui, M., Iwata, S., Nishiyama, A., Mori, K., & Yodoi, J. (1997). AP-1 transcriptional activity is regulated by a direct association between thioredoxin and Ref-1. *Proceedings of the National Academy of Sciences of the United States of America, 94*, 3633–3638.
Hirota, K., Murata, M., Sachi, Y., Nakamura, H., Takeuchi, J., Mori, K., et al. (1999). Distinct roles of thioredoxin in the cytoplasm and in the nucleus. A two-step mechanism of

redox regulation of transcription factor NF-kappaB. *The Journal of Biological Chemistry*, *274*, 27891–27897.

Hirota, K., Nakamura, H., Masutani, H., & Yodoi, J. (2002). Thioredoxin superfamily and thioredoxin-inducing agents. *Annals of the New York Academy of Sciences*, *957*, 189–199.

Holmgren, A. (1968). Thioredoxin. 6. The amino acid sequence of the protein from escherichia coli B. *European Journal of Biochemistry*, *6*, 475–484.

Holmgren, A. (1985). Thioredoxin. *Annual Review of Biochemistry*, *54*, 237–271.

Holmgren, A., Soderberg, B. O., Eklund, H., & Branden, C. I. (1975). Three-dimensional structure of *Escherichia coli* thioredoxin-S2 to 2.8 A resolution. *Proceedings of the National Academy of Sciences of the United States of America*, *72*, 2305–2309.

Huber, K., Patel, P., Zhang, L., Evans, H., Westwell, A. D., Fischer, P. M., et al. (2008). 2-[(1-methylpropyl)dithio]-1H-imidazole inhibits tubulin polymerization through cysteine oxidation. *Molecular Cancer Therapeutics*, *7*, 143–151.

Huh, K. H., Cho, Y., Kim, B. S., Do, J. H., Park, Y. J., Joo, D. J., et al. (2013). The role of thioredoxin 1 in the mycophenolic acid-induced apoptosis of insulin-producing cells. *Cell Death & Disease*, *4*, e721.

Im, J. Y., Lee, K. W., Woo, J. M., Junn, E., & Mouradian, M. M. (2012). DJ-1 induces thioredoxin 1 expression through the Nrf2 pathway. *Human Molecular Genetics*, *21*, 3013–3024.

Imoto, K., Kukidome, D., Nishikawa, T., Matsuhisa, T., Sonoda, K., Fujisawa, K., et al. (2006). Impact of mitochondrial reactive oxygen species and apoptosis signal-regulating kinase 1 on insulin signaling. *Diabetes*, *55*, 1197–1204.

Inomata, Y., Tanihara, H., Tanito, M., Okuyama, H., Hoshino, Y., Kinumi, T., et al. (2008). Suppression of choroidal neovascularization by thioredoxin-1 via interaction with complement factor H. *Investigative Ophthalmology & Visual Science*, *49*, 5118–5125.

Irwin, M. E., Rivera-Del Valle, N., & Chandra, J. (2013). Redox control of leukemia: From molecular mechanisms to therapeutic opportunities. *Antioxidants & Redox Signaling*, *18*, 1349–1383.

Ito, K., Hirao, A., Arai, F., Takubo, K., Matsuoka, S., Miyamoto, K., et al. (2006). Reactive oxygen species act through p38 MAPK to limit the lifespan of hematopoietic stem cells. *Nature Medicine*, *12*, 446–451.

Jakupoglu, C., Przemeck, G. K., Schneider, M., Moreno, S. G., Mayr, N., Hatzopoulos, A. K., et al. (2005). Cytoplasmic thioredoxin reductase is essential for embryogenesis but dispensable for cardiac development. *Molecular and Cellular Biology*, *25*, 1980–1988.

Jang, Y. Y., & Sharkis, S. J. (2007). A low level of reactive oxygen species selects for primitive hematopoietic stem cells that may reside in the low-oxygenic niche. *Blood*, *110*, 3056–3063.

Jarvela, S., Bragge, H., Paunu, N., Jarvela, T., Paljarvi, L., Kalimo, H., et al. (2006). Antioxidant enzymes in oligodendroglial brain tumors: Association with proliferation, apoptotic activity and survival. *Journal of Neuro-Oncology*, *77*, 131–140.

Jimenez, A., Zu, W., Rawe, V. Y., Pelto-Huikko, M., Flickinger, C. J., Sutovsky, P., et al. (2004). Spermatocyte/spermatid-specific thioredoxin-3, a novel Golgi apparatus-associated thioredoxin, is a specific marker of aberrant spermatogenesis. *The Journal of Biological Chemistry*, *279*, 34971–34982.

Jordan, B. F., Runquist, M., Raghunand, N., Gillies, R. J., Tate, W. R., Powis, G., et al. (2005). The thioredoxin-1 inhibitor 1-methylpropyl 2-imidazolyl disulfide (PX-12) decreases vascular permeability in tumor xenografts monitored by dynamic contrast enhanced magnetic resonance imaging. *Clinical Cancer Research*, *11*, 529–536.

Kalinina, E. V., Chernov, N. N., Saprin, A. N., Kotova, Y. N., Gavrilova, Y. A., Chermnykh, N. S., et al. (2007). Expression of genes for thioredoxin 1 and thioredoxin

2 in multidrug resistance ovarian carcinoma cells SKVLB. *Bulletin of Experimental Biology and Medicine, 144*, 301–303.
Kim, Y. C., Masutani, H., Yamaguchi, Y., Itoh, K., Yamamoto, M., & Yodoi, J. (2001). Hemin-induced activation of the thioredoxin gene by Nrf2. A differential regulation of the antioxidant responsive element by a switch of its binding factors. *Journal of Biological Chemistry, 276*, 18399–18406.
Kim, S. J., Miyoshi, Y., Taguchi, T., Tamaki, Y., Nakamura, H., Yodoi, J., et al. (2005). High thioredoxin expression is associated with resistance to docetaxel in primary breast cancer. *Clinical Cancer Research, 11*, 8425–8430.
Kirkpatrick, D. L., Ehrmantraut, G., Stettner, S., Kunkel, M., & Powis, G. (1997). Redox active disulfides: The thioredoxin system as a drug target. *Oncology Research, 9*, 351–356.
Kirkpatrick, D. L., Kuperus, M., Dowdeswell, M., Potier, N., Donald, L. J., Kunkel, M., et al. (1998). Mechanisms of inhibition of the thioredoxin growth factor system by antitumor 2-imidazolyl disulfides. *Biochemical Pharmacology, 55*, 987–994.
Koharyova, M., & Kolarova, M. (2008). Oxidative stress and thioredoxin system. *General Physiology and Biophysics, 27*, 71–84.
Kondo, N., Ishii, Y., Kwon, Y. W., Tanito, M., Horita, H., Nishinaka, Y., et al. (2004). Redox-sensing release of human thioredoxin from T lymphocytes with negative feedback loops. *Journal of Immunology, 172*, 442–448.
Laurent, T. C., Moore, E. C., & Reichard, P. (1964). Enzymatic synthesis of deoxyribonucleotides. Iv. Isolation and characterization of thioredoxin, the hydrogen donor from *Escherichia coli* B. *The Journal of Biological Chemistry, 239*, 3436–3444.
Lee, S. R., Kim, J. R., Kwon, K. S., Yoon, H. W., Levine, R. L., Ginsburg, A., et al. (1999). Molecular cloning and characterization of a mitochondrial selenocysteine-containing thioredoxin reductase from rat liver. *The Journal of Biological Chemistry, 274*, 4722–4734.
Lee, S. R., Yang, K. S., Kwon, J., Lee, C., Jeong, W., & Rhee, S. G. (2002). Reversible inactivation of the tumor suppressor PTEN by H_2O_2. *The Journal of Biological Chemistry, 277*, 20336–20342.
Lemarechal, H., Anract, P., Beaudeux, J. L., Bonnefont-Rousselot, D., Ekindjian, O. G., & Borderie, D. (2007). Expression and extracellular release of Trx80, the truncated form of thioredoxin, by TNF-alpha- and IL-1beta-stimulated human synoviocytes from patients with rheumatoid arthritis. *Clinical Science (London England: 1979), 113*, 149–155.
Li, G. X., Hirabayashi, Y., Yoon, B. I., Kawasaki, Y., Tsuboi, I., Kodama, Y., et al. (2006). Thioredoxin overexpression in mice, model of attenuation of oxidative stress, prevents benzene-induced hemato-lymphoid toxicity and thymic lymphoma. *Experimental Hematology, 34*, 1687–1697.
Li, C., Thompson, M. A., Tamayo, A. T., Zuo, Z., Lee, J., Vega, F., et al. (2012). Overexpression of Thioredoxin-1 mediates growth, survival, and chemoresistance and is a druggable target in diffuse large B-cell lymphoma. *Oncotarget, 3*, 314–326.
Li, L., & Xie, T. (2005). Stem cell niche: Structure and function. *Annual Review of Cell and Developmental Biology, 21*, 605–631.
Lisa, C. F., Flores, M. O., Dube, Sara, Gene, B., Hubbard, Shuko Lee, Salmon, Adam, et al. (2012). Thioredoxin, oxidative stress, cancer and aging. *Longevity & Healthspan, 1*, 4.
Liu, Y., & Min, W. (2002). Thioredoxin promotes ASK1 ubiquitination and degradation to inhibit ASK1-mediated apoptosis in a redox activity-independent manner. *Circulation Research, 90*, 1259–1266.
Mahmood, D. F., Abderrazak, A., Couchie, D., Lunov, O., Diderot, V., Syrovets, T., et al. (2013). Truncated thioredoxin (Trx-80) promotes pro-inflammatory macrophages of the M1 phenotype and enhances atherosclerosis. *Journal of Cellular Physiology, 228*, 1577–1583.
Majmundar, A. J., Wong, W. J., & Simon, M. C. (2010). Hypoxia-inducible factors and the response to hypoxic stress. *Molecular Cell, 40*, 294–309.

Matsui, M., Oshima, M., Oshima, H., Takaku, K., Maruyama, T., Yodoi, J., et al. (1996). Early embryonic lethality caused by targeted disruption of the mouse thioredoxin gene. *Developmental Biology, 178*, 179–185.

Matsukawa, J., Matsuzawa, A., Takeda, K., & Ichijo, H. (2004). The ASK1-MAP kinase cascades in mammalian stress response. *Journal of Biochemistry, 136*, 261–265.

Matsuo, Y., & Yodoi, J. (2013). Extracellular thioredoxin: A therapeutic tool to combat inflammation. *Cytokine & Growth Factor Reviews, 24*, 345–353.

Matthews, J. R., Wakasugi, N., Virelizier, J. L., Yodoi, J., & Hay, R. T. (1992). Thioredoxin regulates the DNA binding activity of NF-kappa B by reduction of a disulphide bond involving cysteine 62. *Nucleic Acids Research, 20*, 3821–3830.

Maurice, M. M., Nakamura, H., van der Voort, E. A., van Vliet, A. I., Staal, F. J., Tak, P. P., et al. (1997). Evidence for the role of an altered redox state in hyporesponsiveness of synovial T cells in rheumatoid arthritis. *Journal of Immunology, 158*, 1458–1465.

Meuillet, E. J., Mahadevan, D., Berggren, M., Coon, A., & Powis, G. (2004). Thioredoxin-1 binds to the C2 domain of PTEN inhibiting PTEN's lipid phosphatase activity and membrane binding: A mechanism for the functional loss of PTEN's tumor suppressor activity. *Archives of Biochemistry and Biophysics, 429*, 123–133.

Miranda-Vizuete, A., Sadek, C. M., Jimenez, A., Krause, W. J., Sutovsky, P., & Oko, R. (2004). The mammalian testis-specific thioredoxin system. *Antioxidants & Redox Signaling, 6*, 25–40.

Mitchell, D. A., Morton, S. U., Fernhoff, N. B., & Marletta, M. A. (2007). Thioredoxin is required for S-nitrosation of procaspase-3 and the inhibition of apoptosis in Jurkat cells. *Proceedings of the National Academy of Sciences of the United States of America, 104*, 11609–11614.

Mitsui, A., Hamuro, J., Nakamura, H., Kondo, N., Hirabayashi, Y., Ishizaki-Koizumi, S., et al. (2002). Overexpression of human thioredoxin in transgenic mice controls oxidative stress and life span. *Antioxidants & Redox Signaling, 4*, 693–696.

Miwa, K., Kishimoto, C., Nakamura, H., Makita, T., Ishii, K., Okuda, N., et al. (2003). Increased oxidative stress with elevated serum thioredoxin level in patients with coronary spastic angina. *Clinical Cardiology, 26*, 177–181.

Miyazaki, K., Noda, N., Okada, S., Hagiwara, Y., Miyata, M., Sakurabayashi, I., et al. (1998). Elevated serum level of thioredoxin in patients with hepatocellular carcinoma. *Biotherapy, 11*, 277–288.

Mougiakakos, D., Johansson, C. C., Jitschin, R., Bottcher, M., & Kiessling, R. (2011). Increased thioredoxin-1 production in human naturally occurring regulatory T cells confers enhanced tolerance to oxidative stress. *Blood, 117*, 857–861.

Mustacich, D., & Powis, G. (2000). Thioredoxin reductase. *The Biochemical Journal, 346*(Pt 1), 1–8.

Nadeau, P. J., Charette, S. J., Toledano, M. B., & Landry, J. (2007). Disulfide Bond-mediated multimerization of Ask1 and its reduction by thioredoxin-1 regulate H(2)O (2)-induced c-Jun NH(2)-terminal kinase activation and apoptosis. *Molecular Biology of the Cell, 18*, 3903–3913.

Nakamura, H. (2004). Thioredoxin as a key molecule in redox signaling. *Antioxidants & Redox Signaling, 6*, 15–17.

Nakamura, H., Bai, J., Nishinaka, Y., Ueda, S., Sasada, T., Ohshio, G., et al. (2000). Expression of thioredoxin and glutaredoxin, redox-regulating proteins, in pancreatic cancer. *Cancer Detection and Prevention, 24*, 53–60.

Nakamura, H., De Rosa, S., Roederer, M., Anderson, M. T., Dubs, J. G., Yodoi, J., et al. (1996). Elevation of plasma thioredoxin levels in HIV-infected individuals. *International Immunology, 8*, 603–611.

Nakamura, H., Herzenberg, L. A., Bai, J., Araya, S., Kondo, N., Nishinaka, Y., et al. (2001). Circulating thioredoxin suppresses lipopolysaccharide-induced neutrophil chemotaxis.

Proceedings of the National Academy of Sciences of the United States of America, 98, 15143–15148.
Nakamura, H., Hoshino, Y., Okuyama, H., Matsuo, Y., & Yodoi, J. (2009). Thioredoxin 1 delivery as new therapeutics. *Advanced Drug Delivery Reviews, 61*, 303–309.
Nakamura, H., Mitsui, A., & Yodoi, J. (2002). Thioredoxin overexpression in transgenic mice. *Methods in Enzymology, 347*, 436–440.
Nakamura, H., Nakamura, K., & Yodoi, J. (1997). Redox regulation of cellular activation. *Annual Review of Immunology, 15*, 351–369.
Nakamura, H., Tamura, S., Watanabe, I., Iwasaki, T., & Yodoi, J. (2002). Enhanced resistancy of thioredoxin-transgenic mice against influenza virus-induced pneumonia. *Immunology Letters, 82*, 165–170.
Nickel, W. (2003). The mystery of nonclassical protein secretion. A current view on cargo proteins and potential export routes. *European Journal of Biochemistry, 270*, 2109–2119.
Nishinaka, Y., Masutani, H., Nakamura, H., & Yodoi, J. (2001). Regulatory roles of thioredoxin in oxidative stress-induced cellular responses. *Redox Report, 6*, 289–295.
Nonn, L., Williams, R. R., Erickson, R. P., & Powis, G. (2003). The absence of mitochondrial thioredoxin 2 causes massive apoptosis, exencephaly, and early embryonic lethality in homozygous mice. *Molecular and Cellular Biology, 23*, 916–922.
Oblong, J. E., Berggren, M., Gasdaska, P. Y., & Powis, G. (1994). Site-directed mutagenesis of active site cysteines in human thioredoxin produces competitive inhibitors of human thioredoxin reductase and elimination of mitogenic properties of thioredoxin. *The Journal of Biological Chemistry, 269*, 11714–11720.
Owusu-Ansah, E., & Banerjee, U. (2009). Reactive oxygen species prime Drosophila haematopoietic progenitors for differentiation. *Nature, 461*, 537–541.
Parmar, K., Mauch, P., Vergilio, J. A., Sackstein, R., & Down, J. D. (2007). Distribution of hematopoietic stem cells in the bone marrow according to regional hypoxia. *Proceedings of the National Academy of Sciences of the United States of America, 104*, 5431–5436.
Pekkari, K., Avila-Carino, J., Bengtsson, A., Gurunath, R., Scheynius, A., & Holmgren, A. (2001). Truncated thioredoxin (Trx80) induces production of interleukin-12 and enhances CD14 expression in human monocytes. *Blood, 97*, 3184–3190.
Pekkari, K., Avila-Carino, J., Gurunath, R., Bengtsson, A., Scheynius, A., & Holmgren, A. (2003). Truncated thioredoxin (Trx80) exerts unique mitogenic cytokine effects via a mechanism independent of thiol oxido-reductase activity. *FEBS Letters, 539*, 143–148.
Pekkari, K., Gurunath, R., Arner, E. S., & Holmgren, A. (2000). Truncated thioredoxin is a mitogenic cytokine for resting human peripheral blood mononuclear cells and is present in human plasma. *The Journal of Biological Chemistry, 275*, 37474–37480.
Pekkari, K., & Holmgren, A. (2004). Truncated thioredoxin: Physiological functions and mechanism. *Antioxidants & Redox Signaling, 6*, 53–61.
Pelicano, H., Carney, D., & Huang, P. (2004). ROS stress in cancer cells and therapeutic implications. *Drug Resistance Updates*, 7, 97–110.
Perez, V. I., Cortez, L. A., Lew, C. M., Rodriguez, M., Webb, C. R., Van Remmen, H., et al. (2011). Thioredoxin 1 overexpression extends mainly the earlier part of life span in mice. *The Journals of Gerontology Series A, Biological Sciences and Medical Sciences, 66*, 1286–1299.
Pileggi, A., Molano, R. D., Berney, T., Cattan, P., Vizzardelli, C., Oliver, R., et al. (2001). Heme oxygenase-1 induction in islet cells results in protection from apoptosis and improved in vivo function after transplantation. *Diabetes, 50*, 1983–1991.
Powis, G., & Montfort, W. R. (2001). Properties and biological activities of thioredoxins. *Annual Review of Biophysics and Biomolecular Structure, 30*, 421–455.
Powis, G., Oblong, J. E., Gasdaska, P. Y., Berggren, M., Hill, S. R., & Kirkpatrick, D. L. (1994). The thioredoxin/thioredoxin reductase redox system and control of cell growth. *Oncology Research, 6*, 539–544.

Raffel, J., Bhattacharyya, A. K., Gallegos, A., Cui, H., Einspahr, J. G., Alberts, D. S., et al. (2003). Increased expression of thioredoxin-1 in human colorectal cancer is associated with decreased patient survival. *The Journal of Laboratory and Clinical Medicine*, *142*, 46–51.

Ramanathan, R. K., Abbruzzese, J., Dragovich, T., Kirkpatrick, L., Guillen, J. M., Baker, A. F., et al. (2011). A randomized phase II study of PX-12, an inhibitor of thioredoxin in patients with advanced cancer of the pancreas following progression after a gemcitabine-containing combination. *Cancer Chemotherapy and Pharmacology*, *67*, 503–509.

Ramanathan, R. K., Kirkpatrick, D. L., Belani, C. P., Friedland, D., Green, S. B., Chow, H. H., et al. (2007). A Phase I pharmacokinetic and pharmacodynamic study of PX-12, a novel inhibitor of thioredoxin-1, in patients with advanced solid tumors. *Clinical Cancer Research*, *13*, 2109–2114.

Rhee, S. G., Kang, S. W., Chang, T. S., Jeong, W., & Kim, K. (2001). Peroxiredoxin, a novel family of peroxidases. *IUBMB Life*, *52*, 35–41.

Rosen, A., Lundman, P., Carlsson, M., Bhavani, K., Srinivasa, B. R., Kjellstrom, G., et al. (1995). A CD4+ T cell line-secreted factor, growth promoting for normal and leukemic B cells, identified as thioredoxin. *International Immunology*, *7*, 625–633.

Rubartelli, A., Bajetto, A., Allavena, G., Wollman, E., & Sitia, R. (1992). Secretion of thioredoxin by normal and neoplastic cells through a leaderless secretory pathway. *The Journal of Biological Chemistry*, *267*, 24161–24164.

Rubartelli, A., & Sitia, R. (1991). Interleukin 1 beta and thioredoxin are secreted through a novel pathway of secretion. *Biochemical Society Transactions*, *19*, 255–259.

Sahaf, B., & Rosen, A. (2000). Secretion of 10-kDa and 12-kDa thioredoxin species from blood monocytes and transformed leukocytes. *Antioxidants & Redox Signaling*, *2*, 717–726.

Sahaf, B., Soderberg, A., Spyrou, G., Barral, A. M., Pekkari, K., Holmgren, A., et al. (1997). Thioredoxin expression and localization in human cell lines: Detection of full-length and truncated species. *Experimental Cell Research*, *236*, 181–192.

Saitoh, M., Nishitoh, H., Fujii, M., Takeda, K., Tobiume, K., Sawada, Y., et al. (1998). Mammalian thioredoxin is a direct inhibitor of apoptosis signal-regulating kinase (ASK) 1. *The EMBO Journal*, *17*, 2596–2606.

Schenk, H., Vogt, M., Droge, W., & Schulze-Osthoff, K. (1996). Thioredoxin as a potent costimulus of cytokine expression. *Journal of Immunology*, *156*, 765–771.

Schofield, R. (1978). The relationship between the spleen colony-forming cell and the haemopoietic stem cell. *Blood Cells*, *4*, 7–25.

Schumacker, P. T. (2006). Reactive oxygen species in cancer cells: Live by the sword, die by the sword. *Cancer Cell*, *10*, 175–176.

Shao, L., Diccianni, M. B., Tanaka, T., Gribi, R., Yu, A. L., Pullen, J. D., et al. (2001). Thioredoxin expression in primary T-cell acute lymphoblastic leukemia and its therapeutic implication. *Cancer Research*, *61*, 7333–7338.

Shaulian, E., & Karin, M. (2001). AP-1 in cell proliferation and survival. *Oncogene*, *20*, 2390–2400.

Shim, Y. K., Kim, J. T., Seong, M. H., Kim, Y. J., Shim, T. J., Kim, S. M., et al. (2012). Serum thioredoxin 1 level has close relation with myocardial damage amount in acute myocardial infarction patients. *Journal of Korean Medical Science*, *27*, 1162–1169.

Shioji, K., Kishimoto, C., Nakamura, H., Masutani, H., Yuan, Z., Oka, S., et al. (2002). Overexpression of thioredoxin-1 in transgenic mice attenuates adriamycin-induced cardiotoxicity. *Circulation*, *106*, 1403–1409.

Silber, J. R., Bobola, M. S., Blank, A., Schoeler, K. D., Haroldson, P. D., Huynh, M. B., et al. (2002). The apurinic/apyrimidinic endonuclease activity of Ape1/Ref-1 contributes to human glioma cell resistance to alkylating agents and is elevated by oxidative stress. *Clinical Cancer Research*, *8*, 3008–3018.

Soejima, H., Suefuji, H., Miyamoto, S., Kajiwaram, I., Kojima, S., Hokamaki, J., et al. (2003). Increased plasma thioredoxin in patients with acute myocardial infarction. *Clinical Cardiology, 26*, 583–587.
Spector, A., Yan, G. Z., Huang, R. R., McDermott, M. J., Gascoyne, P. R., & Pigiet, V. (1988). The effect of H2O2 upon thioredoxin-enriched lens epithelial cells. *The Journal of Biological Chemistry, 263*, 4984–4990.
Stehlik, C., de Martin, R., Kumabashiri, I., Schmid, J. A., Binder, B. R., & Lipp, J. (1998). Nuclear factor (NF)-kappaB-regulated X-chromosome-linked iap gene expression protects endothelial cells from tumor necrosis factor alpha-induced apoptosis. *The Journal of Experimental Medicine, 188*, 211–216.
Suda, T., Arai, F., & Hirao, A. (2005). Hematopoietic stem cells and their niche. *Trends in Immunology, 26*, 426–433.
Sumida, Y., Nakashima, T., Yoh, T., Nakajima, Y., Ishikawa, H., Mitsuyoshi, H., et al. (2000). Serum thioredoxin levels as an indicator of oxidative stress in patients with hepatitis C virus infection. *Journal of Hepatology, 33*, 616–622.
Sun, Q. A., Kirnarsky, L., Sherman, S., & Gladyshev, V. N. (2001). Selenoprotein oxidoreductase with specificity for thioredoxin and glutathione systems. *Proceedings of the National Academy of Sciences of the United States of America, 98*, 3673–3678.
Tada-Oikawa, S., Hiraku, Y., Kawanishi, M., & Kawanishi, S. (2003). Mechanism for generation of hydrogen peroxide and change of mitochondrial membrane potential during rotenone-induced apoptosis. *Life Sciences, 73*, 3277–3288.
Tagaya, Y., Maeda, Y., Mitsui, A., Kondo, N., Matsui, H., Hamuro, J., et al. (1989). ATL-derived factor (ADF), an IL-2 receptor/Tac inducer homologous to thioredoxin; possible involvement of dithiol-reduction in the IL-2 receptor induction. *The EMBO Journal, 8*, 757–764.
Takagi, Y., Mitsui, A., Nishiyama, A., Nozaki, K., Sono, H., Gon, Y., et al. (1999). Overexpression of thioredoxin in transgenic mice attenuates focal ischemic brain damage. *Proceedings of the National Academy of Sciences of the United States of America, 96*, 4131–4136.
Tamaki, H., Nakamura, H., Nishio, A., Nakase, H., Ueno, S., Uza, N., et al. (2006). Human thioredoxin-1 ameliorates experimental murine colitis in association with suppressed macrophage inhibitory factor production. *Gastroenterology, 131*, 1110–1121.
Tanaka, T., Hosoi, F., Yamaguchi-Iwai, Y., Nakamura, H., Masutani, H., Ueda, S., et al. (2002). Thioredoxin-2 (TRX-2) is an essential gene regulating mitochondria-dependent apoptosis. *The EMBO Journal, 21*, 1695–1703.
Tangye, S. G., & Raison, R. L. (1997). Human cytokines suppress apoptosis of leukaemic CD5+ B cells and preserve expression of bcl-2. *Immunology and Cell Biology, 75*, 127–135.
Tanito, M., Masutani, H., Kim, Y. C., Nishikawa, M., Ohira, A., & Yodoi, J. (2005). Sulforaphane induces thioredoxin through the antioxidant-responsive element and attenuates retinal light damage in mice. *Investigative Ophthalmology & Visual Science, 46*, 979–987.
Teshigawara, K., Maeda, M., Nishino, K., Nikaido, T., Uchiyama, T., Tsudo, M., et al. (1985). Adult T leukemia cells produce a lymphokine that augments interleukin 2 receptor expression. *The Journal of Molecular and Cellular Immunology, 2*, 17–26.
Tome, M. E., Johnson, D. B., Rimsza, L. M., Roberts, R. A., Grogan, T. M., Miller, T. P., et al. (2005). A redox signature score identifies diffuse large B-cell lymphoma patients with a poor prognosis. *Blood, 106*, 3594–3601.
Tothova, Z., Kollipara, R., Huntly, B. J., Lee, B. H., Castrillon, D. H., Cullen, D. E., et al. (2007). FoxOs are critical mediators of hematopoietic stem cell resistance to physiologic oxidative stress. *Cell, 128*, 325–339.
Trachootham, D., Lu, W., Ogasawara, M. A., Nilsa, R. D., & Huang, P. (2008). Redox regulation of cell survival. *Antioxidants & Redox Signaling, 10*, 1343–1374.

Ueno, M., Masutani, H., Arai, R. J., Yamauchi, A., Hirota, K., Sakai, T., et al. (1999). Thioredoxin-dependent redox regulation of p53-mediated p21 activation. *The Journal of Biological Chemistry*, *274*, 35809–35815.

Vakkila, J., & Lotze, M. T. (2004). Inflammation and necrosis promote tumour growth. *Nature Reviews Immunology*, *4*, 641–648.

Valko, M., Rhodes, C. J., Moncol, J., Izakovic, M., & Mazur, M. (2006). Free radicals, metals and antioxidants in oxidative stress-induced cancer. *Chemico-Biological Interactions*, *160*, 1–40.

van Os, R. P., Dethmers-Ausema, B., & de Haan, G. (2008). In vitro assays for cobblestone area-forming cells, LTC-IC, and CFU-C. *Methods in Molecular Biology*, *430*, 143–157.

Wakasugi, N., Tagaya, Y., Wakasugi, H., Mitsui, A., Maeda, M., Yodoi, J., et al. (1990). Adult T-cell leukemia-derived factor/thioredoxin, produced by both human T-lymphotropic virus type I- and Epstein-Barr virus-transformed lymphocytes, acts as an autocrine growth factor and synergizes with interleukin 1 and interleukin 2. *Proceedings of the National Academy of Sciences of the United States of America*, *87*, 8282–8286.

Wang, J., Kobayashi, M., Sakurada, K., Imamura, M., Moriuchi, T., & Hosokawa, M. (1997). Possible roles of an adult T-cell leukemia (ATL)-derived factor/thioredoxin in the drug resistance of ATL to adriamycin. *Blood*, *89*, 2480–2487.

Wang, Y., Liu, L., Pazhanisamy, S. K., Li, H., Meng, A., & Zhou, D. (2010). Total body irradiation causes residual bone marrow injury by induction of persistent oxidative stress in murine hematopoietic stem cells. *Free Radical Biology & Medicine*, *48*, 348–356.

Wang, C. Y., Mayo, M. W., Korneluk, R. G., Goeddel, D. V., & Baldwin, A. S., Jr. (1998). NF-kappaB antiapoptosis: Induction of TRAF1 and TRAF2 and c-IAP1 and c-IAP2 to suppress caspase-8 activation. *Science*, *281*, 1680–1683.

Wang, K., Zhang, T., Dong, Q., Nice, E. C., Huang, C., & Wei, Y. (2013). Redox homeostasis: The linchpin in stem cell self-renewal and differentiation. *Cell Death& Disease*, *4*, e537.

Welsh, S. J., Bellamy, W. T., Briehl, M. M., & Powis, G. (2002). The redox protein thioredoxin-1 (Trx-1) increases hypoxia-inducible factor 1alpha protein expression: Trx-1 overexpression results in increased vascular endothelial growth factor production and enhanced tumor angiogenesis. *Cancer Research*, *62*, 5089–5095.

Welsh, S. J., Williams, R. R., Birmingham, A., Newman, D. J., Kirkpatrick, D. L., & Powis, G. (2003). The thioredoxin redox inhibitors 1-methylpropyl 2-imidazolyl disulfide and pleurotin inhibit hypoxia-induced factor 1alpha and vascular endothelial growth factor formation. *Molecular Cancer Therapeutics*, *2*, 235–243.

Widder, J. D., Fraccarollo, D., Galuppo, P., Hansen, J. M., Jones, D. P., Ertl, G., et al. (2009). Attenuation of angiotensin II-induced vascular dysfunction and hypertension by overexpression of Thioredoxin 2. *Hypertension*, *54*, 338–344.

Wiesel, P., Foster, L. C., Pellacani, A., Layne, M. D., Hsieh, C. M., Huggins, G. S., et al. (2000). Thioredoxin facilitates the induction of heme oxygenase-1 in response to inflammatory mediators. *The Journal of Biological Chemistry*, *275*, 24840–24846.

Wu, W. S. (2006). The signaling mechanism of ROS in tumor progression. *Cancer Metastasis Reviews*, *25*, 695–705.

Yahata, T., Muguruma, Y., Yumino, S., Sheng, Y., Uno, T., Matsuzawa, H., et al. (2008). Quiescent human hematopoietic stem cells in the bone marrow niches organize the hierarchical structure of hematopoiesis. *Stem Cells*, *26*, 3228–3236.

Yahata, T., Takanashi, T., Muguruma, Y., Ibrahim, A. A., Matsuzawa, H., Uno, T., et al. (2011). Accumulation of oxidative DNA damage restricts the self-renewal capacity of human hematopoietic stem cells. *Blood*, *118*, 2941–2950.

Yamada, M., Tomida, A., Yoshikawa, H., Taketani, Y., & Tsuruo, T. (1996). Increased expression of thioredoxin/adult T-cell leukemia-derived factor in cisplatin-resistant human cancer cell lines. *Clinical Cancer Research*, *2*, 427–432.

Yamada, M., Tomida, A., Yoshikawa, H., Taketani, Y., & Tsuruo, T. (1997). Overexpression of thioredoxin does not confer resistance to cisplatin in transfected human ovarian and colon cancer cell lines. *Cancer Chemotherapy and Pharmacology, 40*, 31–37.
Yodoi, J., Takatsuki, K., & Masuda, T. (1974). Letter: Two cases of T-cell chronic lymphocytic leukemia in Japan. *The New England Journal of Medicine, 290*, 572–573.
Yodoi, J., & Tursz, T. (1991). ADF, a growth-promoting factor derived from adult T cell leukemia and homologous to thioredoxin: Involvement in lymphocyte immortalization by HTLV-I and EBV. *Advances in Cancer Research, 57*, 381–411.
Yokomizo, A., Ono, M., Nanri, H., Makino, Y., Ohga, T., Wada, M., et al. (1995). Cellular levels of thioredoxin associated with drug sensitivity to cisplatin, mitomycin C, doxorubicin, and etoposide. *Cancer Research, 55*, 4293–4296.
Yoshida, S., Katoh, T., Tetsuka, T., Uno, K., Matsui, N., & Okamoto, T. (1999). Involvement of thioredoxin in rheumatoid arthritis: Its costimulatory roles in the TNF-alpha-induced production of IL-6 and IL-8 from cultured synovial fibroblasts. *Journal of Immunology, 163*, 351–358.
You, B. R., Shin, H. R., & Park, W. H. (2014). PX-12 inhibits the growth of A549 lung cancer cells via G2/M phase arrest and ROS-dependent apoptosis. *International Journal of Oncology, 44*, 301–308.
Zhang, R., Al-Lamki, R., Bai, L., Streb, J. W., Miano, J. M., Bradley, J., et al. (2004). Thioredoxin-2 inhibits mitochondria-located ASK1-mediated apoptosis in a JNK-independent manner. *Circulation Research, 94*, 1483–1491.
Zhang, H., Go, Y. M., & Jones, D. P. (2007). Mitochondrial thioredoxin-2/peroxiredoxin-3 system functions in parallel with mitochondrial GSH system in protection against oxidative stress. *Archives of Biochemistry and Biophysics, 465*, 119–126.
Zhang, H., Luo, Y., Zhang, W., He, Y., Dai, S., Zhang, R., et al. (2007). Endothelial-specific expression of mitochondrial thioredoxin improves endothelial cell function and reduces atherosclerotic lesions. *The American Journal of Pathology, 170*, 1108–1120.
Zhang, X., Zheng, Y., Fried, L. E., Du, Y., Montano, S. J., Sohn, A., et al. (2011). Disruption of the mitochondrial thioredoxin system as a cell death mechanism of cationic triphenylmethanes. *Free Radical Biology & Medicine, 50*, 811–820.
Zhou, F. L., Zhang, W. G., Wei, Y. C., Meng, S., Bai, G. G., Wang, B. Y., et al. (2010). Involvement of oxidative stress in the relapse of acute myeloid leukemia. *The Journal of Biological Chemistry, 285*, 15010–15015.

CHAPTER EIGHT

Role of the Keap1–Nrf2 Pathway in Cancer

Hanna M. Leinonen*, Emilia Kansanen*, Petri Pölönen*,†, Merja Heinäniemi*,†, Anna-Liisa Levonen*,1

*Department of Biotechnology and Molecular Medicine, A.I. Virtanen Institute for Molecular Sciences, University of Eastern Finland, P.O. Box 1627, Kuopio, Finland
†Institute of Biomedicine, School of Medicine, University of Eastern Finland, P.O. Box 1627, Kuopio, Finland
[1]Corresponding author: e-mail address: anna-liisa.levonen@uef.fi

Contents

Abstract

The Kelch-like ECH-associated protein 1 (Keap1)–nuclear factor E2-related factor 2 (Nrf2) pathway is one of the major signaling cascades involved in cell defense and survival against endogenous and exogenous stress. While Nrf2 and its target genes provide protection against various age-related diseases including tumorigenesis, constitutively active Nrf2 in cancer cells increases the expression of cytoprotective genes and, consequently, enhances proliferation via metabolic reprogramming and inhibition of apoptosis. Herein, we review the current understanding of the regulation of Nrf2 in normal cells as well as its dual role in cancer. Furthermore, the mechanisms of Nrf2 dysregulation in cancer, consequences of unchecked Nrf2 activity, and therapies targeting the Keap1–Nrf2 system are discussed.

Advances in Cancer Research, Volume 122
ISSN 0065-230X
http://dx.doi.org/10.1016/B978-0-12-420117-0.00008-6

1. INTRODUCTION

Oxidative stress, an imbalance between the production and disposal of reactive oxygen species (ROS), plays an important role in various diseases including cancer (Valko et al., 2007). The Kelch-like ECH-associated protein 1 (Keap1)–nuclear factor E2-related factor 2 (Nrf2) pathway is one of the most important signaling cascades involved in cell defense and survival against xenobiotics and oxidative stress. Nrf2 protects against the development of several diseases, such as cancer and neurodegenerative diseases (Hybertson, Gao, Bose, & McCord, 2011; Ma & He, 2012), and various activators of the Keap1–Nrf2 pathway have been extensively studied for chemoprevention (Zhao, Gao, & Qu, 2010). Recently, it has become evident that Nrf2 has a protective role not only in normal cells but also in cancer cells, in which aberrant signaling via the Keap1-Nrf2 pathway leads to constitutive activation of Nrf2 and upregulation of target genes resulting in enhanced cancer cell survival. (Padmanabhan et al., 2006; Shibata, Kokubu, et al., 2008; Wang, Sun, Villeneuve, et al., 2008). In this chapter, we will summarize the current knowledge on the regulation of Nrf2 in normal cells as well as the mechanisms of dysregulation in cancer. In addition, we will discuss the clinical ramifications of Nrf2 overactivity and how the Keap1–Nrf2 system can be targeted in cancer therapy.

2. THE KEAP1–NRF2 PATHWAY

Cells have evolved highly efficient protective mechanisms that enable their survival under conditions of stress. One of the major cytoprotective signaling cascades is the Keap1–Nrf2 pathway, which is activated in response to endogenous or exogenous stress. Under basal conditions, Nrf2 is bound to its cytosolic inhibitor Keap1, which functions as an adaptor protein in the cullin 3 (Cul3)-based E3 ligase complex that ubiquitinates Nrf2 resulting in proteasomal degradation (Kobayashi et al., 2004). In the presence of oxidative or electrophilic stress, ubiquitination of Nrf2 is disturbed allowing nuclear accumulation of Nrf2. In the nucleus, Nrf2 binds to the antioxidant response element (ARE) (Rushmore, Morton, & Pickett, 1991) in the regulatory regions of the target genes and drives their expression. Target genes of Nrf2 have been identified by gene expression profiling of Nrf2 knockout ($Nrf2^{-/-}$) mice (Kwak et al., 2003; Lee, Calkins, Chan, Kan, & Johnson, 2003). These genes include detoxification enzymes, such as NAD(P)H quinone oxidoreductase 1 (*NQO1*) and glutathione *S*-transferases (*GSTs*); antioxidant enzymes, including heme oxygenase-1 (*HMOX1*); and peroxiredoxins, and enzymes involved in glutathione (GSH) synthesis and metabolism such as glutamate–cysteine ligase (*GCL*) subunits. Furthermore, Nrf2 regulates proteasomal and chaperone proteins suggesting that Nrf2 has an important role also in the reparation and removal of damaged proteins (reviewed in Digaleh, Kiaei, & Khodagholi, 2013).

The inhibitor protein of the pathway, Keap1, consists of five domains, the N-terminal domain, Broad complex; Tramtrack and Bric-à-Brac (BTB) domain; the intervening region (IVR); the Kelch domain, also known as double glycine repeat; and C-terminal domain (Fig. 8.1A). Keap1 is a homodimer that dimerizes through the BTB domain and forms a structure that resembles a "cherry bob" (Ogura et al., 2010) (Fig. 8.1C). In the cell, Keap1 is tethered to the cytoskeletal actin through the Kelch domain, which binds Nrf2 (Itoh et al., 1999). Nrf2 protein consists of seven Neh (Nrf2–ECH homology) domains (Neh1–Neh7) (Fig. 8.1B). Neh1 contains CNC basic region for DNA binding and a leucine zipper structure for dimerization with small Maf proteins (Itoh et al., 1997). Neh2 is the domain that negatively regulates the transcriptional activity of Nrf2 via binding to Keap1 (Itoh et al., 1999). The Neh2 domain has two different motifs that bind Keap1, the ETGE and DLG

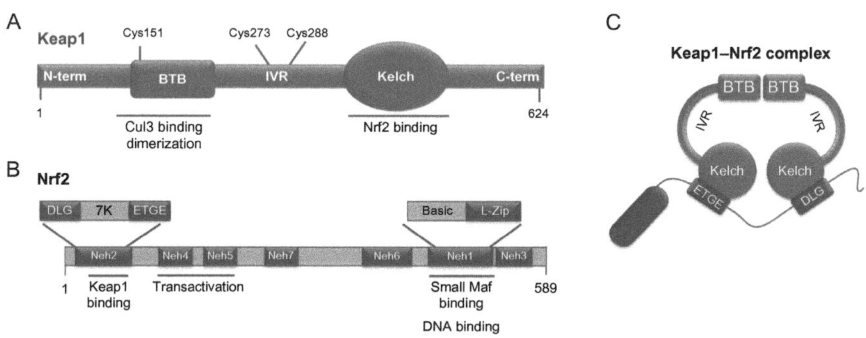

Figure 8.1 *Structures of Keap1 and Nrf2 proteins.* (A) Keap1 consists of 624 amino acid residues and has five domains. (B) Nrf2 consists of 589 amino acids and has 7 evolutionarily highly conserved domains, Neh1–7. Neh2 domain contains ETGE and DLG motifs, which are required for the interaction with Keap1, and a hydrophilic region of lysine residues (7 K), which is indispensable for the Keap1-dependent polyubiquitination and degradation of Nrf2. (C) Nrf2 interacts with two molecules of Keap1 through its Neh2 ETGE and DLG motifs. (See the color plate.)

motifs, and this results in a Keap1–Nrf2 complex with 1:2 stoichiometry with two binding sites in Nrf2 (Tong, Kobayashi, Katsuoka, & Yamamoto, 2006) (Fig. 8.1C). The two-site binding is beneficial for ubiquitination of Nrf2, because the lysine residues are aligned between the Keap1 binding ETGE and DLG motifs (Fig. 8.1A). Neh3 (Nioi, Nguyen, Sherratt, & Pickett, 2005), Neh4 (Itoh et al., 1997), and Neh5 (Zhang et al., 2007) are transactivation domains. The most recently defined Neh domain is Neh7 comprising amino acids 209–316 in Nrf2. Neh7 is postulated to bind RXR, which inhibits Nrf2–ARE signaling (Wang et al., 2013).

The Keap1–Nrf2 pathway-activating compounds were originally screened based on their ability to activate ARE and increase NQO1 activity (Prestera, Holtzclaw, Zhang, & Talalay, 1993). This and subsequent studies revealed that ARE inducers belong to several different chemical classes. The common feature of these diverse chemical compounds is that they react with sulfhydryls such as cysteine thiols within proteins (reviewed in Itoh, Mimura, & Yamamoto, 2010). The discovery of the cysteine rich Keap1 in 1999 (Itoh et al., 1999) and the finding that ARE activating chemicals bind Keap1 (Dinkova-Kostova et al., 2002) led to the notion that Keap1 is the key sensor of the electrophile stress within cells.

2.1. Keap1-dependent regulation of Nrf2

Under basal conditions, Keap1 functions as an adaptor protein in the Cul3-based E3 ligase complex resulting in rapid ubiquitination and subsequent degradation of Nrf2 (Kobayashi et al., 2004) (Fig. 8.2A). The half-life of the Nrf2 protein is short, 10–20 min (Itoh et al., 2003), which prevents the constitutive activation of antioxidant response. However, during oxidative or electrophilic stress, the cysteine residues in Keap1 are modified leading to the conformational change in the Keap1–Nrf2 complex disrupting the ubiquitination and degradation of Nrf2 (Fig. 8.2B). This allows the *de novo* synthesized Nrf2 to translocate into the nucleus, bind to ARE, and induce the expression of target genes. In proteomic analyses, several of the 27 cysteines in human Keap1 have been found to be modified in response to different electrophiles (reviewed in Kansanen, Jyrkkänen, & Levonen, 2012). However, only three of the Keap1 cysteines, Cys151, Cys273, and Cys288, are found to be functionally important for the Keap1–Nrf2 system (Yamamoto et al., 2008). While Cys273 and Cys288 are critical for Keap1 to inhibit Nrf2 under basal conditions, a subset of Nrf2 activators target

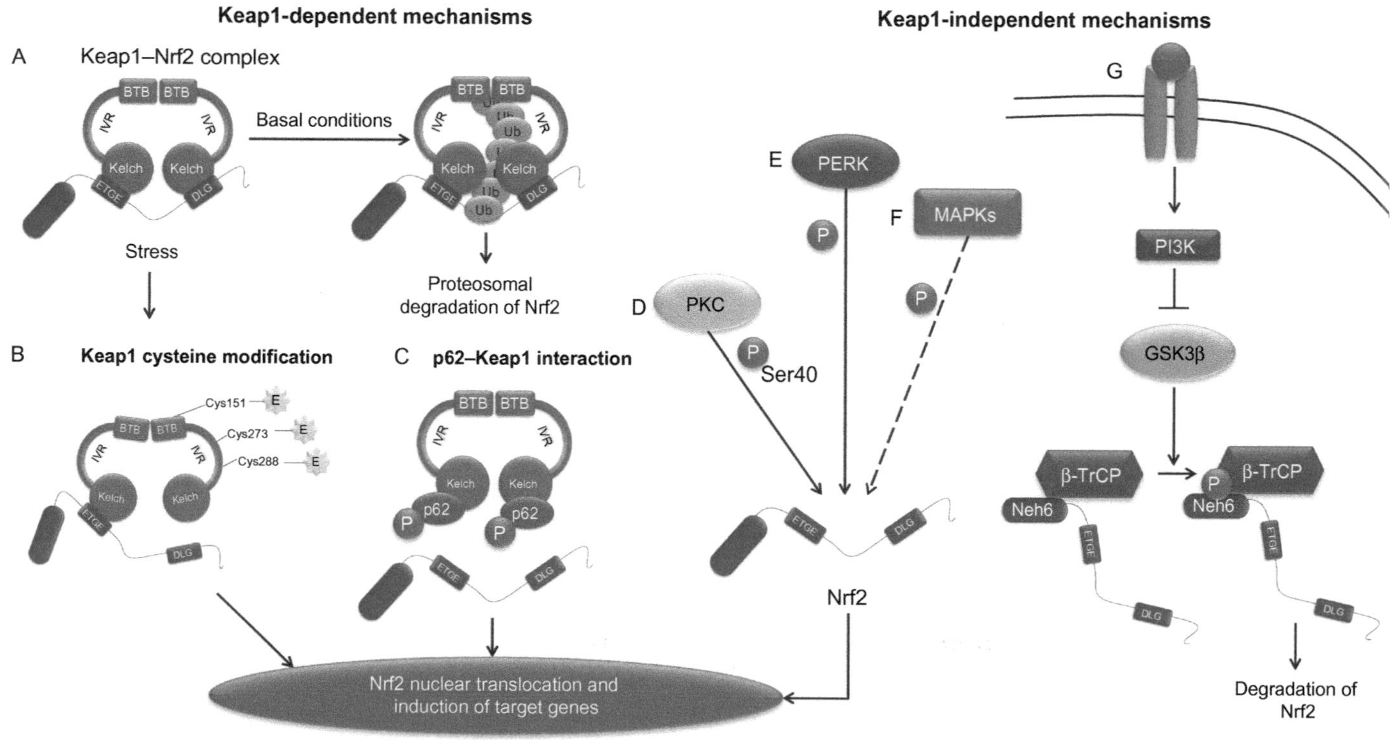
Keap1-dependent mechanisms
A
Keap1–Nrf2 complex
BTB
IVR
Kelch
ETGE
DLG
Basal conditions
Ub
Proteosomal degradation of Nrf2
Stress
B
Keap1 cysteine modification
Cys151
Cys273
Cys288
E
C
p62–Keap1 interaction
p62
P
Nrf2 nuclear translocation and induction of target genes
Keap1-independent mechanisms
D
PKC
Ser40
E
PERK
F
MAPKs
Nrf2
G
PI3K
GSK3β
β-TrCP
Neh6
Degradation of Nrf2

Cys151 (Kobayashi et al., 2009; McMahon, Lamont, Beattie, & Hayes, 2010). However, lipid derived electrophilic species nitro-oleic acid (OA-NO_2) (Kansanen et al., 2011), prostaglandin A_2 (PGA_2), 15-deoxy-$\Delta^{12,14}$-prostaglandin J_2 (15d-PGJ_2) (Kobayashi et al., 2009), and the heavy metal compound As(III) (Wang, Sun, Chen, et al., 2008) can activate Nrf2 independent of Keap1 Cys151.

In addition to Keap1 cysteine modifications, other Keap1-dependent mechanisms for Nrf2 activation have been found. There are recent reports identifying proteins that compete with the binding of Nrf2 to Keap1 (Hast et al., 2013). Sequestosome 1/p62 (SQSTM1/p62) activates the Keap1–Nrf2 pathway via an interaction with Keap1 through a motif within the C-terminal region resembling the ETGE motif in Nrf2 (Jain et al., 2010; Komatsu et al., 2010; Taguchi et al., 2012) (Fig. 8.2C). Phosphorylation of p62 at Ser351 increases the interaction of p62 and Keap1 leading to nuclear accumulation of Nrf2 and subsequent Nrf2 target gene expression (Ichimura et al., 2013).

2.2. Keap1-independent regulation of Nrf2

Based on the data published to date, Keap1 cysteine modifications appear to be the main mechanism for the activation of the Keap1–Nrf2 pathway. However, there is a body of evidence showing that alternative, Keap1-independent, mechanisms for Nrf2 activation exist.

Various kinases are suggested to activate the Keap1–Nrf2 pathway. For example, protein kinase C (PKC) can directly phosphorylate Nrf2 at Ser40 leading to the induction of Nrf2 target genes (Bloom & Jaiswal, 2003;

Figure 8.2 *Keap1–Nrf2 pathway activation.* Keap1-dependent mechanisms: (A) In basal conditions, two Keap1 molecules bind to Nrf2 and Nrf2 is polyubiquitylated by the Cul3-based E3 ligase complex. The polyubiquitilation results in rapid Nrf2 degradation by the proteasome. (B) Under stress conditions, cysteine residues in Keap1 are modified leading to the inhibition of Nrf2 polyubiquitylation. (C) In p62-dependent activation, p62 binds to the Kelch domain of Keap1 leading to attenuated Nrf2 ubiquitination, which leads to Nrf2 nuclear accumulation. In these models, the newly synthesized Nrf2 proteins translocate into the nucleus, bind to the antioxidant response element (ARE), and drive the expression of Nrf2 target genes such as *NQO1, HMOX1, GCL,* and GSTs. Keap1-independent mechanisms include phosphorylation of Nrf2 by PKC (D), PERK (E), or MAP kinases (F). PI3K may regulate Nrf2 via indirect mechanism by inhibiting GSK-3β activity. Active GSK-3β phosphorylates Nrf2 Neh6 domain-binding protein β-TrCP, which results in Cul1-dependent ubiquitination and degradation of Nrf2 (G). (See the color plate.)

Huang, Nguyen, & Pickett, 2000, 2002) (Fig. 8.2D). In addition, double-stranded RNA-activated protein kinase-like endoplasmic reticulum (ER) kinase (PERK), activated by ER stress, phosphorylates Nrf2 at an unknown residue resulting in nuclear accumulation of Nrf2 (Cullinan et al., 2003) (Fig. 8.2E). Furthermore, MAP kinases (MAPKs), such as p38 MAPK, c-Jun NH_2-terminal kinase (JNK), and extracellular signal-regulated kinase (ERK), as well as phosphatidylinositol 3-kinase (PI3K) (Kang, Ryu, & Kim, 2000; Nguyen, Sherratt, & Pickett, 2003; Sun, Huang, & Zhang, 2009), are suggested to play a role in Nrf2 activation (Fig. 8.2F). However, whether Nrf2 is directly phosphorylated by these kinases remains to be investigated.

In addition to Keap1-dependent degradation of Nrf2, there are alternative pathways that can degrade Nrf2. Nrf2 can bind β-transducin repeat-containing protein (β-TrCP) via the Neh6 domain that contains several serine residues, which can be phosphorylated by glycogen synthase kinase 3β (GSK-3β) leading to cullin 1 (Cul1)-dependent ubiquitination and degradation of Nrf2 (Chowdhry et al., 2013; McMahon, Thomas, Itoh, Yamamoto, & Hayes, 2004; Rada et al., 2012). The activity of GSK-3β is regulated by various kinases. For example, PI3K and Akt inhibitors increase the activity of GSK-3β and thus downregulate Nrf2 (Chowdhry et al., 2013) suggesting that PI3K pathway inhibits the activity of GSK-3β leading to increase in Nrf2 (Fig. 8.2G). GSK-3β-mediated phosphorylation occurs after electrophile-induced Keap1-dependent activation of Nrf2 target genes (Rojo et al., 2012). Thus, it may serve as the mechanism for Nrf2 degradation post-induction.

3. DUAL ROLES OF NRF2 IN CANCER

3.1. The role of oxidative stress in cancer

Oxidative stress plays an important role in carcinogenesis (Valko et al., 2007). Reactive oxygen species (ROS) are involved in both the initiation and progression of cancer and are generated by cancer cells via increased mitochondrial activity and accelerated metabolism needed for tumor growth and cellular proliferation (Hanahan & Weinberg, 2011; Storz, 2005; Szatrowski & Nathan, 1991; Valko et al., 2007). ROS can cause mutations in DNA, including sugar and base modifications, single- or double-strand breaks, and DNA cross-links (Dizdaroglu, Jaruga, Birincioglu, & Rodriguez, 2002). ROS can affect not only nuclear but also mitochondrial DNA, which is more prone to mutations due to the lack of histones (Birch-Machin, 2006). Moreover, activation of signaling cascades, genomic instability, errors in replication, and gene arrest or transcription of tumor suppressor genes and oncogenes may occur when oxidative stress lasts for a longer period (Valko et al., 2007). One of the most studied both mutagenic and carcinogenic DNA adducts caused by ROS is 8-hydroxydeoxyguanine (OH8dG) (Shigenaga, Gimeno, & Ames, 1989), which triggers guanine to thymine transversions (Moriya, 1993) that are common in oncogenes and tumor suppressor genes (Harris & Hollstein, 1993; Hruban et al., 1993). In addition, ROS are involved in protein oxidation as well as lipid peroxidation reactions, thus producing mutagenic and carcinogenic reactive aldehydes such as 4-hydroxy-2-nonenal (4-HNE) and malondialdehyde (MDA) (Dalle-Donne et al., 2005; Feng, Hu, & Tang, 2004; Wang et al., 1996). ROS also affect epigenetic regulation mechanisms by modifying DNA methylation patterns, thus likely inducing the expression of otherwise quiescent genes such as oncogenes (Cerda & Weitzman, 1997; Counts & Goodman, 1995).

3.2. The use of the Keap1–Nrf2-inducing compounds in chemoprevention

A variety of agents, such as naturally existing phytochemicals and synthetic compounds that are used, for example, as additives of cosmetics or food, have been used to prevent or suppress carcinogenesis (Wattenberg, 1985). Many of these compounds have the ability to increase the expression of cytoprotective genes, and this was hypothesized to be the mechanism of protection (Talalay, De Long, & Prochaska, 1988). Several of the genes induced by these compounds were later identified as Nrf2 target genes (Hayes & McMahon, 2001). Nrf2$^{-/-}$ mice are more susceptible to toxicity DNA adduct formation, and carcinogenesis caused by different chemicals, such as acetaminophen, cigarette smoke, diesel exhaust, benzo[a]pyrene, urethane, and hyperoxia (Aoki et al., 2001; Cho, Reddy, & Kleeberger, 2006; Enomoto, 2001; Fahey et al., 2002; Satoh, Moriguchi, Takai, Ebina, & Yamamoto, 2013). In addition, in a mouse model of Braf^{V600E}-induced lung cancer, Nrf2 ablation accelerated tumor growth (Strohecker et al., 2013). Activation of ARE-driven genes by chemopreventive compounds, such as ethoxyquin, oltipraz, and isothiocyanates (e.g., sulforaphane, SFN), is almost completely lost in Nrf2$^{-/-}$ mice, and thus, they cannot be protected against carcinogenesis by these compounds (Fahey et al., 2002; Hayes et al., 2000; Ramos-Gomez et al., 2001). Therefore, several inducers of the Keap1–Nrf2 pathway are suggested to be potent compounds for chemoprevention.

The most prototypic Keap1–Nrf2 pathway-activating chemopreventive agents are SFN, oltipraz, and CDDO-imidazolide (1[2-cyano-3,12-dioxooleana-1,9(11)-dien-28-oyl]imidazole, CDDO-Im) (reviewed in Kwak & Kensler, 2010), and they all have already undergone clinical trials (http://www.clinicaltrials.gov). A phytochemical SFN is a naturally occurring isothiocyanate found in cruciferous vegetables such as broccoli (Dinkova-Kostova et al., 2002; Zhang, Kensler, Cho, Posner, & Talalay, 1994). SFN has a chemopreventive effect in various cancers, such as colon, skin, lung, and stomach cancers (Zhang & Tang, 2007). Several clinical trials studying SFN as a chemopreventive agent for various cancers, administered usually in the form of broccoli extract, are underway (http://www.clinicaltrials.gov).

Another commonly used chemopreventive agent, oltipraz (4-methyl-5-[2-pyrazinyl]-1,2-dithiole-3-thione), a synthetic dithiolethione, protects against the development of various cancers in rodents (Zhang & Munday, 2008). From the beginning of 1995, clinical trials of oltipraz in the protection of aflatoxin-induced hepatotoxicity have been conducted in Qidong, China, where people are at high risk for exposure of dietary aflatoxin (Glintborg, Weimann, Kensler, & Poulsen, 2006; Jacobson et al., 1997; Kensler et al., 1998). However, due to significant side effects, the clinical trials of oltipraz as a chemopreventive compound have been halted. CDDO is a synthetic oleanolic triterpenoid, which inhibits inflammation, and because of its antiproliferative and proapoptotic effects, it also acts as a chemopreventive compound *in vivo* (Place et al., 2003). However, recently, a phase 3 study using bardoxolone methyl, a CDDO-derivative, to treat chronic kidney disease in type 2 diabetics was stopped because of a higher rate of cardiovascular events and mortality (De Zeeuw et al., 2013), and currently, all clinical trials of CDDO derivatives listed on clinical trials registry are either terminated or withdrawn.

3.3. Oncogenic role of Nrf2

Traditionally, Nrf2 has been considered as a transcription factor that protects cells from aging and age-related diseases such as cancer and cardiovascular diseases (reviewed in Hybertson et al., 2011). However, several Nrf2 target genes induced by chemopreventive compounds, such as *NQO1* and *GST* isoenzymes are overexpressed in cancer (Galloway, Blake, Shepherd, & McLellan, 1997; Traver et al., 1992). Cancer cells with resistance to chemotherapeutic agents usually contain high GSH levels and upregulated expression of various drug-metabolizing enzymes and drug efflux pumps that are coordinately regulated by Nrf2 (reviewed in Meijerman, Beijnen, & Schellens, 2008). Indeed, it was reported in 2006 that in some lung cancer patient samples and lung cancer cell lines, Nrf2 is constitutively localized into the nucleus as a result of missense mutations in *KEAP1* (Padmanabhan et al., 2006; Singh et al., 2006). Thereafter, several papers have been published describing elevated activity of Nrf2 in pancreatic, gallbladder, and prostate cancers, among others, and multiple mechanisms for Keap1–Nrf2 pathway hyperactivity beyond *KEAP1* mutations have been discovered (Lister et al., 2011; Shibata, Kokubu, et al., 2008; Zhang et al., 2010) (Fig. 8.3). It was also soon observed that overactivity of the Keap1–Nrf2 pathway leads to significantly poorer prognosis of patients as a result of constitutive expression of cytoprotective genes in cancer cells (Shibata, Kokubu, et al., 2008; Solis et al., 2010). Importantly, although Nrf2 hyperactivity has been shown to be advantageous to cancer progression, as has been demonstrated in the studies with Keap1 knockdown mice, accumulation of Nrf2 has not been proved to be sufficient for spontaneous cancer formation (Taguchi et al., 2010). Therefore, it is suggested that once tumor is initiated, cancer cells hijack the Keap1–Nrf2 system for acquiring stress resistance and growth advantage.

3.4. Mechanisms of Nrf2 overactivity in cancer

3.4.1 Disrupted protein–protein interaction of Keap1–Nrf2

Initially, loss-of-function mutations in *KEAP1* gene were identified in human adenocarcinoma cell lines (Padmanabhan et al., 2006) (Fig. 8.3). These mutations substituted Gly to Cys in the Kelch domain of Keap1, which reduces the affinity of Nrf2 binding. Consequently, nuclear accumulation of Nrf2 is constitutively increased. Mutations have also been observed in many other cancer types including breast cancer (Sjöblom et al., 2006), gallbladder cancer (Shibata, Kokubu, et al., 2008), liver cancer (Yoo, Kim, Kim, An, & Lee, 2012), ovarian cancer (Konstantinopoulos et al., 2011), clear renal cell carcinoma (De Zeeuw et al., 2013), and lung papillary carcinomas (Li, Singh, Biswal, Askin, & Gabrielson, 2011), and most of them are located within the Kelch domain of *KEAP1* (Ohta et al., 2008; Padmanabhan et al., 2006). However, mutations in other regions of *KEAP1* have also been found (Ohta et al., 2008; Singh et al., 2006). Importantly, a heteroallelic mutation in *KEAP1* is sufficient to increase Nrf2 activity, as mutated Keap1 functions in a dominant-negative fashion (Suzuki,

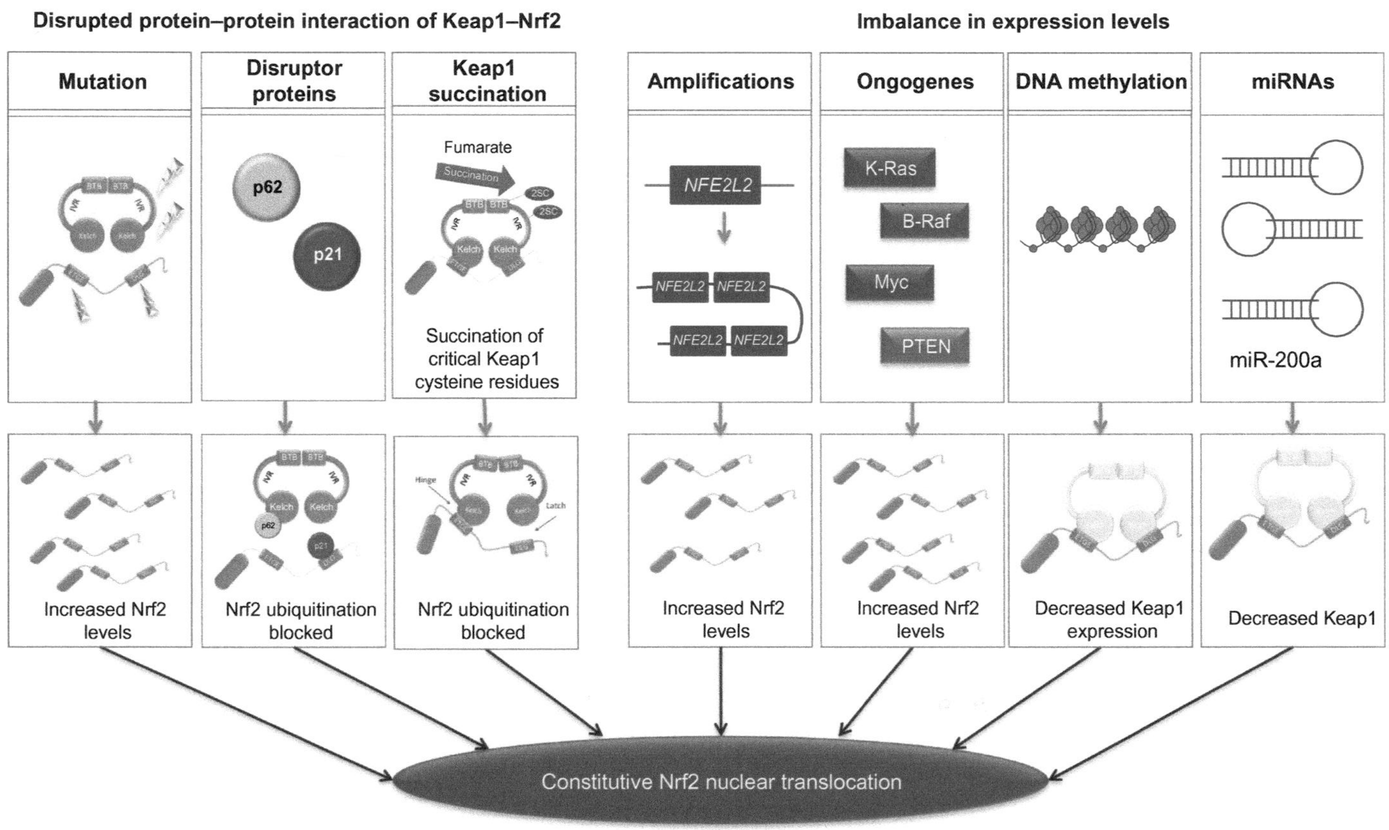

Disrupted protein–protein interaction of Keap1–Nrf2
Imbalance in expression levels
Mutation
Disruptor proteins
Keap1 succination
Amplifications
Ongogenes
DNA methylation
miRNAs
p62
p21
Fumarate
Succination
2SC
Succination of critical Keap1 cysteine residues
NFE2L2
K-Ras
B-Raf
Myc
PTEN
miR-200a
Increased Nrf2 levels
Nrf2 ubiquitination blocked
Nrf2 ubiquitination blocked
Increased Nrf2 levels
Increased Nrf2 levels
Decreased Keap1 expression
Decreased Keap1
Constitutive Nrf2 nuclear translocation

Maher, & Yamamoto, 2011). In addition to point mutations, truncated mRNA transcripts of Keap1 have been observed in prostate cancer cells. These differentially spliced transcripts are nonfunctional and therefore unable to suppress Nrf2 activity (Zhang et al., 2010).

In Nrf2 gene (*NFE2L2*), gain-of-function mutations are frequent (Fig. 8.3). These mutations have been reported in several cancers, including lung, head and neck, and esophageal carcinomas (Kim et al., 2010; Shibata et al., 2011; Shibata, Ohta, et al., 2008). These mutations are found exclusively within the ETGE and DLG motifs of Nrf2 leading to decreased Keap1 binding affinity and inhibition of the Nrf2 degradation. Comprehensive genomic characterization has provided the genomic landscape of common human cancers and identified mutations in *NFE2L2* gene as one of the ~140 driver mutations (Vogelstein et al., 2013).

Somatic mutations of *CUL3* gene have been identified in squamous cell lung cancer and papillary renal cell carcinoma (Cancer Genome Atlas Research Network, 2012; Ooi et al., 2013). Cul3 is the molecular scaffold binding Keap1, Rbx1, and E2 ligase in the ubiquitin ligase complex. However, mutations in *CUL3* are likely to influence also many other pathways, as Cul3 binds to numerous other BTB-containing proteins (Genschik, Sumara, & Lechner, 2013).

Recently, large-scale multi-omics projects have begun to decipher the molecular complexity across multiple cancer types, providing a unique opportunity to assess the prevalence and clinical relevance of aberrations across pathways. One of the large-scale efforts, The Cancer Genome Atlas (TCGA) consortium, collects tumor samples from patient cohorts and measures gene expression, copy number alteration, protein levels, and mutation

Figure 8.3 *Mechanisms for constitutive Nrf2 activity in cancer. Disrupted protein–protein interaction of Keap1–Nrf2*: Mutations in Nrf2 ETGE or DLG motifs or in Keap1 disrupt the interaction between the two proteins promoting Nrf2 nuclear accumulation. Disruptor proteins, such as p62 and p21, compete with Nrf2 and Keap1 disturbing their interaction leading to impaired ubiquitination and degradation of Nrf2. Loss-of-function mutations in fumarate hydratase cause accumulation of fumarate and subsequent succination of Keap1 cysteine residues, which disrupts the ubiquitination and degradation of Nrf2. *Imbalance in expression levels*: Increased copy number of Nrf2-gene (*NFE2L2*) can amplify the basal expression levels of Nrf2. Furthermore, the transcriptional induction of Nrf2 by oncogenes, such as K-Ras, B-Raf, Myc, and PTEN, can increase Nrf2. The hypermethylation of Keap1 promoter decreases Keap1 mRNA expression causing imbalance in the Keap1–Nrf2 ratio. High expression of miR-200a reduces Keap1 levels, which increases Nrf2. (See the color plate.)

profiles from several cancer types. The ongoing project is committed to profile over 10,000 tumor samples from at least 20 tumor types, and it has become a useful resource for integrating cancer data and for characterizing the genetic and epigenetic landscape of tumors.

In accordance with previous reports, a high somatic mutation rate (34%) in key components of the Keap1–Nrf2 pathway, *NFE2L2*, *KEAP1*, and *CUL3* was observed in TCGA squamous cell lung carcinoma clinical samples (in total 178) (Cancer Genome Atlas Research Network, 2012). Furthermore, these mutations are concordantly found in the classical cancer subtype, suggesting that Nrf2 pathway alterations are a hallmark of this subtype. The high rate of *NFE2L2–KEAP1–CUL3* mutations has also been observed in the TCGA head and neck carcinoma, lung adenocarcinoma, uterine corpus endometrial carcinoma, and bladder urothelial carcinoma (Kandoth et al., 2013). These studies highlight the role of Nrf2 as a cancer driver gene with potential clinical ramifications.

Among nongenetic changes that can lead to high Nrf2 activity, several competing protein–protein interaction partners have been identified (Fig. 8.3). p62 has been shown to promote Keap1 degradation via autophagy (Copple et al., 2010; Jain et al., 2010; Komatsu et al., 2010; Lau et al., 2010). p62 is a multidomain protein that binds to various signaling proteins and forms protein aggregates in the cell. It associates with ubiquitinated proteins and localizes to autophagosome via binding to LC3 (Bjørkøy et al., 2005). Therefore, p62 is considered to induce autophagosomal degradation of ubiquitinated proteins, damaged organelles, and microbes (Johansen & Lamark, 2011). p62 contains STGE motif that is similar to the ETGE motif in Nrf2. Therefore, p62 can compete with Nrf2 to bind Keap1 and promote Keap1 degradation. Interestingly, p62 is also a direct target of Nrf2 forming a positive feedback loop (Jain et al., 2010).Typically, p62-mediated degradation of Keap1 occurs in autophagy-deficient cells that have excessive p62 levels (Copple et al., 2010; Lau et al., 2010, 2013). It has been also shown that p62 can be phosphorylated which increases its affinity to Keap1 and promotes Keap1 autophagy (Ichimura et al., 2013).

In addition to p62, various Keap1-interacting proteins have been recently discovered to increase nuclear accumulation of Nrf2. The common property of these disruptor proteins is the ETGE motif that can be used to compete with Keap1 binding and therefore suppress Nrf2 degradation and promote nuclear accumulation of Nrf2. In addition to p62, PALB2, DPP3, and WTX have been shown to inhibit Nrf2 ubiquitination (Camp et al., 2012; Hast et al., 2013; Ma et al., 2012).

P21 is a downstream target of p53, and it is induced in response to oxidative stress (O'Reilly, 2005). P21 contains a KRR motif that can bind to both the ETGE and DLG motifs in Nrf2, thus competing with Keap1 (Chen et al., 2009).

Moreover, Nrf2 overactivity has been reported in the hereditary form of type 2 papillary renal cell carcinoma, where mutated fumarate hydratase (FH) was shown to cause succination of Keap1 cysteines (Adam et al., 2011; Ooi et al., 2011) (Fig. 8.3). FH is an enzyme that reversibly catalyzes fumarate to malate in the tricarboxylic acid cycle (TCA cycle). Homozygous loss-of-function mutations in FH decrease FH activity and cause accumulation of fumarate, succination of Keap1 cysteines, and reduction of functional Keap1. Increased Nrf2 activity was considered to promote tumor progression in hereditary papillary renal cell carcinoma (Adam et al., 2011; Ooi et al., 2011).

3.4.2 Imbalance in expression levels

Based on numerous observations in normal cells, Nrf2 activity is mainly regulated posttranslationally without affecting Nrf2 mRNA levels. However, in cancer, also transcriptional regulation of Nrf2 has been shown to promote Nrf2 overactivation. Duplications and deletions of chromosomal segments are typically observed in cancer. These alterations cause abnormal copy numbers for certain genes and can result in cancer promoting effects. TCGA has observed amplification of chromosomal segments containing *NFE2L2*, which is likely to increase the basal expression of Nrf2 and promote constitutive activation (The Cancer Genome Atlas) (Fig. 8.3).

DNA methylation is an epigenetic mechanism of gene regulation occurring in CpG sites. Usually, most human CpG sites are methylated (60–80%), but promoters of the ubiquitously expressed genes are often methylation-free areas, whereas methylation of promoter regions silences gene expression (Smith & Meissner, 2013). *KEAP1* promoter has been found to be hypermethylated in lung (Wang, An, Ji, et al., 2008), prostate (Zhang et al., 2010), brain (Muscarella et al., 2011), colorectal (Hanada et al., 2012), and breast cancers (Barbano et al., 2013), which reduces basal Keap1 expression levels (Fig. 8.3). Consequently, transcription factor binding affinity is reduced, which decreases gene transcription. In addition, at the posttranscriptional level, miR-200a was shown to decrease Keap1 mRNA and protein levels in breast cancer cells resulting in nuclear accumulation of Nrf2 (Eades, Yang, Yao, Zhang, & Zhou, 2011) (Fig. 8.3).

K-Ras, B-Raf, and Myc oncogenes have been shown to elevate Nrf2 and its target gene levels independently of Keap1 via MAP kinases JNK and ERK (DeNicola et al., 2011) (Fig. 8.3). In addition, disruption of PTEN tumor suppressor and mutated constitutively active epithelial growth factor receptor (EGFR) induces Nrf2 gene expression and promotes tumor progression (Mitsuishi et al., 2012; Yamadori et al., 2012). Similarly, NF-κB induction of Nrf2 expression may lead to constitutive activity of Nrf2, as was reported in acute myeloid leukemia (Rushworth et al., 2012). Moreover, Nrf2 hyperactivity has been indicated in several hormone-related cancers, such as prostate (Zhang et al., 2010), endometrial (Jiang et al., 2010), breast (Nioi & Nguyen, 2007), and ovarian cancers (Konstantinopoulos et al., 2011; Liao et al., 2012). In some cases, mutations or epigenetic mechanisms cause the constitutive Nrf2 activity (Nioi & Nguyen, 2007; Zhang et al., 2010). Also, gonadotropins and sex steroids can transcriptionally induce Nrf2 (Gorrini et al., 2013; Liao et al., 2012).

3.5. Consequences of Nrf2 hyperactivity

Typically, during cancer development, cancer cells acquire new characteristics including continuous proliferative signaling, avoidance of antigrowth signals and cell death, induction of angiogenesis, reprogrammed energy metabolism, invasion, and metastasis (Hanahan & Weinberg, 2011). Overactive Nrf2 has been shown to contribute to the progression of several of these cancer hallmarks.

In Nrf2 overexpressing cancers, the elevated expression of cytoprotective genes confers growth and survival advantages (Shibata, Kokubu, et al., 2008; Singh, Bodas, Wakabayashi, Bunz, & Biswal, 2010; Wang, Sun, Villeneuve, et al., 2008). Cancers with elevated Nrf2 activity are more resistant to chemo- and radiotherapy. Moreover, these cancer cells are able to evade apoptosis and proliferate faster due to metabolic reprogramming. In addition, hyperactivation of Nrf2 affects angiogenesis (Fig. 8.4).

3.5.1 Chemo- and radioresistance

Resistance to chemotherapy and/or radiotherapy is the major obstacle in the successful treatment of cancer. There are numerous factors that contribute to cancer cell drug resistance such as increase in drug efflux, decreased drug accumulation due to drug resistance-associated proteins, increased detoxification due to increased cellular thiols, methylation of the promoter regions of tumor suppressor genes, mutations in DNA mismatch repair enzymes, mutations in p53, and avoidance of apoptosis (Wilson, Longley, & Johnston, 2006). Similar

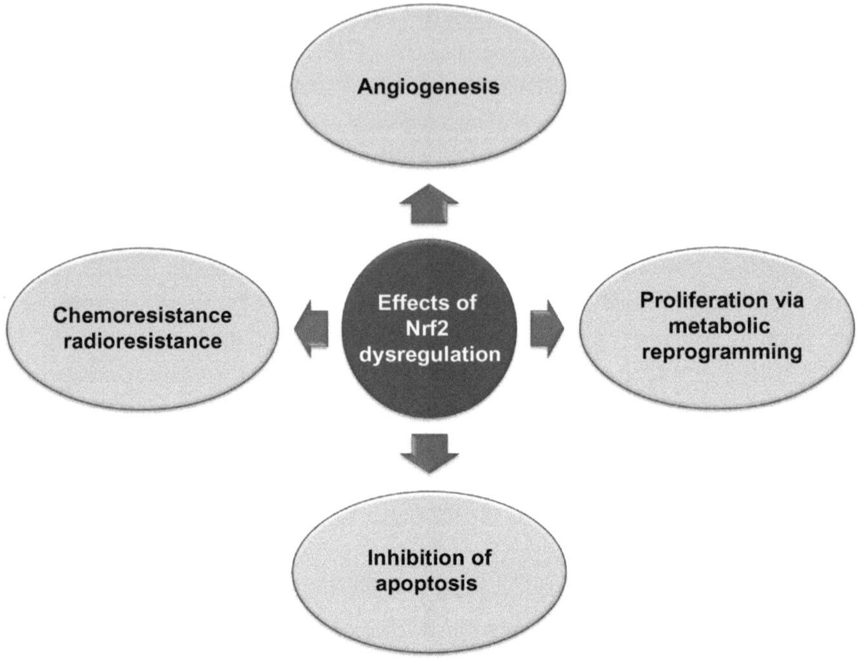

Figure 8.4 Consequences of the constitutive activation of Nrf2 in cancer.

mechanisms also lead to radioresistance (Borek, 2004; Zhivotovsky, Joseph, & Orrenius, 1999). Therefore, finding specific target genes behind these mechanisms would be important for discovering strategies to overcome the resistance. Intriguingly, several genes reported to be the cause of drug- and radioresistance are either direct or indirect Nrf2 target genes or are otherwise functionally linked to Nrf2 (Hayes & McMahon, 2009).

The main mechanism behind the therapeutic effect of radiation therapy is the generation of ROS via ionization of water molecules (Borek, 2004). Similarly, several compounds used in cancer treatment, such as cisplatin (Casares et al., 2012), doxorubicin (Tsang, Chau, Kong, Fung, & Kwok, 2003), carmustine (An et al., 2011), and tamoxifen (Lee, Kang, Lee, & Kim, 2000), function by inducing the formation of ROS leading to cancer cell death. Moreover, both ionizing radiation and many chemotherapeutics are known to induce ARE activity (McDonald et al., 2010; Wang, Hayes, & Wolf, 2006) indicating that Nrf2 is involved also in acquired resistance and may serve as an important nodal point in the regulation of sensitivity to these therapies. Indeed, a common consequence of high constitutive Nrf2 activity is chemo- and radioresistance of cancer cells.

After discovering the upregulation of the Keap1–Nrf2 pathway in cancer, it was soon also noted that constitutive activity of Nrf2 is at least in part responsible for drug- and radioresistance of cancer cells (Shibata, Kokubu, et al., 2008; Solis et al., 2010; Zhang et al., 2010). The activation of Nrf2 results in enhanced chemoresistance of lung carcinoma, breast adenocarcinoma, and neuroblastoma cells to cisplatin, doxorubicin, and etoposide, whereas silencing Nrf2 with siRNA treatment sensitizes cells to chemotherapeutic agents (Wang, Sun, Villeneuve, et al., 2008). Furthermore, constitutive Nrf2 activity in response to alterations in Keap1 expression causes enhanced resistance of gallbladder cancer to 5-fluorouracil (Shibata, Kokubu, et al., 2008). Increased activity of Nrf2 is shown to be associated not only with resistance to radiotherapy and chemotherapeutics, which directly target redox balance of the cells, but also with resistance to drugs that do not function via ROS formation. This mechanism is mainly through the regulation of the drug efflux pump family (ABC transporters) and multidrug resistance-associated protein genes, which currently are known targets of Nrf2 (Ji et al., 2013; Singh, Wu, et al., 2010; Vollrath, Wielandt, Iruretagoyena, & Chianale, 2006).

3.5.2 Inhibition of cancer cell apoptosis

Apoptosis serves as a natural barrier of tumor formation and resisting cell death is one of the hallmarks of cancer (Adams & Cory, 2007; Hanahan & Weinberg, 2011). With respect to Nrf2, two mechanisms by which it can protect against cell death have been proposed: increased expression of antioxidant enzymes and interaction with proteins involved in apoptosis.

During malignant transformation, cells progressively increase the formation of ROS (Gorrini et al., 2013; Trachootham, Alexandre, & Huang, 2009). Given that Nrf2 is the key regulator of cellular antioxidant and cytoprotective genes, the most important function by which Nrf2 is able to cause resistance to cellular death is via upregulation of the antioxidant system, counterbalancing increased ROS. Of the numerous Nrf2 targets that have been shown to affect the sensitivity to ROS-dependent cellular killing, those involved in GSH metabolism are likely to be central (Gorrini et al., 2013; Trachootham et al., 2009). Nrf2 regulates the expression of GCL subunits that are required for synthesis of GSH, the key antioxidant in prevention of damaging effects caused by ROS on cellular macromolecules (Mulcahy, Wartman, Bailey, & Gipp, 1997; Wild, Moinova, & Mulcahy, 1999). Inhibition of GCL activity by buthionine sulfoximine reversed the protective effect of an Nrf2 activator against apoptosis caused by toxic lipid aldehyde 4-hydroxynonenal (Levonen et al., 2001). In addition to traditional antioxidant enzymes, Nrf2 also regulates the expression of ferritin subunits (Chorley et al., 2012), which via sequestration of free Fe^{2+} inhibits its participation in free radical reactions, thus inhibiting Fe^{2+}-mediated toxicity and sensitivity to ROS-mediated chemotherapy (Arosio, Ingrassia, & Cavadini, 2009; Liu et al., 2011).

In addition to direct antioxidant function, Nrf2 may intersect with other pathways affecting apoptosis within the cell. The most important tumor suppressor p53 is activated upon DNA damage causing growth arrest or triggering apoptosis, which is at least partly dependent on the formation of ROS. Notably, p53 is the most frequently mutated or inactivated tumor suppressor found in human cancer (reviewed in Muller & Vousden, 2013). Apoptosis triggered by p53 is dependent on decreased Nrf2 activity (Faraonio et al., 2006), and thus, increased Nrf2 expression interferes with p53-dependent apoptosis leading to cancer cell survival. In addition, when mice with low p53 levels were treated with a genotoxic *N*-nitrosobutyl(4-hydroxybutyl)amine, Nrf2 target gene expression was increased compared to mice with normal p53 levels (Iida et al., 2007). However, it has also been proposed that $p53^{R273H}$ mutant is able to attenuate Nrf2 target gene expression resulting in increased ROS formation (Kalo et al., 2012), casting doubt that p53-dependent repression of Nrf2 signaling plays a major role in its function as a tumor repressor. Conversely, it was recently shown that caveolin-1-dependent repression of Nrf2 signaling activates the p53/p21 pathway, thereby promoting premature senescence and inhibiting malignant transformation (Volonte et al., 2013).

Normally wt-p53 upregulates the expression of its downstream target p21 in response to DNA damage, thus controlling cell growth, whereas mutated p53 leads to decreased p21 expression (Elbendary et al., 1996). In addition, oxidative stress upregulates p21 expression both p53 dependently and p53 independently (O'Reilly, 2005). Interestingly, p21 is able to compete with Keap1 and thus stabilizes Nrf2 by direct binding to Neh2 domain leading to upregulated Nrf2 target gene expression under both basal and induced conditions (Chen et al., 2009).

3.5.3 Cancer cell proliferation through metabolic reprogramming

Cell proliferation is required in embryogenesis, as well as in growth and proper function of several adult tissues, whereas aberrant cell proliferation is another hallmark of cancer (Hanahan & Weinberg, 2011). In proliferating cells, the need of energy and building blocks, such as amino acids, fatty acids, and monosaccharides, for generation of macromolecules increases. In aerobic conditions, most of the glucose taken up by the quiescent cells is converted to pyruvate in glycolysis, and TCA cycle, together with oxidative phosphorylation, is the major route of ATP formation. In contrast, in anaerobic conditions, glycolysis is favored, majority of the resulting pyruvate is converted to lactate, and only a small amount of pyruvate enters TCA cycle. It was observed already in the 1920s by Otto Warburg that cancer cells tend to reprogram their metabolism and thus consume considerably higher amount of glucose compared to quiescent cells. In addition, in cancer cells even in the presence of oxygen, most of the glucose is converted to lactate in aerobic glycolysis instead of oxidizing it completely in oxidative phosphorylation. This phenomenon is also known as "Warburg effect" (Hanahan & Weinberg, 2011; Vander Heiden, Cantley, & Thompson, 2009; Ward & Thompson, 2012). Alternative to glycolysis, glucose can be metabolized via pentose phosphate pathway (PPP) in which ribose 5-phosphate, essential for proliferating cells, is generated for purine synthesis, and erythrose 4-phosphate is produced for the use in the synthesis of aromatic amino acids. In addition, NADPH, an important cofactor of anabolism (i.e., the biosynthesis of cellular macromolecules), is formed (Buchakjian & Kornbluth, 2010; Kruger & von Schaewen, 2003).

It was recently observed that several activated oncogenes and mutated tumor suppressor genes, such as K-Ras, Myc, and p53, directly promote metabolism and nutrient uptake, resulting in similar features as in proliferating cells (Ward & Thompson, 2012). Oncogenes are able to increase the basal Nrf2 expression, activate Nrf2-mediated antioxidant response, and lower the cellular ROS levels in primary murine cells (DeNicola et al., 2011). In addition, $Nrf2^{-/-}$ cells are significantly less proliferative. Therefore, Nrf2 promotes oncogene-initiated carcinogenesis and proliferation (DeNicola et al., 2011).

It was shown later that in cancer cells, Nrf2 promotes the generation of building blocks for macromolecule synthesis and cancer cell proliferation by either directly or indirectly regulating genes involved in GSH biosynthesis and recycling, NADPH production, and PPP activation (Mitsuishi et al., 2012; Singh et al., 2013). The expression of glucose-6-phosphate dehydrogenase (*G6PD*), transketolase (*TKT*), transaldolase (*TALDO1*), phosphoribosylpyrophosphate amidotransferase (*PPAT*), methylenetetrahydrofolate dehydrogenase 2 (*MTHFD2*), isocitrate dehydrogenase (*IDH1*), and malic enzyme 1 (*ME1*) was decreased when Nrf2 overexpressing A549 lung cancer cells were treated with siNrf2 (Mitsuishi et al., 2012). Moreover, all other genes except *PPAT* and *MTHFFD2* were shown to exhibit well-conserved AREs in the regulatory regions, therefore being direct Nrf2 targets. Similar results were also published by another group; however, they only found a strong Nrf2 binding site in *TALDO1* gene and hence suggested an indirect mechanism of regulation of PPP genes by Nrf2. The authors sought for miRNAs and found that Nrf2 is able to attenuate miR-1 and miR-206 expression, which in turn leads to increased expression of PPP genes (Singh et al., 2013). Therefore, the effect of Nrf2 on glucose metabolism was proposed at least partially to be mediated through miR-1 and miR-206, both of which have previously been associated with cancer (Hudson et al., 2012; Nasser et al., 2008; Singh et al., 2013; Wang, Ling, Bai, & Zhao, 2011). In addition, a positive feedback loop was shown to exist between the PI3K–Akt pathway, most commonly mutated signaling cascades in cancer, and the Keap1–Nrf2 pathway (Buzzai et al., 2005; Elstrom et al., 2004; Mitsuishi et al., 2012).

Moreover, in its reduced form, GSH is important for cell proliferation. Nrf2 regulates both genes involved in GSH synthesis and PPP genes needed for the formation of NADPH, the main reducing agent of GSH (Lu, 2009). Nrf2 is also able to regulate genes involved in metabolism of fatty acids and other lipids (Kitteringham et al., 2010).

3.5.4 The role of Nrf2 in angiogenesis

Yet another hallmark of cancer is the induction of angiogenesis (Hanahan & Weinberg, 2011). Similar to normal cells, blood flow is needed in cancer cells for nutrient and oxygen supply and for depleting metabolic wastes and carbon dioxide. In adults, angiogenesis occurs usually only during certain physiological processes, such as wound healing, and is otherwise turned off. However, during cancer progression, angiogenesis is almost always permanently active and is induced already in the early phase of cancer development (Raica, Cimpean, & Ribatti, 2009). Several factors contribute to the regulation of angiogenesis either directly or indirectly. One of the most prototypic proangiogenic factors is vascular endothelial growth factor A (VEGF-A), which regulates both physiological and pathological angiogenesis (Cao, Baeriswyl, & Christofori, 2009).

Nrf2 and its target genes have been shown to be involved in the regulation of cancer cell angiogenesis. The expression of Nrf2 target gene, *HMOX1*, is involved in the angiogenesis of glioma, pancreatic cancer, and melanoma (Nishie et al., 1999; Sunamura et al., 2003; Torisu-Itakura,

Furue, Kuwano, & Ono, 2000), and the product of heme oxygenase reaction, carbon monoxide (CO), increases VEGF-A expression via HIF-1α (Choi et al., 2010). Furthermore, stable knockdown of Nrf2 using lentiviral RNAi technology inhibits HIF-1α–VEGF signaling likely via reduced mitochondrial O_2 consumption and suppresses cancer growth and angiogenesis in a mouse xenograft model of colon cancer (Kim et al., 2011). Recently, Nrf2 has been shown to promote sprouting angiogenesis during vascular development (Wei et al., 2013), but whether this has a role in tumor angiogenesis remains to be determined. Nevertheless, there is an emerging link between Nrf2 and tumor angiogenesis that may add to the list of adverse effects of Nrf2 hyperactivity in cancer.

4. APPLICATIONS OF THE KEAP1–NRF2 SYSTEM IN CANCER TREATMENT

4.1. Nrf2 as a prognostic marker of cancer

Inasmuch as patients with overactive Nrf2 usually have chemo- and radioresistance and poor prognosis, Nrf2 expression could be used as a novel biomarker for personalized molecular therapies. In addition to analyzing mutations of *NFE2L2*, *KEAP1*, or other factors involved in the signaling pathway, other markers, such as *KEAP1* promoter methylation, Nrf2 protein, or the expression of its target genes, could be analyzed, as the mechanisms of Nrf2 overactivity are diverse. However, while the association of high Nrf2 activity with poor clinical outcome in certain cancers, such as NSCLC, is well established (Hu et al., 2012; Yang et al., 2011), opposite associations have been reported. In malignant glioma, *KEAP1* promoter methylation resulting in decreased Keap1 mRNA expression is associated with a better prognosis (Muscarella et al., 2011), and in acute myelogenous leukemia patients, high Nrf2 mRNA measured from peripheral blood mononuclear cells was associated with longer survival (Garcia-Manero et al., 2012). Given the dichotomous role of Nrf2 and the multitude of mechanisms of Nrf2 dysregulation in cancer, a more thorough characterization of potential markers with larger patient groups is warranted.

In addition to Nrf2 expression status, other prognostic factors related to Nrf2 are small nucleotide polymorphisms (SNPs) associated with the *NFE2L2* gene. So far, several SNPs have been found in the promoter region of the gene and are suggested to serve as indicators of risk for developing certain diseases. These SNPs include rs6721961C/A (−617C/A), rs35652124A/G (−653A/G) (Marzec et al., 2007), −686A/G, −684G/A,

−650C/A (Yamamoto et al., 2004), and rs6721961 T/T and rs2706110 A/A (Hartikainen et al., 2012). In addition, one triplet repeat polymorphism has been found in the regulatory region of *NFE2L2* (Yamamoto et al., 2004). People carrying SNPs −686/−684 A/G are associated with a reduced risk for gastric malignancies (Arisawa et al., 2007). SNP −617 (rs6721961) is important predictor of various diseases, as SNP −617A is associated with the increased risk of acute lung injury in response to major trauma or sepsis (Marzec et al., 2007) as well as the development of lung adenocarcinoma in female nonsmokers (Okano et al., 2013), and rare homozygous SNPs −617A/A and T/T are associated with higher risk of breast and lung cancers, respectively (Hartikainen et al., 2012; Suzuki et al., 2013). In addition, lung adenocarcinoma patients with homozygous −617A/A allele together with wt-MDM2 (mouse double minute 2 homolog also known as E3 ubiquitin protein ligase) displayed exceptional survival for over 1700 days after surgery (Okano et al., 2013). Moreover, rs2706110 AA is associated with poor survival of breast cancer (Hartikainen et al., 2012).

4.2. Enhancing chemo- and radiotherapy with small molecular compounds

Consistent with the notion that Nrf2 upregulation contributes to chemo- and radioresistance of cancer cells, inhibition of Nrf2 enhances the effect of chemotherapy and/or radiotherapy (Lee, Lim, et al., 2012; Manandhar, Lee, & Kwak, 2010; Ren et al., 2011; Singh et al., 2008; Tang et al., 2011). Inhibition of Nrf2 reduces the expression of antioxidant and xenobiotic enzymes resulting oxidative damage of cancer cells and sensitizing them to apoptosis. Moreover, as Nrf2 target genes include multidrug resistance proteins and drug efflux proteins, inhibition of Nrf2 also increases sensitivity to ROS-independent chemotherapeutic agents (Taguchi, Motohashi, & Yamamoto, 2011).

So far, the most commonly used method for Nrf2 inhibition is RNA interference (RNAi) (Fig. 8.5A). Because Nrf2-dependent response to chemotherapy and radiotherapy is not limited to certain cancer types or to a specific chemotherapeutic agent, Nrf2 knockdown could be used as a general approach for sensitization to therapy (Homma et al., 2009; Manandhar et al., 2010; Singh, Bodas, et al., 2010; Singh et al., 2008). However, although RNAi has enormous potential in cancer therapy, its transfer to the clinic is challenging because of the obstacles related to the delivery issues (Pecot, Calin, Coleman, Lopez-Berestein, & Sood, 2011).

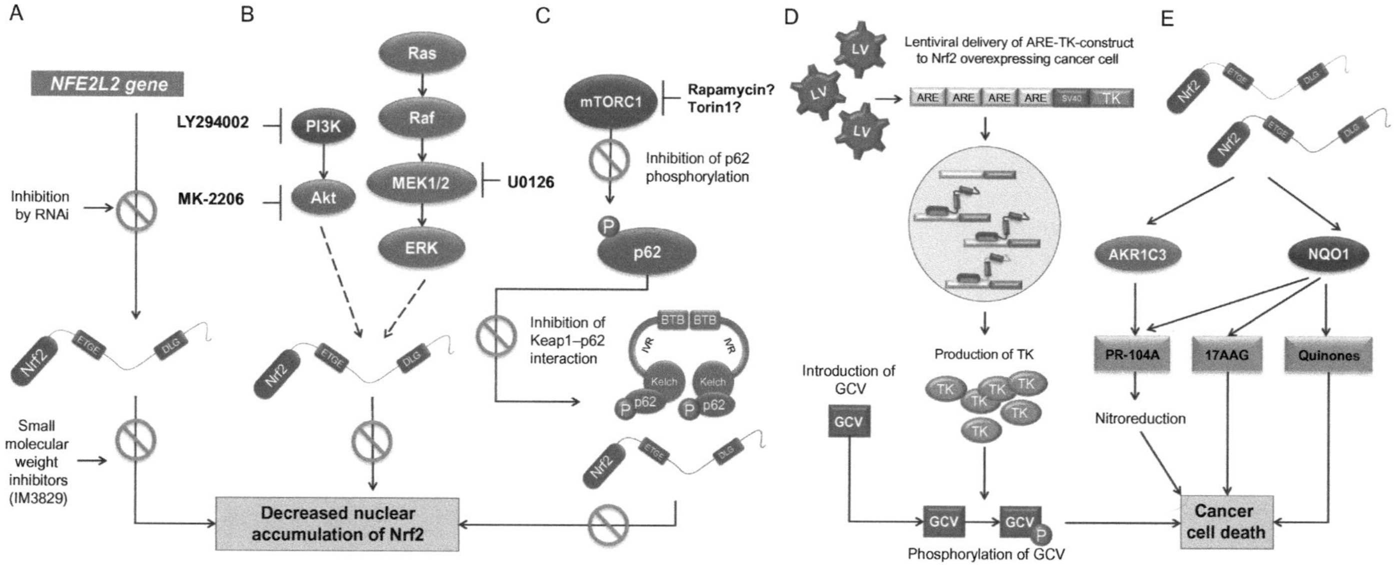

Figure 8.5 *Therapeutic targeting of Nrf2 overexpressing cancers*. Proposed approaches to target Nrf2 overexpressing cancer cells include (A) direct inhibition of Nrf2 by RNAi or small molecular weight inhibitors; (B) inhibition of upstream kinases, such as MEK1/2 and PI3K; (C) inhibition of members of autophagy pathway, such as mTORC1; (D) harnessing overactive Nrf2 for suicide gene therapy utilizing Nrf2-driven gene therapy vectors; and (E) using bioreductive prodrugs in Nrf2 overexpressing cancers. (See the color plate.)

Another way to inhibit Nrf2 expression is to use low-molecular-weight inhibitors (Fig. 8.5A). In the study of Lee et al., altogether 8000 synthetic compounds were screened to identify small molecular inhibitors of antioxidant responses, which increase the rate of cancer cell apoptosis induced by radiotherapy (Lee, Lim, et al., 2012). IM3829 (4-(2-cyclohexylethoxy)aniline) was found to inhibit Nrf2 expression and nuclear translocation, and when combined with radiotherapy, it significantly inhibited the survival of H1299, A549, and H460 lung cancer cells both *in vitro* and *in vivo* in xenograft-bearing mice (Lee, Lim, et al., 2012). In addition, brusatol, a quasinoid compound purified from the *Brucea javanica* shrub plant extract, and luteolin, a dietary flavonoid, have shown to inhibit Nrf2 activity, and both of them have been used to enhance the efficacy of A549 lung cancer cell chemotherapy (Ren et al., 2011; Tang et al., 2011). However, because of the structural similarity of other members of the CNC protein family (such as NF-E2 p45, Nrf1, and Bach1; Motohashi, O'Connor, Katsuoka, Engel, & Yamamoto, 2002) and thus risk of off-target toxic effects, the development of specific Nrf2 inhibitors has been challenging.

Certain small molecular compounds have been shown to have different effects on Nrf2 in malignant and nonmalignant cells. Ginsenoside Rg3 enhanced the sensitivity of cancer cells to cisplatin by reducing Nrf2-mediated HMOX1 and NQO1 activity in cancer cells with high Nrf2 activity, but protecting normal cells from cytotoxic effects of cisplatin (Lee, Park, Chung, & Chung, 2012). Similarly, parthenolide was demonstrated to contribute to radiosensitization of prostate cancer cells by increasing the cellular ROS levels and causing thioredoxin oxidation, leading to reduction of Keap1 thiols and degradation of PGAM5 and Bcl-XL and finally to cancer cell apoptosis. In healthy cells, parthenolide had an opposite effect; by increasing Keap1 oxidation and Nrf2-dependent target gene expression, parthenolide treatment resulted in cellular protection (Xu et al., 2013).

In addition to direct targeting of Nrf2, kinase inhibitors may be used to inhibit Nrf2 as several protein kinases, such as PI3K, PKC, and MAPK, are known to regulate Nrf2 (Hayes & McMahon, 2009) (Fig. 8.5B). Specifically, MEK1/2 inhibitor U0126 inhibited Nrf2 activation by EGFR in lung cancer cells, and PI3K–Akt inhibitor LY294002 and PKB/Akt inhibitor MK-2206 reduced Nrf2 protein and target gene induction and increased the sensitivity of Nrf2 overexpressing A549 cells to acrolein, chlorambucil, and cisplatin (Chowdhry et al., 2013).PKB/Akt inhibitor MK-2206 has also been used in combination with the thioredoxin reductase inhibitor auranofin in non-small-cell lung carcinoma (Dai et al., 2013). In addition, selective PI3K inhibitor NVP-BKM120 decreased Nrf2 protein and sensitized radioresistant Nrf2 overexpressing squamous cell lung carcinoma cell lines to radiation therapy (Abazeed et al., 2013). Inasmuch as both MAPK/ERK pathway inhibitors (Roberts & Der, 2007) and PI3K/Akt pathway (Courtney, Corcoran, & Engelman, 2010) inhibitors are under active clinical development for the treatment of various cancers, these approaches may provide a viable option for clinical use.

In addition to upstream kinase inhibitors affecting Nrf2 activation as p62, the autophagy pathway that intersects with the Keap1–Nrf2 pathway via p62-dependent degradation of Keap1 could be a target of therapy (Fig. 8.5C). Specifically, the recently identified S351-phosphorylation site in p62 that increases its affinity to Keap1 is a potential target (Ichimura et al., 2013). However, while mTORC1 was shown to be upstream of pS351 phosphorylation caused by arsenic in primary hepatocytes assessed by using mTOR kinase inhibitor rapamycin, mTOR kinase inhibitors were ineffective in hepatocellular carcinoma cells having high levels of phosphorylated p62, indicating that other kinase pathways are involved (Ichimura et al., 2013). Clearly, a better understanding of the role of autophagy in the regulation and dysregulation of the Keap1–Nrf2 pathway is warranted in order to identify potential targets for therapy.

4.3. Taking advantage of high Nrf2 activity: Nrf2-regulated lentiviral suicide gene therapy and the use of bioreductive prodrugs

Opposite to Nrf2 inhibition, another application of the use Keap1–Nrf2 system in cancer treatment is to exploit increased Nrf2 activity for targeting gene therapy to cancer cells (Fig. 8.5D). We have previously designed a lentiviral gene therapy vector that responds to Nrf2 via multiple AREs derived from human *GCLM* gene cloned to a vector backbone with minimal SV40 promoter (Hurttila et al., 2008). This vector was utilized for suicide cancer gene therapy by cloning of herpes simplex virus thymidine kinase (HSV-TK) gene under the control of AREs (Leinonen et al., 2012). In this approach, a prodrug ganciclovir (GCV) is converted to its toxic phosphorylated form (Moolten, 1986). We have demonstrated that ARE-regulated suicide gene therapy based on HSV-TK and GCV is effective in A549 lung cancer cells both *in vitro* and *in vivo* in a mouse xenograft model, providing a proof-of-principle that this approach is viable in Nrf2 overexpressing cancers (Leinonen et al., 2012).

Cytochrome P450 is a key enzyme in a phase I-dependent xenobiotic metabolism. It is expressed in several tumors, thus increasing drug resistance. High CYP450 expression has been utilized in cancer therapy because it is able to catalyze the activation of anticancer agents such as cyclophosphamide and ifosfamide (Rodriguez-Antona & Ingelman-Sundberg, 2006). A similar approach was employed in tumors with high Keap1–Nrf2 pathway activity by using bioreductive prodrug, dinitrobenzamide mustard PR-104A, which, in response to oxygenated conditions, is converted into cytotoxic metabolites by nitroreduction (Guise et al., 2010; Patterson et al., 2007) (Fig. 8.5E). In clinical trials, PR-104A has already been used to treat patients with solid tumors and refractory/relapsed acute leukemia (http://www.cancer.gov). In the study by Guise and colleagues, it was reported that Keap1 knockdown in MDA-MB-231 cells results in induced *AKR1C3* expression and elevated PR104A metabolism, suggesting that the metabolism and cytotoxic effects of PR-104A are at least partially dependent on Nrf2 expression levels (Guise et al., 2010). Correspondingly, siNrf2 treatment of Nrf2 overexpressing A549 lung cancer cells suppressed PR-104A metabolism. Another, well-characterized Nrf2 target gene, *NQO1*, is highly expressed in many solid tumors and able to reduce several quinone derivatives to their toxic forms that either alkylate DNA or generate ROS via redox cycling (Parkinson, Bair, Cismesia, & Hergenrother, 2013). The HSP90 inhibitor 17-allylamino-17-demethoxygeldanamycin (17-AAG), which is in clinical development for breast cancer and NSCLC (Neckers & Workman, 2012), is metablized to its active form by NQO1 (Guo et al., 2005) (Fig. 8.5E).

5. CONCLUDING REMARKS AND FUTURE PERSPECTIVES

Within the past few years, it has become evident that Nrf2 not only plays a role in prevention of malignant transformation in healthy tissues but also is important for protecting cancer cells from chemotherapy and radiotherapy and facilitating cancer growth via, for example, metabolic reprogramming. Just as large-scale genomics endeavors like the TCGA project have revolutionized cancer research, they have provided important insights into the dysregulation of the Keap1–Nrf2 pathway in cancer. As the regulation of Nrf2 occurs largely at the posttranslational level, large-scale proteomics analyses likely significantly add to the current understanding of the role of Keap1–Nrf2 system in cancer and the pathway is now acknowledged as one of the drivers of tumorigenesis (Vogelstein et al., 2013). In addition, the development of bioinformatics analyses is crucially important for distilling meaningful insights from unbiased omics data. The notion that intelligent data analysis can yield unexpected connections between previously unrelated phenomena is beautifully exemplified in a paper by Ooi et al. (2011), in which the observation that the gene expression profile of FH-deficient cells largely overlapped with Nrf2 activation signature led to the identification of a novel Nrf2 dysregulation mechanism in papillary renal cell carcinoma.

With respect to targeting overactive Nrf2 in cancer therapy, it is important not only to focus on direct Nrf2 inhibition but also to explore indirect methods for inhibition such as inhibition of upstream protein kinases. In addition, harnessing Nrf2 overactivity for therapeutic use by Nrf2-regulated gene therapy or bioreductive prodrugs is an option, as is targeting key Nrf2 regulated genes such as GSH synthetic enzymes.

ACKNOWLEDGMENTS

This work was supported by the Academy of Finland, the Sigrid Juselius Foundation, Finnish Cultural Organization, Emil Aaltonen Foundation, and the Finnish Cancer Organisations.

REFERENCES

Abazeed, M. E., Adams, D. J., Hurov, K. E., Tamayo, P., Creighton, C. J., Sonkin, D., et al. (2013). Integrative radiogenomic profiling of squamous cell lung cancer. *Cancer Research, 73*(20), 6289–6298.

Adam, J., Hatipoglu, E., O'Flaherty, L., Ternette, N., Sahgal, N., Lockstone, H., et al. (2011). Renal cyst formation in Fh1-deficient mice is independent of the Hif/Phd pathway: Roles for fumarate in KEAP1 succination and Nrf2 signaling. *Cancer Cell, 20*(4), 524–537.

Adams, J. M., & Cory, S. (2007). The Bcl-2 apoptotic switch in cancer development and therapy. *Oncogene, 26*(9), 1324–1337.

An, J. M., Kim, S. S., Rhie, J. H., Shin, D. M., Seo, S. R., & Seo, J. T. (2011). Carmustine induces ERK- and JNK-dependent cell death of neuronally-differentiated PC12 cells via generation of reactive oxygen species. *Toxicology In Vitro: An International Journal Published in Association with BIBRA, 25*(7), 1359–1365.

Aoki, Y., Sato, H., Nishimura, N., Takahashi, S., Itoh, K., & Yamamoto, M. (2001). Accelerated DNA adduct formation in the lung of the Nrf2 knockout mouse exposed to diesel exhaust. *Toxicology and Applied Pharmacology, 173*(3), 154–160.

Arisawa, T., Tahara, T., Shibata, T., Nagasaka, M., Nakamura, M., Kamiya, Y., et al. (2007). The relationship between Helicobacter pylori infection and promoter polymorphism of the Nrf2 gene in chronic gastritis. *International Journal of Molecular Medicine, 19*(1), 143–148.

Arosio, P., Ingrassia, R., & Cavadini, P. (2009). Ferritins: A family of molecules for iron storage, antioxidation and more. *Biochimica et Biophysica Acta, 1790*(7), 589–599.

Barbano, R., Muscarella, L. A., Pasculli, B., Valori, V. M., Fontana, A., Coco, M., et al. (2013). Aberrant Keap1 methylation in breast cancer and association with clinicopathological features. *Epigenetics: Official Journal of the DNA Methylation Society, 8*(1), 105–112.

Birch-Machin, M. A. (2006). The role of mitochondria in ageing and carcinogenesis. *Clinical and Experimental Dermatology, 31*(4), 548–552.

Bjørkøy, G., Lamark, T., Brech, A., Outzen, H., Perander, M., Overvatn, A., et al. (2005). p62/SQSTM1 forms protein aggregates degraded by autophagy and has a protective effect on huntingtin-induced cell death. *The Journal of Cell Biology, 171*(4), 603–614.

Bloom, D. A., & Jaiswal, A. K. (2003). Phosphorylation of Nrf2 at Ser40 by protein kinase C in response to antioxidants leads to the release of Nrf2 from INrf2, but is not required for Nrf2 stabilization/accumulation in the nucleus and transcriptional activation of antioxidant response element. *The Journal of Biological Chemistry, 278*(45), 44675–44682.

Borek, C. (2004). Antioxidants and radiation therapy. *The Journal of Nutrition, 134*(11), 3207S–3209S.

Buchakjian, M. R., & Kornbluth, S. (2010). The engine driving the ship: Metabolic steering of cell proliferation and death. *Nature Reviews. Molecular Cell Biology, 11*(10), 715–727.

Buzzai, M., Bauer, D. E., Jones, R. G., Deberardinis, R. J., Hatzivassiliou, G., Elstrom, R. L., et al. (2005). The glucose dependence of Akt-transformed cells can be reversed by pharmacologic activation of fatty acid beta-oxidation. *Oncogene, 24*(26), 4165–4173.

Camp, N. D., James, R. G., Dawson, D. W., Yan, F., Davison, J. M., Houck, S. A., et al. (2012). Wilms tumor gene on X chromosome (WTX) inhibits degradation of NRF2 protein through competitive binding to KEAP1 protein. *The Journal of Biological Chemistry, 287*(9), 6539–6550.

Cancer Genome Atlas Research Network. (2012). Comprehensive genomic characterization of squamous cell lung cancers. *Nature, 489*(7417), 519–525.

Cao, Y., Baeriswyl, V., & Christofori, G. (2009). The angiogenic switch in carcinogenesis. *Seminars in Cancer Biology, 19*(5), 329–337.

Casares, C., Ramírez-Camacho, R., Trinidad, A., Roldán, A., Jorge, E., & García-Berrocal, J. R. (2012). Reactive oxygen species in apoptosis induced by cisplatin: Review of physiopathological mechanisms in animal models. *European Archives of Oto-Rhino-Laryngology: Official Journal of the European Federation of Oto-Rhino-Laryngological Societies (EUFOS): Affiliated with the German Society for Oto-Rhino-Laryngology—Head and Neck Surgery, 269*(12), 2455–2459.

Cerda, S., & Weitzman, S. A. (1997). Influence of oxygen radical injury on DNA methylation. *Mutation Research, 386*(2), 141–152.

Chen, W., Sun, Z., Wang, X.-J., Jiang, T., Huang, Z., Fang, D., et al. (2009). Direct interaction between Nrf2 and p21(Cip1/WAF1) upregulates the Nrf2-mediated antioxidant response. *Molecular Cell, 34*(6), 663–673.

Cho, H.-Y., Reddy, S. P., & Kleeberger, S. R. (2006). Nrf2 defends the lung from oxidative stress. *Antioxidants & Redox Signaling, 8*(1–2), 76–87.

Choi, Y. K., Kim, C.-K., Lee, H., Jeoung, D., Ha, K.-S., Kwon, Y.-G., et al. (2010). Carbon monoxide promotes VEGF expression by increasing HIF-1alpha protein level via two distinct mechanisms, translational activation and stabilization of HIF-1alpha protein. *The Journal of Biological Chemistry, 285*(42), 32116–32125.

Chorley, B. N., Campbell, M. R., Wang, X., Karaca, M., Sambandan, D., Bangura, F., et al. (2012). Identification of novel NRF2-regulated genes by ChIP-Seq: Influence on retinoid X receptor alpha. *Nucleic Acids Research, 40*(15), 7416–7429.

Chowdhry, S., Zhang, Y., McMahon, M., Sutherland, C., Cuadrado, A., & Hayes, J. D. (2013). Nrf2 is controlled by two distinct β-TrCP recognition motifs in its Neh6 domain, one of which can be modulated by GSK-3 activity. *Oncogene, 32*(32), 3765–3781.

Copple, I. M., Lister, A., Obeng, A. D., Kitteringham, N. R., Jenkins, R. E., Layfield, R., et al. (2010). Physical and functional interaction of sequestosome 1 with Keap1 regulates the Keap1-Nrf2 cell defense pathway. *The Journal of Biological Chemistry, 285*(22), 16782–16788.

Counts, J. L., & Goodman, J. I. (1995). Hypomethylation of DNA: A nongenotoxic mechanism involved in tumor promotion. *Toxicology Letters, 82–83*, 663–672.

Courtney, K. D., Corcoran, R. B., & Engelman, J. A. (2010). The PI3K pathway as drug target in human cancer. *Journal of Clinical Oncology: Official Journal of the American Society of Clinical Oncology, 28*(6), 1075–1083.

Cullinan, S. B., Zhang, D., Hannink, M., Arvisais, E., Kaufman, R. J., & Diehl, J. A. (2003). Nrf2 is a direct PERK substrate and effector of PERK-dependent cell survival. *Molecular and Cellular Biology, 23*(20), 7198–7209.

Dai, B., Yoo, S.-Y., Bartholomeusz, G., Graham, R. A., Majidi, M., Yan, S., et al. (2013). KEAP1-dependent synthetic lethality induced by AKT and TXNRD1 inhibitors in lung cancer. *Cancer Research, 73*, 5532–5543.

Dalle-Donne, I., Scaloni, A., Giustarini, D., Cavarra, E., Tell, G., Lungarella, G., et al. (2005). Proteins as biomarkers of oxidative/nitrosative stress in diseases: The contribution of redox proteomics. *Mass Spectrometry Reviews, 24*(1), 55–99.

DeNicola, G. M., Karreth, F. A., Humpton, T. J., Gopinathan, A., Wei, C., Frese, K., et al. (2011). Oncogene-induced Nrf2 transcription promotes ROS detoxification and tumorigenesis. *Nature, 475*(7354), 106–109.

De Zeeuw, D., Akizawa, T., Audhya, P., Bakris, G. L., Chin, M., Christ-Schmidt, H., et al. (2013). Bardoxolone methyl in type 2 diabetes and stage 4 chronic kidney disease. *New England Journal of Medicine, 369*, 2492–2503.

Digaleh, H., Kiaei, M., & Khodagholi, F. (2013). Nrf2 and Nrf1 signaling and ER stress crosstalk: Implication for proteasomal degradation and autophagy. *Cellular and Molecular Life Sciences, 70*(24), 4681–4694.

Dinkova-Kostova, A. T., Holtzclaw, W. D., Cole, R. N., Itoh, K., Wakabayashi, N., Katoh, Y., et al. (2002). Direct evidence that sulfhydryl groups of Keap1 are the sensors regulating induction of phase 2 enzymes that protect against carcinogens and oxidants. *Proceedings of the National Academy of Sciences of the United States of America, 99*(18), 11908–11913.

Dizdaroglu, M., Jaruga, P., Birincioglu, M., & Rodriguez, H. (2002). Free radical-induced damage to DNA: Mechanisms and measurement. *Free Radical Biology and Medicine, 32*(11), 1102–1115.

Eades, G., Yang, M., Yao, Y., Zhang, Y., & Zhou, Q. (2011). miR-200a regulates Nrf2 activation by targeting Keap1 mRNA in breast cancer cells. *The Journal of Biological Chemistry, 286*(47), 40725–40733.

Elbendary, A. A., Cirisano, F. D., Evans, A. C., Davis, P. L., Iglehart, J. D., Marks, J. R., et al. (1996). Relationship between p21 expression and mutation of the p53 tumor suppressor gene in normal and malignant ovarian epithelial cells. *Clinical Cancer Research: An Official Journal of the American Association for Cancer Research, 2*(9), 1571–1575.

Elstrom, R. L., Bauer, D. E., Buzzai, M., Karnauskas, R., Harris, M. H., Plas, D. R., et al. (2004). Akt stimulates aerobic glycolysis in cancer cells. *Cancer Research, 64*(11), 3892–3899.

Enomoto, A. (2001). High sensitivity of Nrf2 knockout mice to acetaminophen hepatotoxicity associated with decreased expression of ARE-regulated drug metabolizing enzymes and antioxidant genes. *Toxicological Sciences, 59*(1), 169–177.

Fahey, J. W., Haristoy, X., Dolan, P. M., Kensler, T. W., Scholtus, I., Stephenson, K. K., et al. (2002). Sulforaphane inhibits extracellular, intracellular, and antibiotic-resistant strains of Helicobacter pylori and prevents benzo[a]pyrene-induced stomach tumors. *Proceedings of the National Academy of Sciences of the United States of America, 99*(11), 7610–7615.

Faraonio, R., Vergara, P., Di Marzo, D., Pierantoni, M. G., Napolitano, M., Russo, T., et al. (2006). p53 suppresses the Nrf2-dependent transcription of antioxidant response genes. *The Journal of Biological Chemistry, 281*(52), 39776–39784.

Feng, Z., Hu, W., & Tang, M.-S. (2004). Trans-4-hydroxy-2-nonenal inhibits nucleotide excision repair in human cells: A possible mechanism for lipid peroxidation-induced carcinogenesis. *Proceedings of the National Academy of Sciences of the United States of America, 101*(23), 8598–8602.

Galloway, D. C., Blake, D. G., Shepherd, A. G., & McLellan, L. I. (1997). Regulation of human gamma-glutamylcysteine synthetase: Co-ordinate induction of the catalytic and regulatory subunits in HepG2 cells. *The Biochemical Journal, 328*(Pt 1), 99–104.

Garcia-Manero, G., Tambaro, F. P., Bekele, N. B., Yang, H., Ravandi, F., Jabbour, E., et al. (2012). Phase II trial of vorinostat with idarubicin and cytarabine for patients with newly diagnosed acute myelogenous leukemia or myelodysplastic syndrome. *Journal of Clinical Oncology: Official Journal of the American Society of Clinical Oncology, 30*(18), 2204–2210.

Genschik, P., Sumara, I., & Lechner, E. (2013). The emerging family of CULLIN3-RING ubiquitin ligases (CRL3s): Cellular functions and disease implications. *The EMBO Journal, 32*(17), 2307–2320.

Glintborg, B., Weimann, A., Kensler, T. W., & Poulsen, H. E. (2006). Oltipraz chemoprevention trial in Qidong, People's Republic of China: Unaltered oxidative biomarkers. *Free Radical Biology & Medicine, 41*(6), 1010–1014.

Gorrini, C., Baniasadi, P. S., Harris, I. S., Silvester, J., Inoue, S., Snow, B., et al. (2013). BRCA1 interacts with Nrf2 to regulate antioxidant signaling and cell survival. *The Journal of Experimental Medicine, 210*(8), 1529–1544.

Guise, C. P., Abbattista, M. R., Singleton, R. S., Holford, S. D., Connolly, J., Dachs, G. U., et al. (2010). The bioreductive prodrug PR-104A is activated under aerobic conditions by human aldo-keto reductase 1C3. *Cancer Research, 70*(4), 1573–1584.

Guo, W., Reigan, P., Siegel, D., Zirrolli, J., Gustafson, D., & Ross, D. (2005). Formation of 17-allylamino-demethoxygeldanamycin (17-AAG) hydroquinone by NAD(P)H:quinone oxidoreductase 1: Role of 17-AAG hydroquinone in heat shock protein 90 inhibition. *Cancer Research, 65*(21), 10006–10015.

Hanada, N., Takahata, T., Zhou, Q., Ye, X., Sun, R., Itoh, J., et al. (2012). Methylation of the KEAP1 gene promoter region in human colorectal cancer. *BMC Cancer, 12*, 66.

Hanahan, D., & Weinberg, R. A. (2011). Hallmarks of cancer: The next generation. *Cell, 144*(5), 646–674.

Harris, C. C., & Hollstein, M. (1993). Clinical implications of the p53 tumor-suppressor gene. *The New England Journal of Medicine, 329*(18), 1318–1327.

Hartikainen, J. M., Tengström, M., Kosma, V.-M., Kinnula, V. L., Mannermaa, A., & Soini, Y. (2012). Genetic polymorphisms and protein expression of NRF2 and Sulfiredoxin predict survival outcomes in breast cancer. *Cancer Research, 72*(21), 5537–5546.

Hast, B. E., Goldfarb, D., Mulvaney, K. M., Hast, M. A., Siesser, P. F., Yan, F., et al. (2013). Proteomic analysis of ubiquitin ligase KEAP1 reveals associated proteins that inhibit NRF2 ubiquitination. *Cancer Research, 73*(7), 2199–2210.

Hayes, J. D., Chanas, S. A., Henderson, C. J., McMahon, M., Sun, C., Moffat, G. J., et al. (2000). The Nrf2 transcription factor contributes both to the basal expression of glutathione S-transferases in mouse liver and to their induction by the chemopreventive synthetic antioxidants, butylated hydroxyanisole and ethoxyquin. *Biochemical Society Transactions, 28*(2), 33–41.

Hayes, J. D., & McMahon, M. (2001). Molecular basis for the contribution of the antioxidant responsive element to cancer chemoprevention. *Cancer Letters, 174*(2), 103–113.

Hayes, J. D., & McMahon, M. (2009). NRF2 and KEAP1 mutations: Permanent activation of an adaptive response in cancer. *Trends in Biochemical Sciences, 34*(4), 176–188.

Homma, S., Ishii, Y., Morishima, Y., Yamadori, T., Matsuno, Y., Haraguchi, N., et al. (2009). Nrf2 enhances cell proliferation and resistance to anticancer drugs in human lung cancer. *Clinical Cancer Research: An Official Journal of the American Association for Cancer Research, 15*(10), 3423–3432.

Hruban, R. H., van Mansfeld, A. D., Offerhaus, G. J., van Weering, D. H., Allison, D. C., Goodman, S. N., et al. (1993). K-ras oncogene activation in adenocarcinoma of the human pancreas. A study of 82 carcinomas using a combination of mutant-enriched polymerase chain reaction analysis and allele-specific oligonucleotide hybridization. *The American Journal of Pathology*, *143*(2), 545–554.

Hu, Y., Ju, Y., Lin, D., Wang, Z., Huang, Y., Zhang, S., et al. (2012). Mutation of the Nrf2 gene in non-small cell lung cancer. *Molecular Biology Reports*, *39*(4), 4743–4747.

Huang, H. C., Nguyen, T., & Pickett, C. B. (2000). Regulation of the antioxidant response element by protein kinase C-mediated phosphorylation of NF-E2-related factor 2. *Proceedings of the National Academy of Sciences of the United States of America*, *97*(23), 12475–12480.

Huang, H.-C., Nguyen, T., & Pickett, C. B. (2002). Phosphorylation of Nrf2 at Ser-40 by protein kinase C regulates antioxidant response element-mediated transcription. *The Journal of Biological Chemistry*, *277*(45), 42769–42774.

Hudson, R. S., Yi, M., Esposito, D., Watkins, S. K., Hurwitz, A. A., Yfantis, H. G., et al. (2012). MicroRNA-1 is a candidate tumor suppressor and prognostic marker in human prostate cancer. *Nucleic Acids Research*, *40*(8), 3689–3703.

Hurttila, H., Koponen, J. K., Kansanen, E., Jyrkkänen, H.-K., Kivelä, A., Kylätie, R., et al. (2008). Oxidative stress-inducible lentiviral vectors for gene therapy. *Gene Therapy*, *15*(18), 1271–1279.

Hybertson, B. M., Gao, B., Bose, S. K., & McCord, J. M. (2011). Oxidative stress in health and disease: The therapeutic potential of Nrf2 activation. *Molecular Aspects of Medicine*, *32*(4–6), 234–246.

Ichimura, Y., Waguri, S., Sou, Y.-S., Kageyama, S., Hasegawa, J., Ishimura, R., et al. (2013). Phosphorylation of p62 activates the Keap1-Nrf2 pathway during selective autophagy. *Molecular Cell*, *51*(5), 618–631.

Iida, K., Itoh, K., Maher, J. M., Kumagai, Y., Oyasu, R., Mori, Y., et al. (2007). Nrf2 and p53 cooperatively protect against BBN-induced urinary bladder carcinogenesis. *Carcinogenesis*, *28*(11), 2398–2403.

Itoh, K., Chiba, T., Takahashi, S., Ishii, T., Igarashi, K., Katoh, Y., et al. (1997). An Nrf2/small Maf heterodimer mediates the induction of phase II detoxifying enzyme genes through antioxidant response elements. *Biochemical and Biophysical Research Communications*, *236*(2), 313–322.

Itoh, K., Mimura, J., & Yamamoto, M. (2010). Discovery of the negative regulator of Nrf2, Keap1: A historical overview. *Antioxidants & Redox Signaling*, *13*(11), 1665–1678.

Itoh, K., Wakabayashi, N., Katoh, Y., Ishii, T., Igarashi, K., Engel, J. D., et al. (1999). Keap1 represses nuclear activation of antioxidant responsive elements by Nrf2 through binding to the amino-terminal Neh2 domain. *Genes & Development*, *13*(1), 76–86.

Itoh, K., Wakabayashi, N., Katoh, Y., Ishii, T., O'Connor, T., & Yamamoto, M. (2003). Keap1 regulates both cytoplasmic-nuclear shuttling and degradation of Nrf2 in response to electrophiles. *Genes to Cells: Devoted to Molecular & Cellular Mechanisms*, *8*(4), 379–391.

Jacobson, L. P., Zhang, B. C., Zhu, Y. R., Wang, J. B., Wu, Y., Zhang, Q. N., et al. (1997). Oltipraz chemoprevention trial in Qidong, People's Republic of China: Study design and clinical outcomes. *Cancer Epidemiology, Biomarkers & Prevention: A Publication of the American Association for Cancer Research, Cosponsored by the American Society of Preventive Oncology*, *6*(4), 257–265.

Jain, A., Lamark, T., Sjøttem, E., Larsen, K. B., Awuh, J. A., Øvervatn, A., et al. (2010). p62/SQSTM1 is a target gene for transcription factor NRF2 and creates a positive feedback loop by inducing antioxidant response element-driven gene transcription. *The Journal of Biological Chemistry*, *285*(29), 22576–22591.

Ji, L., Li, H., Gao, P., Shang, G., Zhang, D. D., Zhang, N., et al. (2013). Nrf2 pathway regulates multidrug-resistance-associated protein 1 in small cell lung cancer. *PLoS One*, *8*(5), e63404.

Jiang, T., Chen, N., Zhao, F., Wang, X.-J., Kong, B., Zheng, W., et al. (2010). High levels of Nrf2 determine chemoresistance in type II endometrial cancer. *Cancer Research*, *70*(13), 5486–5496.

Johansen, T., & Lamark, T. (2011). Selective autophagy mediated by autophagic adapter proteins. *Autophagy*, 7(3), 279–296.

Kalo, E., Kogan-Sakin, I., Solomon, H., Bar-Nathan, E., Shay, M., Shetzer, Y., et al. (2012). Mutant p53R273H attenuates the expression of phase 2 detoxifying enzymes and promotes the survival of cells with high levels of reactive oxygen species. *Journal of Cell Science*, *125*(Pt 22), 5578–5586.

Kandoth, C., McLellan, M. D., Vandin, F., Ye, K., Niu, B., Lu, C., et al. (2013). Mutational landscape and significance across 12 major cancer types. *Nature*, *502*(7471), 333–339.

Kang, K. W., Ryu, J. H., & Kim, S. G. (2000). The essential role of phosphatidylinositol 3-kinase and of p38 mitogen-activated protein kinase activation in the antioxidant response element-mediated rGSTA2 induction by decreased glutathione in H4IIE hepatoma cells. *Molecular Pharmacology*, *58*(5), 1017–1025.

Kansanen, E., Bonacci, G., Schopfer, F. J., Kuosmanen, S. M., Tong, K. I., Leinonen, H., et al. (2011). Electrophilic nitro-fatty acids activate NRF2 by a KEAP1 cysteine 151-independent mechanism. *The Journal of Biological Chemistry*, *286*(16), 14019–14027.

Kansanen, E., Jyrkkänen, H.-K., & Levonen, A.-L. (2012). Activation of stress signaling pathways by electrophilic oxidized and nitrated lipids. *Free Radical Biology & Medicine*, *52*(6), 973–982.

Kensler, T. W., He, X., Otieno, M., Egner, P. A., Jacobson, L. P., Chen, B., et al. (1998). Oltipraz chemoprevention trial in Qidong, People's Republic of China: Modulation of serum aflatoxin albumin adduct biomarkers. *Cancer Epidemiology, Biomarkers & Prevention: A Publication of the American Association for Cancer Research, Cosponsored by the American Society of Preventive Oncology*, 7(2), 127–134.

Kim, T.-H., Hur, E., Kang, S.-J., Kim, J.-A., Thapa, D., Lee, Y. M., et al. (2011). NRF2 blockade suppresses colon tumor angiogenesis by inhibiting hypoxia-induced activation of HIF-1α. *Cancer Research*, *71*(6), 2260–2275.

Kim, Y. R., Oh, J. E., Kim, M. S., Kang, M. R., Park, S. W., Han, J. Y., et al. (2010). Oncogenic NRF2 mutations in squamous cell carcinomas of oesophagus and skin. *The Journal of Pathology*, *220*(4), 446–451.

Kitteringham, N. R., Abdullah, A., Walsh, J., Randle, L., Jenkins, R. E., Sison, R., et al. (2010). Proteomic analysis of Nrf2 deficient transgenic mice reveals cellular defence and lipid metabolism as primary Nrf2-dependent pathways in the liver. *Journal of Proteomics*, *73*(8), 1612–1631.

Kobayashi, A., Kang, M.-I., Okawa, H., Ohtsuji, M., Zenke, Y., Chiba, T., et al. (2004). Oxidative stress sensor Keap1 functions as an adaptor for Cul3-based E3 ligase to regulate proteasomal degradation of Nrf2. *Molecular and Cellular Biology*, *24*(16), 7130–7139.

Kobayashi, M., Li, L., Iwamoto, N., Nakajima-Takagi, Y., Kaneko, H., Nakayama, Y., et al. (2009). The antioxidant defense system Keap1-Nrf2 comprises a multiple sensing mechanism for responding to a wide range of chemical compounds. *Molecular and Cellular Biology*, *29*(2), 493–502.

Komatsu, M., Kurokawa, H., Waguri, S., Taguchi, K., Kobayashi, A., Ichimura, Y., et al. (2010). The selective autophagy substrate p62 activates the stress responsive transcription factor Nrf2 through inactivation of Keap1. *Nature Cell Biology*, *12*(3), 213–223.

Konstantinopoulos, P. A., Spentzos, D., Fountzilas, E., Francoeur, N., Sanisetty, S., Grammatikos, A. P., et al. (2011). Keap1 mutations and Nrf2 pathway activation in epithelial ovarian cancer. *Cancer Research*, *71*(15), 5081–5089.

Kruger, N. J., & von Schaewen, A. (2003). The oxidative pentose phosphate pathway: Structure and organisation. *Current Opinion in Plant Biology*, *6*(3), 236–246.
Kwak, M.-K., & Kensler, T. W. (2010). Targeting NRF2 signaling for cancer chemoprevention. *Toxicology and Applied Pharmacology*, *244*(1), 66–76.
Kwak, M.-K., Wakabayashi, N., Itoh, K., Motohashi, H., Yamamoto, M., & Kensler, T. W. (2003). Modulation of gene expression by cancer chemopreventive dithiolethiones through the Keap1-Nrf2 pathway. Identification of novel gene clusters for cell survival. *The Journal of Biological Chemistry*, *278*(10), 8135–8145.
Lau, A., Wang, X.-J., Zhao, F., Villeneuve, N. F., Wu, T., Jiang, T., et al. (2010). A noncanonical mechanism of Nrf2 activation by autophagy deficiency: Direct interaction between Keap1 and p62. *Molecular and Cellular Biology*, *30*(13), 3275–3285.
Lau, A., Zheng, Y., Tao, S., Wang, H., Whitman, S. A., White, E., et al. (2013). Arsenic inhibits autophagic flux, activating the Nrf2-Keap1 pathway in a p62-dependent manner. *Molecular and Cellular Biology*, *33*(12), 2436–2446.
Lee, J.-M., Calkins, M. J., Chan, K., Kan, Y. W., & Johnson, J. A. (2003). Identification of the NF-E2-related factor-2-dependent genes conferring protection against oxidative stress in primary cortical astrocytes using oligonucleotide microarray analysis. *The Journal of Biological Chemistry*, *278*(14), 12029–12038.
Lee, Y. S., Kang, Y. S., Lee, S. H., & Kim, J. A. (2000). Role of NAD(P)H oxidase in the tamoxifen-induced generation of reactive oxygen species and apoptosis in HepG2 human hepatoblastoma cells. *Cell Death and Differentiation*, 7(10), 925–932.
Lee, S., Lim, M.-J., Kim, M.-H., Yu, C.-H., Yun, Y.-S., Ahn, J., et al. (2012). An effective strategy for increasing the radiosensitivity of human lung cancer cells by blocking Nrf2-dependent antioxidant responses. *Free Radical Biology & Medicine*, *53*(4), 807–816.
Lee, C. K., Park, K.-K., Chung, A.-S., & Chung, W.-Y. (2012). Ginsenoside Rg3 enhances the chemosensitivity of tumors to cisplatin by reducing the basal level of nuclear factor erythroid 2-related factor 2-mediated heme oxygenase-1/NAD(P)H quinone oxidoreductase-1 and prevents normal tissue damage by scavenging cisp. *Food and Chemical Toxicology*, *50*(7), 2565–2574.
Leinonen, H. M., Ruotsalainen, A.-K., Määttä, A.-M., Laitinen, H. M., Kuosmanen, S. M., Kansanen, E., et al. (2012). Oxidative stress-regulated lentiviral TK/GCV gene therapy for lung cancer treatment. *Cancer Research*, 72(23), 6227–6235.
Levonen, A. L., Dickinson, D. A., Moellering, D. R., Mulcahy, R. T., Forman, H. J., & Darley-Usmar, V. M. (2001). Biphasic effects of 15-deoxy-delta(12,14)-prostaglandin J(2) on glutathione induction and apoptosis in human endothelial cells. *Arteriosclerosis, Thrombosis, and Vascular Biology*, *21*(11), 1846–1851.
Li, Q. K., Singh, A., Biswal, S., Askin, F., & Gabrielson, E. (2011). KEAP1 gene mutations and NRF2 activation are common in pulmonary papillary adenocarcinoma. *Journal of Human Genetics*, *56*(3), 230–234.
Liao, H., Zhou, Q., Zhang, Z., Wang, Q., Sun, Y., Yi, X., et al. (2012). NRF2 is overexpressed in ovarian epithelial carcinoma and is regulated by gonadotrophin and sex-steroid hormones. *Oncology Reports*, *27*(6), 1918–1924.
Lister, A., Nedjadi, T., Kitteringham, N. R., Campbell, F., Costello, E., Lloyd, B., et al. (2011). Nrf2 is overexpressed in pancreatic cancer: Implications for cell proliferation and therapy. *Molecular Cancer*, *10*, 37.
Liu, X., Madhankumar, A. B., Slagle-Webb, B., Sheehan, J. M., Surguladze, N., & Connor, J. R. (2011). Heavy chain ferritin siRNA delivered by cationic liposomes increases sensitivity of cancer cells to chemotherapeutic agents. *Cancer Research*, *71*(6), 2240–2249.
Lu, S. C. (2009). Regulation of glutathione synthesis. *Molecular Aspects of Medicine*, *30*(1–2), 42–59.

Ma, J., Cai, H., Wu, T., Sobhian, B., Huo, Y., Alcivar, A., et al. (2012). PALB2 interacts with KEAP1 to promote NRF2 nuclear accumulation and function. *Molecular and Cellular Biology, 32*(8), 1506–1517.

Ma, Q., & He, X. (2012). Molecular basis of electrophilic and oxidative defense: Promises and perils of Nrf2. *Pharmacological Reviews, 64*(4), 1055–1081.

Manandhar, S., Lee, S., & Kwak, M.-K. (2010). Effect of stable inhibition of NRF2 on doxorubicin sensitivity in human ovarian carcinoma OV90 cells. *Archives of Pharmacal Research, 33*(5), 717–726.

Marzec, J. M., Christie, J. D., Reddy, S. P., Jedlicka, A. E., Vuong, H., Lanken, P. N., et al. (2007). Functional polymorphisms in the transcription factor NRF2 in humans increase the risk of acute lung injury. *FASEB Journal: Official Publication of the Federation of American Societies for Experimental Biology, 21*(9), 2237–2246.

McDonald, J. T., Kim, K., Norris, A. J., Vlashi, E., Phillips, T. M., Lagadec, C., et al. (2010). Ionizing radiation activates the Nrf2 antioxidant response. *Cancer Research, 70*(21), 8886–8895.

McMahon, M., Lamont, D. J., Beattie, K. A., & Hayes, J. D. (2010). Keap1 perceives stress via three sensors for the endogenous signaling molecules nitric oxide, zinc, and alkenals. *Proceedings of the National Academy of Sciences of the United States of America, 107*(44), 18838–18843.

McMahon, M., Thomas, N., Itoh, K., Yamamoto, M., & Hayes, J. D. (2004). Redox-regulated turnover of Nrf2 is determined by at least two separate protein domains, the redox-sensitive Neh2 degron and the redox-insensitive Neh6 degron. *The Journal of Biological Chemistry, 279*(30), 31556–31567.

Meijerman, I., Beijnen, J. H., & Schellens, J. H. M. (2008). Combined action and regulation of phase II enzymes and multidrug resistance proteins in multidrug resistance in cancer. *Cancer Treatment Reviews, 34*(6), 505–520.

Mitsuishi, Y., Taguchi, K., Kawatani, Y., Shibata, T., Nukiwa, T., Aburatani, H., et al. (2012). Nrf2 redirects glucose and glutamine into anabolic pathways in metabolic reprogramming. *Cancer Cell, 22*(1), 66–79.

Moolten, F. L. (1986). Tumor chemosensitivity conferred by inserted herpes thymidine kinase genes: Paradigm for a prospective cancer control strategy. *Cancer Research, 46*(10), 5276–5281.

Moriya, M. (1993). Single-stranded shuttle phagemid for mutagenesis studies in mammalian cells: 8-Oxoguanine in DNA induces targeted G.C–>T.A transversions in simian kidney cells. *Proceedings of the National Academy of Sciences of the United States of America, 90*(3), 1122–1126.

Motohashi, H., O'Connor, T., Katsuoka, F., Engel, J. D., & Yamamoto, M. (2002). Integration and diversity of the regulatory network composed of Maf and CNC families of transcription factors. *Gene, 294*(1–2), 1–12.

Mulcahy, R. T., Wartman, M. A., Bailey, H. H., & Gipp, J. J. (1997). Constitutive and beta-naphthoflavone-induced expression of the human gamma-glutamylcysteine synthetase heavy subunit gene is regulated by a distal antioxidant response element/TRE sequence. *The Journal of Biological Chemistry, 272*(11), 7445–7454.

Muller, P. A. J., & Vousden, K. H. (2013). p53 mutations in cancer. *Nature Cell Biology, 15*(1), 2–8.

Muscarella, L. A., Barbano, R., D'Angelo, V., Copetti, M., Coco, M., Balsamo, T., et al. (2011). Regulation of KEAP1 expression by promoter methylation in malignant gliomas and association with patient's outcome. *Epigenetics: Official Journal of the DNA Methylation Society, 6*(3), 317–325.

Nasser, M. W., Datta, J., Nuovo, G., Kutay, H., Motiwala, T., Majumder, S., et al. (2008). Down-regulation of micro-RNA-1 (miR-1) in lung cancer. Suppression of tumorigenic

property of lung cancer cells and their sensitization to doxorubicin-induced apoptosis by miR-1. *The Journal of Biological Chemistry*, *283*(48), 33394–33405.
Neckers, L., & Workman, P. (2012). Hsp90 molecular chaperone inhibitors: Are we there yet? *Clinical Cancer Research: An Official Journal of the American Association for Cancer Research*, *18*(1), 64–76.
Nguyen, T., Sherratt, P. J., & Pickett, C. B. (2003). Regulatory mechanisms controlling gene expression mediated by the antioxidant response element. *Annual Review of Pharmacology and Toxicology*, *43*, 233–260.
Nioi, P., & Nguyen, T. (2007). A mutation of Keap1 found in breast cancer impairs its ability to repress Nrf2 activity. *Biochemical and Biophysical Research Communications*, *362*(4), 816–821.
Nioi, P., Nguyen, T., Sherratt, P. J., & Pickett, C. B. (2005). The carboxy-terminal Neh3 domain of Nrf2 is required for transcriptional activation. *Molecular and Cellular Biology*, *25*(24), 10895–10906.
Nishie, A., Ono, M., Shono, T., Fukushi, J., Otsubo, M., Onoue, H., et al. (1999). Macrophage infiltration and heme oxygenase-1 expression correlate with angiogenesis in human gliomas. *Clinical Cancer Research: An Official Journal of the American Association for Cancer Research*, *5*(5), 1107–1113.
O'Reilly, M. A. (2005). Redox activation of p21Cip1/WAF1/Sdi1: A multifunctional regulator of cell survival and death. *Antioxidants & Redox Signaling*, 7(1–2), 108–118.
Ogura, T., Tong, K. I., Mio, K., Maruyama, Y., Kurokawa, H., Sato, C., et al. (2010). Keap1 is a forked-stem dimer structure with two large spheres enclosing the intervening, double glycine repeat, and C-terminal domains. *Proceedings of the National Academy of Sciences of the United States of America*, *107*(7), 2842–2847.
Ohta, T., Iijima, K., Miyamoto, M., Nakahara, I., Tanaka, H., Ohtsuji, M., et al. (2008). Loss of Keap1 function activates Nrf2 and provides advantages for lung cancer cell growth. *Cancer Research*, *68*(5), 1303–1309.
Okano, Y., Nezu, U., Enokida, Y., Lee, M. T. M., Kinoshita, H., Lezhava, A., et al. (2013). SNP (−617C>A) in ARE-like loci of the NRF2 gene: A new biomarker for prognosis of lung adenocarcinoma in Japanese non-smoking women. *PLoS One*, *8*(9), e73794.
Ooi, A., Dykema, K., Ansari, A., Petillo, D., Snider, J., Kahnoski, R., et al. (2013). CUL3 and NRF2 mutations confer an NRF2 activation phenotype in a sporadic form of papillary renal cell carcinoma. *Cancer Research*, *73*(7), 2044–2051.
Ooi, A., Wong, J.-C., Petillo, D., Roossien, D., Perrier-Trudova, V., Whitten, D., et al. (2011). An antioxidant response phenotype shared between hereditary and sporadic type 2 papillary renal cell carcinoma. *Cancer Cell*, *20*(4), 511–523.
Padmanabhan, B., Tong, K. I., Ohta, T., Nakamura, Y., Scharlock, M., Ohtsuji, M., et al. (2006). Structural basis for defects of Keap1 activity provoked by its point mutations in lung cancer. *Molecular Cell*, *21*(5), 689–700.
Parkinson, E. I., Bair, J. S., Cismesia, M., & Hergenrother, P. J. (2013). Efficient NQO1 substrates are potent and selective anticancer agents. *ACS Chemical Biology*, *8*(10), 2173–2183.
Patterson, A. V., Ferry, D. M., Edmunds, S. J., Gu, Y., Singleton, R. S., Patel, K., et al. (2007). Mechanism of action and preclinical antitumor activity of the novel hypoxia-activated DNA cross-linking agent PR-104. *Clinical Cancer Research: An Official Journal of the American Association for Cancer Research*, *13*(13), 3922–3932.
Pecot, C. V., Calin, G. A., Coleman, R. L., Lopez-Berestein, G., & Sood, A. K. (2011). RNA interference in the clinic: Challenges and future directions. *Nature Reviews. Cancer*, *11*(1), 59–67.
Place, A. E., Suh, N., Williams, C. R., Risingsong, R., Honda, T., Honda, Y., et al. (2003). The novel synthetic triterpenoid, CDDO-imidazolide, inhibits inflammatory response

and tumor growth in vivo. *Clinical Cancer Research: An Official Journal of the American Association for Cancer Research, 9*(7), 2798–2806.

Prestera, T., Holtzclaw, W. D., Zhang, Y., & Talalay, P. (1993). Chemical and molecular regulation of enzymes that detoxify carcinogens. *Proceedings of the National Academy of Sciences of the United States of America, 90*(7), 2965–2969.

Rada, P., Rojo, A. I., Evrard-Todeschi, N., Innamorato, N. G., Cotte, A., Jaworski, T., et al. (2012). Structural and functional characterization of Nrf2 degradation by the glycogen synthase kinase 3/β-TrCP axis. *Molecular and Cellular Biology, 32*(17), 3486–3499.

Raica, M., Cimpean, A. M., & Ribatti, D. (2009). Angiogenesis in pre-malignant conditions. *European Journal of Cancer, 45*(11), 1924–1934.

Ramos-Gomez, M., Kwak, M. K., Dolan, P. M., Itoh, K., Yamamoto, M., Talalay, P., et al. (2001). Sensitivity to carcinogenesis is increased and chemoprotective efficacy of enzyme inducers is lost in nrf2 transcription factor-deficient mice. *Proceedings of the National Academy of Sciences of the United States of America, 98*(6), 3410–3415.

Ren, D., Villeneuve, N. F., Jiang, T., Wu, T., Lau, A., Toppin, H. A., et al. (2011). Brusatol enhances the efficacy of chemotherapy by inhibiting the Nrf2-mediated defense mechanism. *Proceedings of the National Academy of Sciences of the United States of America, 108*(4), 1433–1438.

Roberts, P. J., & Der, C. J. (2007). Targeting the Raf-MEK-ERK mitogen-activated protein kinase cascade for the treatment of cancer. *Oncogene, 26*(22), 3291–3310.

Rodriguez-Antona, C., & Ingelman-Sundberg, M. (2006). Cytochrome P450 pharmacogenetics and cancer. *Oncogene, 25*(11), 1679–1691.

Rojo, A. I., Medina-Campos, O. N., Rada, P., Zúñiga-Toalá, A., López-Gazcón, A., Espada, S., et al. (2012). Signaling pathways activated by the phytochemical nordihydroguaiaretic acid contribute to a Keap1-independent regulation of Nrf2 stability: Role of glycogen synthase kinase-3. *Free Radical Biology & Medicine, 52*(2), 473–487.

Rushmore, T. H., Morton, M. R., & Pickett, C. B. (1991). The antioxidant responsive element. Activation by oxidative stress and identification of the DNA consensus sequence required for functional activity. *The Journal of Biological Chemistry, 266*(18), 11632–11639.

Rushworth, S. A., Zaitseva, L., Murray, M. Y., Shah, N. M., Bowles, K. M., & MacEwan, D. J. (2012). The high Nrf2 expression in human acute myeloid leukemia is driven by NF-κB and underlies its chemo-resistance. *Blood, 120*(26), 5188–5198.

Satoh, H., Moriguchi, T., Takai, J., Ebina, M., & Yamamoto, M. (2013). Nrf2 prevents initiation but accelerates progression through the Kras signaling pathway during lung carcinogenesis. *Cancer Research, 73*(13), 4158–4168.

Shibata, T., Kokubu, A., Gotoh, M., Ojima, H., Ohta, T., Yamamoto, M., et al. (2008). Genetic alteration of Keap1 confers constitutive Nrf2 activation and resistance to chemotherapy in gallbladder cancer. *Gastroenterology, 135*(4), 1358–1368.

Shibata, T., Kokubu, A., Saito, S., Narisawa-Saito, M., Sasaki, H., Aoyagi, K., et al. (2011). NRF2 mutation confers malignant potential and resistance to chemoradiation therapy in advanced esophageal squamous cancer. *Neoplasia (New York, NY), 13*(9), 864–873.

Shibata, T., Ohta, T., Tong, K. I., Kokubu, A., Odogawa, R., Tsuta, K., et al. (2008). Cancer related mutations in NRF2 impair its recognition by Keap1-Cul3 E3 ligase and promote malignancy. *Proceedings of the National Academy of Sciences of the United States of America, 105*(36), 13568–13573.

Shigenaga, M. K., Gimeno, C. J., & Ames, B. N. (1989). Urinary 8-hydroxy-2′-deoxyguanosine as a biological marker of in vivo oxidative DNA damage. *Proceedings of the National Academy of Sciences of the United States of America, 86*(24), 9697–9701.

Singh, A., Bodas, M., Wakabayashi, N., Bunz, F., & Biswal, S. (2010). Gain of Nrf2 function in non-small-cell lung cancer cells confers radioresistance. *Antioxidants & Redox Signaling, 13*(11), 1627–1637.

Singh, A., Boldin-Adamsky, S., Thimmulappa, R. K., Rath, S. K., Ashush, H., Coulter, J., et al. (2008). RNAi-mediated silencing of nuclear factor erythroid-2-related factor 2 gene expression in non-small cell lung cancer inhibits tumor growth and increases efficacy of chemotherapy. *Cancer Research*, *68*(19), 7975–7984.

Singh, A., Happel, C., Manna, S. K., Acquaah-Mensah, G., Carrerero, J., Kumar, S., et al. (2013). Transcription factor NRF2 regulates miR-1 and miR-206 to drive tumorigenesis. *The Journal of Clinical Investigation*, *123*(7), 2921–2934.

Singh, A., Misra, V., Thimmulappa, R. K., Lee, H., Ames, S., Hoque, M. O., et al. (2006). Dysfunctional KEAP1-NRF2 interaction in non-small-cell lung cancer. *PLoS Medicine*, *3*(10), e420.

Singh, A., Wu, H., Zhang, P., Happel, C., Ma, J., & Biswal, S. (2010). Expression of ABCG2 (BCRP) is regulated by Nrf2 in cancer cells that confers side population and chemoresistance phenotype. *Molecular Cancer Therapeutics*, *9*(8), 2365–2376.

Sjöblom, T., Jones, S., Wood, L. D., Parsons, D. W., Lin, J., Barber, T. D., et al. (2006). The consensus coding sequences of human breast and colorectal cancers. *Science (New York, NY)*, *314*(5797), 268–274.

Smith, Z. D., & Meissner, A. (2013). DNA methylation: Roles in mammalian development. *Nature Reviews. Genetics*, *14*(3), 204–220.

Solis, L. M., Behrens, C., Dong, W., Suraokar, M., Ozburn, N. C., Moran, C. A., et al. (2010). Nrf2 and Keap1 abnormalities in non-small cell lung carcinoma and association with clinicopathologic features. *Clinical Cancer Research: An Official Journal of the American Association for Cancer Research*, *16*(14), 3743–3753.

Storz, P. (2005). Reactive oxygen species in tumor progression. *Frontiers in Bioscience: A Journal and Virtual Library*, *10*, 1881–1896.

Strohecker, A. M., Guo, J. Y., Karsli-Uzunbas, G., Price, S. M., Chen, G. J., Mathew, R., et al. (2013). Autophagy sustains mitochondrial glutamine metabolism and growth of BrafV600E-driven lung tumors. *Cancer Discovery*, *3*(11), 1272–1285.

Sun, Z., Huang, Z., & Zhang, D. D. (2009). Phosphorylation of Nrf2 at multiple sites by MAP kinases has a limited contribution in modulating the Nrf2-dependent antioxidant response. *PLoS One*, *4*(8), e6588.

Sunamura, M., Duda, D. G., Ghattas, M. H., Lozonschi, L., Motoi, F., Yamauchi, J.-I., et al. (2003). Heme oxygenase-1 accelerates tumor angiogenesis of human pancreatic cancer. *Angiogenesis*, *6*(1), 15–24.

Suzuki, T., Maher, J., & Yamamoto, M. (2011). Select heterozygous Keap1 mutations have a dominant-negative effect on wild-type Keap1 in vivo. *Cancer Research*, *71*(5), 1700–1709.

Suzuki, T., Shibata, T., Takaya, K., Shiraishi, K., Kohno, T., Kunitoh, H., et al. (2013). Regulatory nexus of synthesis and degradation deciphers cellular Nrf2 expression levels. *Molecular and Cellular Biology*, *33*(12), 2402–2412. http://dx.doi.org/10.1128/MCB.00065-13.

Szatrowski, T. P., & Nathan, C. F. (1991). Production of large amounts of hydrogen peroxide by human tumor cells. *Cancer Research*, *51*(3), 794–798.

Taguchi, K., Fujikawa, N., Komatsu, M., Ishii, T., Unno, M., Akaike, T., et al. (2012). Keap1 degradation by autophagy for the maintenance of redox homeostasis. *Proceedings of the National Academy of Sciences of the United States of America*, *109*(34), 13561–13566.

Taguchi, K., Maher, J. M., Suzuki, T., Kawatani, Y., Motohashi, H., & Yamamoto, M. (2010). Genetic analysis of cytoprotective functions supported by graded expression of Keap1. *Molecular and Cellular Biology*, *30*(12), 3016–3026.

Taguchi, K., Motohashi, H., & Yamamoto, M. (2011). Molecular mechanisms of the Keap1–Nrf2 pathway in stress response and cancer evolution. *Genes to Cells: Devoted to Molecular & Cellular Mechanisms*, *16*(2), 123–140.

Talalay, P., De Long, M. J., & Prochaska, H. J. (1988). Identification of a common chemical signal regulating the induction of enzymes that protect against chemical carcinogenesis. *Proceedings of the National Academy of Sciences of the United States of America, 85*(21), 8261–8265.

Tang, X., Wang, H., Fan, L., Wu, X., Xin, A., Ren, H., et al. (2011). Luteolin inhibits Nrf2 leading to negative regulation of the Nrf2/ARE pathway and sensitization of human lung carcinoma A549 cells to therapeutic drugs. *Free Radical Biology & Medicine, 50*(11), 1599–1609. http://dx.doi.org/10.1016/j.freeradbiomed.2011.03.008.

Tong, K. I., Kobayashi, A., Katsuoka, F., & Yamamoto, M. (2006). Two-site substrate recognition model for the Keap1-Nrf2 system: A hinge and latch mechanism. *Biological Chemistry, 387*(10–11), 1311–1320.

Torisu-Itakura, H., Furue, M., Kuwano, M., & Ono, M. (2000). Co-expression of thymidine phosphorylase and heme oxygenase-1 in macrophages in human malignant vertical growth melanomas. *Cancer Science, 91*(9), 906–910.

Trachootham, D., Alexandre, J., & Huang, P. (2009). Targeting cancer cells by ROS-mediated mechanisms: A radical therapeutic approach? *Nature Reviews. Drug Discovery, 8*(7), 579–591.

Traver, R. D., Horikoshi, T., Danenberg, K. D., Stadlbauer, T. H., Danenberg, P. V., Ross, D., et al. (1992). NAD(P)H:quinone oxidoreductase gene expression in human colon carcinoma cells: Characterization of a mutation which modulates DT-diaphorase activity and mitomycin sensitivity. *Cancer Research, 52*(4), 797–802.

Tsang, W., Chau, S. P., Kong, S., Fung, K., & Kwok, T. (2003). Reactive oxygen species mediate doxorubicin induced p53-independent apoptosis. *Life Sciences, 73*(16), 2047–2058.

Valko, M., Leibfritz, D., Moncol, J., Cronin, M. T. D., Mazur, M., & Telser, J. (2007). Free radicals and antioxidants in normal physiological functions and human disease. *The International Journal of Biochemistry & Cell Biology, 39*(1), 44–84.

Vander Heiden, M. G., Cantley, L. C., & Thompson, C. B. (2009). Understanding the Warburg effect: The metabolic requirements of cell proliferation. *Science (New York, NY), 324*(5930), 1029–1033.

Vogelstein, B., Papadopoulos, N., Velculescu, V. E., Zhou, S., Diaz, L. A., & Kinzler, K. W. (2013). Cancer genome landscapes. *Science (New York, NY), 339*(6127), 1546–1558.

Vollrath, V., Wielandt, A. M., Iruretagoyena, M., & Chianale, J. (2006). Role of Nrf2 in the regulation of the Mrp2 (ABCC2) gene. *The Biochemical Journal, 395*(3), 599–609.

Volonte, D., Liu, Z., Musille, P. M., Stoppani, E., Wakabayashi, N., Di, Y.-P., et al. (2013). Inhibition of nuclear factor-erythroid 2-related factor (Nrf2) by caveolin-1 promotes stress-induced premature senescence. *Molecular Biology of the Cell, 24*(12), 1852–1862.

Wang, R., An, J., Ji, F., Jiao, H., Sun, H., & Zhou, D. (2008). Hypermethylation of the Keap1 gene in human lung cancer cell lines and lung cancer tissues. *Biochemical and Biophysical Research Communications, 373*(1), 151–154.

Wang, M., Dhingra, K., Hittelman, W. N., Liehr, J. G., de Andrade, M., & Li, D. (1996). Lipid peroxidation-induced putative malondialdehyde-DNA adducts in human breast tissues. *Cancer Epidemiology, Biomarkers & Prevention: A Publication of the American Association for Cancer Research, Cosponsored by the American Society of Preventive Oncology, 5*(9), 705–710.

Wang, X. J., Hayes, J. D., & Wolf, C. R. (2006). Generation of a stable antioxidant response element-driven reporter gene cell line and its use to show redox-dependent activation of nrf2 by cancer chemotherapeutic agents. *Cancer Research, 66*(22), 10983–10994.

Wang, X., Ling, C., Bai, Y., & Zhao, J. (2011). MicroRNA-206 is associated with invasion and metastasis of lung cancer. *Anatomical Record (Hoboken, NJ: 2007)*, *294*(1), 88–92.

Wang, H., Liu, K., Geng, M., Gao, P., Wu, X., Hai, Y., et al. (2013). RXRα inhibits the NRF2-ARE signaling pathway through a direct interaction with the Neh7 domain of NRF2. *Cancer Research*, *73*(10), 3097–3108.

Wang, X.-J., Sun, Z., Chen, W., Li, Y., Villeneuve, N. F., & Zhang, D. D. (2008). Activation of Nrf2 by arsenite and monomethylarsonous acid is independent of Keap1-C151: Enhanced Keap1-Cul3 interaction. *Toxicology and Applied Pharmacology*, *230*(3), 383–389.

Wang, X.-J., Sun, Z., Villeneuve, N. F., Zhang, S., Zhao, F., Li, Y., et al. (2008). Nrf2 enhances resistance of cancer cells to chemotherapeutic drugs, the dark side of Nrf2. *Carcinogenesis*, *29*(6), 1235–1243.

Ward, P. S., & Thompson, C. B. (2012). Metabolic reprogramming: A cancer hallmark even Warburg did not anticipate. *Cancer Cell*, *21*(3), 297–308.

Wattenberg, L. W. (1985). Chemoprevention of cancer. *Cancer Research*, *45*(1), 1–8.

Wei, Y., Gong, J., Thimmulappa, R. K., Kosmider, B., Biswal, S., & Duh, E. J. (2013). Nrf2 acts cell-autonomously in endothelium to regulate tip cell formation and vascular branching. *Proceedings of the National Academy of Sciences of the United States of America*, *110*(41), E3910–E3918.

Wild, A. C., Moinova, H. R., & Mulcahy, R. T. (1999). Regulation of gamma-glutamylcysteine synthetase subunit gene expression by the transcription factor Nrf2. *The Journal of Biological Chemistry*, *274*(47), 33627–33636.

Wilson, T. R., Longley, D. B., & Johnston, P. G. (2006). Chemoresistance in solid tumours. *Annals of Oncology: Official Journal of the European Society for Medical Oncology/ESMO*, *17*(Suppl. 1), x315–x324.

Xu, Y., Fang, F., Miriyala, S., Crooks, P. A., Oberley, T. D., Chaiswing, L., et al. (2013). KEAP1 is a redox sensitive target that arbitrates the opposing radiosensitive effects of parthenolide in normal and cancer cells. *Cancer Research*, *73*(14), 4406–4417.

Yamadori, T., Ishii, Y., Homma, S., Morishima, Y., Kurishima, K., Itoh, K., et al. (2012). Molecular mechanisms for the regulation of Nrf2-mediated cell proliferation in non-small-cell lung cancers. *Oncogene*, *31*(45), 4768–4777. http://dx.doi.org/10.1038/onc.2011.628.

Yamamoto, T., Suzuki, T., Kobayashi, A., Wakabayashi, J., Maher, J., Motohashi, H., et al. (2008). Physiological significance of reactive cysteine residues of Keap1 in determining Nrf2 activity. *Molecular and Cellular Biology*, *28*(8), 2758–2770.

Yamamoto, T., Yoh, K., Kobayashi, A., Ishii, Y., Kure, S., Koyama, A., et al. (2004). Identification of polymorphisms in the promoter region of the human NRF2 gene. *Biochemical and Biophysical Research Communications*, *321*(1), 72–79.

Yang, H., Wang, W., Zhang, Y., Zhao, J., Lin, E., Gao, J., et al. (2011). The role of NF-E2-related factor 2 in predicting chemoresistance and prognosis in advanced non-small-cell lung cancer. *Clinical Lung Cancer*, *12*(3), 166–171.

Yoo, N. J., Kim, H. R., Kim, Y. R., An, C. H., & Lee, S. H. (2012). Somatic mutations of the KEAP1 gene in common solid cancers. *Histopathology*, *60*(6), 943–952.

Zhang, J., Hosoya, T., Maruyama, A., Nishikawa, K., Maher, J. M., Ohta, T., et al. (2007). Nrf2 Neh5 domain is differentially utilized in the transactivation of cytoprotective genes. *The Biochemical Journal*, *404*(3), 459–466.

Zhang, Y., Kensler, T. W., Cho, C. G., Posner, G. H., & Talalay, P. (1994). Anticarcinogenic activities of sulforaphane and structurally related synthetic norbornyl isothiocyanates. *Proceedings of the National Academy of Sciences of the United States of America*, *91*(8), 3147–3150.

Zhang, Yuesheng, & Munday, R. (2008). Dithiolethiones for cancer chemoprevention: Where do we stand? *Molecular Cancer Therapeutics*, 7(11), 3470–3479.

Zhang, P., Singh, A., Yegnasubramanian, S., Esopi, D., Kombairaju, P., Bodas, M., et al. (2010). Loss of Kelch-like ECH-associated protein 1 function in prostate cancer cells causes chemoresistance and radioresistance and promotes tumor growth. *Molecular Cancer Therapeutics*, *9*(2), 336–346.

Zhang, Y., & Tang, L. (2007). Discovery and development of sulforaphane as a cancer chemopreventive phytochemical. *Acta Pharmacologica Sinica*, *28*(9), 1343–1354.

Zhao, C. R., Gao, Z. H., & Qu, X. J. (2010). Nrf2-ARE signaling pathway and natural products for cancer chemoprevention. *Cancer Epidemiology*, *34*(5), 523–533.

Zhivotovsky, B., Joseph, B., & Orrenius, S. (1999). Tumor radiosensitivity and apoptosis. *Experimental Cell Research*, *248*(1), 10–17.

INDEX

Note: Page numbers followed by "*f*" indicate figures and "*t*" indicate tables.

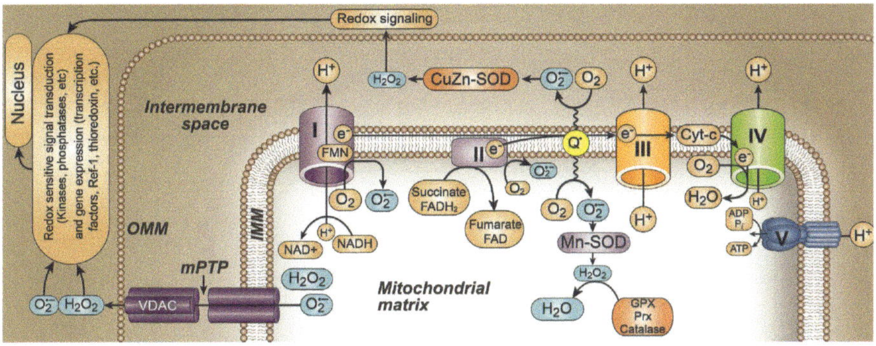

Figure 1.1, Daohong Zhou *et al.* (See Page 5 of this volume.)

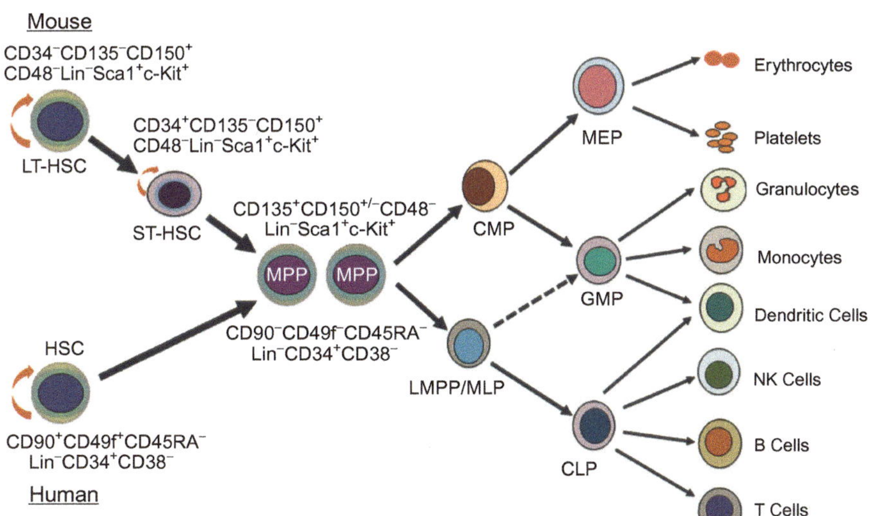

Figure 1.5, Daohong Zhou *et al.* (See Page 20 of this volume.)

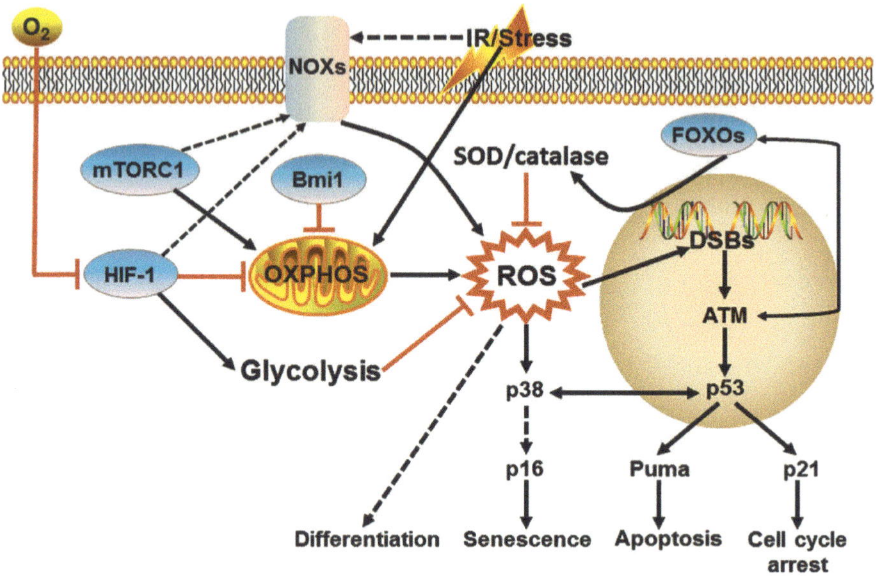

Figure 1.6, Daohong Zhou *et al.* (See Page 27 of this volume.)

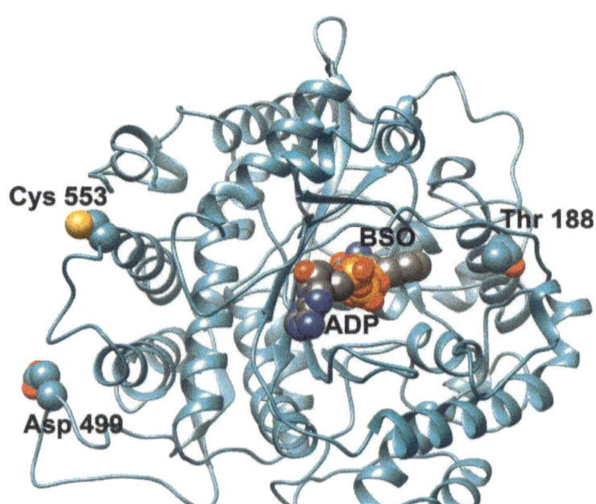

Figure 2.4, Yilin Liu *et al.* (See Page 76 of this volume.)

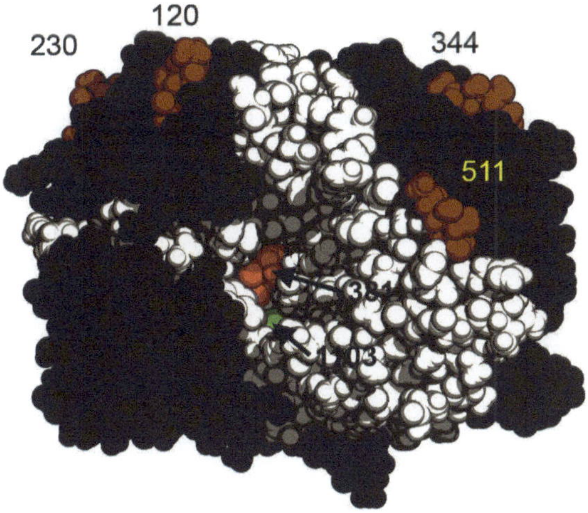

Figure 3.1, Marie H. Hanigan (See Page 108 of this volume.)

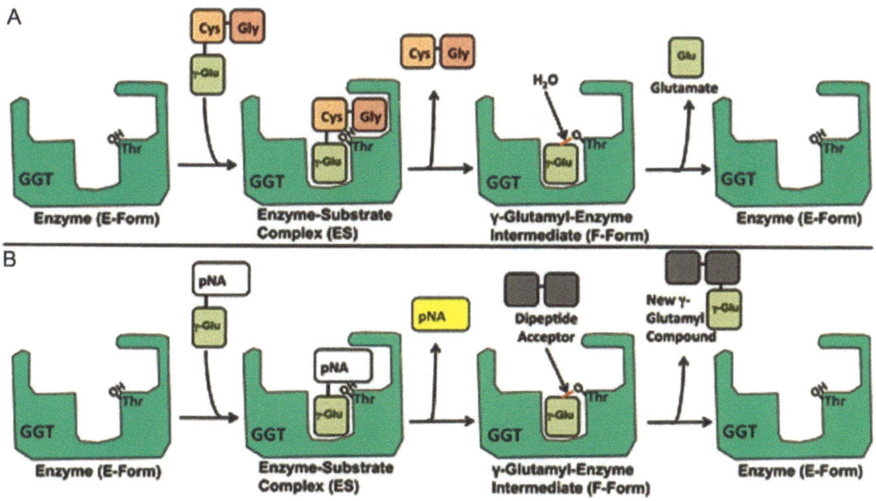

Figure 3.3, Marie H. Hanigan (See Page 110 of this volume.)

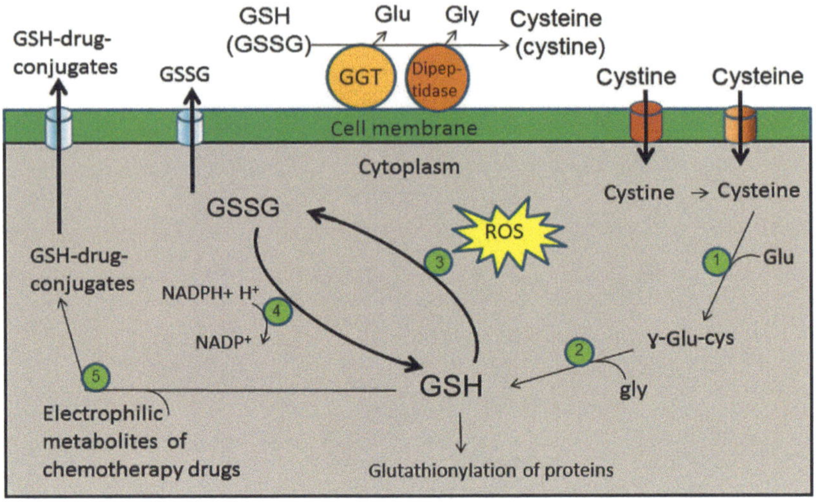

Figure 3.4, Marie H. Hanigan (See Page 115 of this volume.)

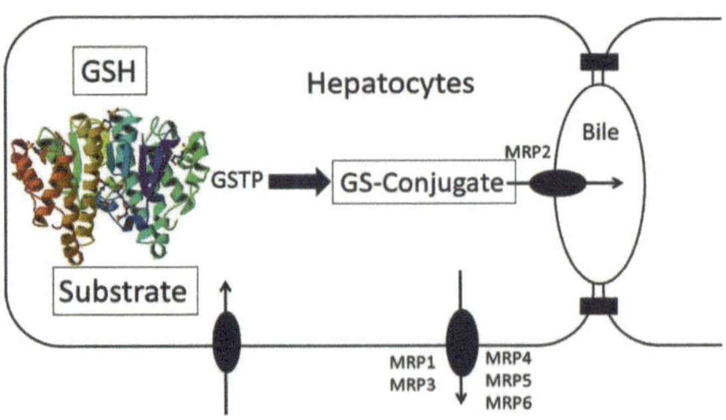

Figure 4.1, Jie Zhang *et al.* (See Page 144 of this volume.)

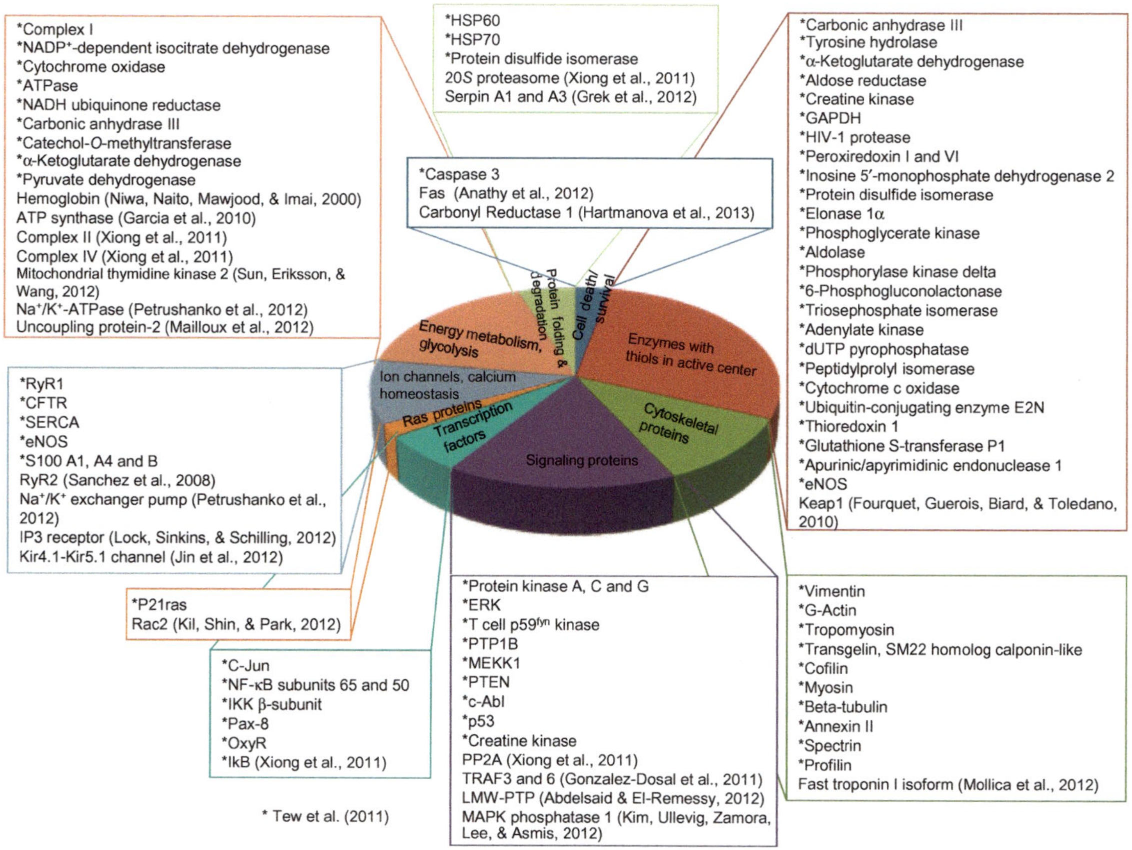

Figure 4.3, Jie Zhang *et al.* (See Page 153 of this volume.)

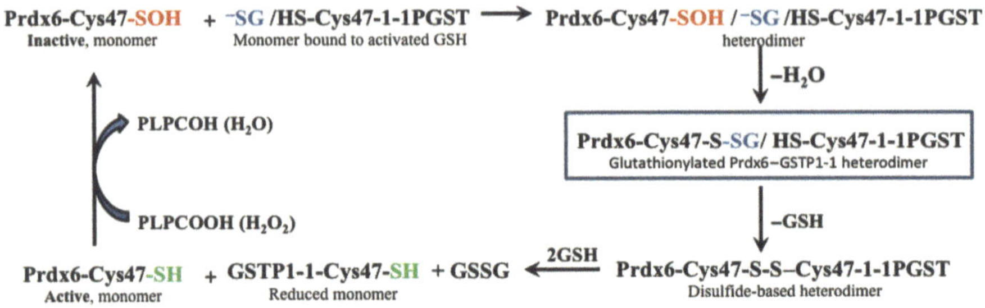

Figure 4.5, Jie Zhang *et al.* (See Page 158 of this volume.)

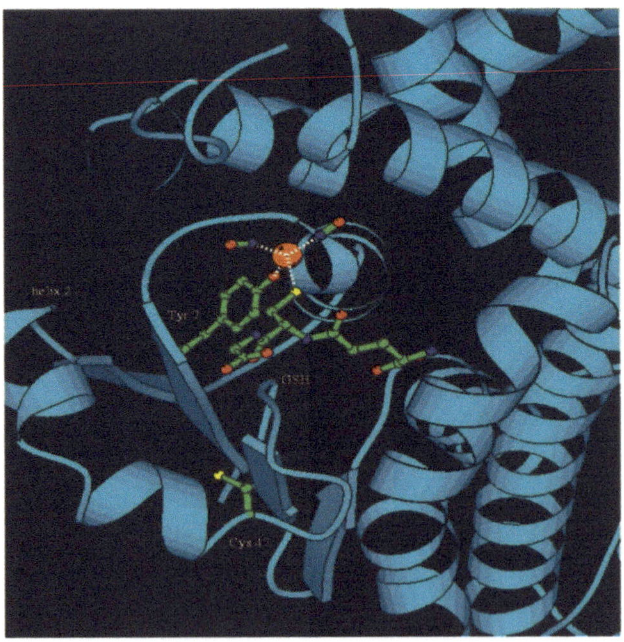

Figure 4.6, Jie Zhang *et al.* (See Page 161 of this volume.)

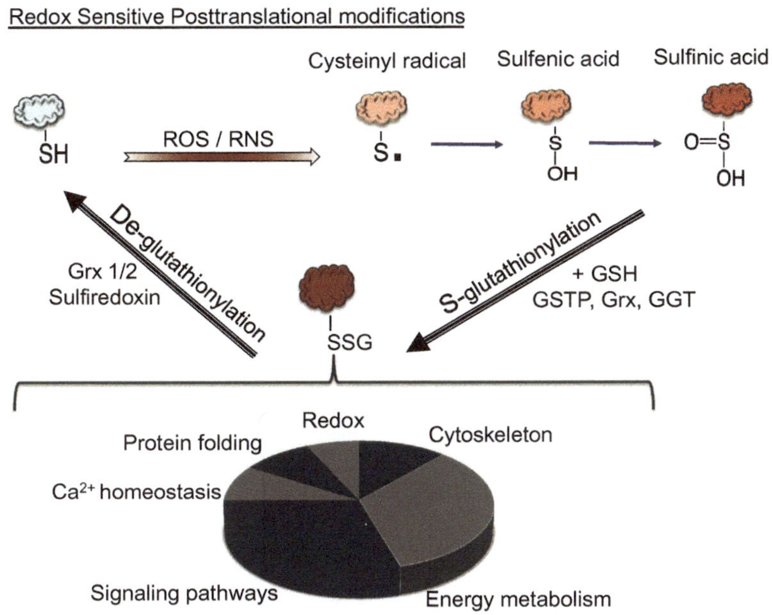

Figure 5.3, Danyelle M. Townsend *et al.* (See Page 184 of this volume.)

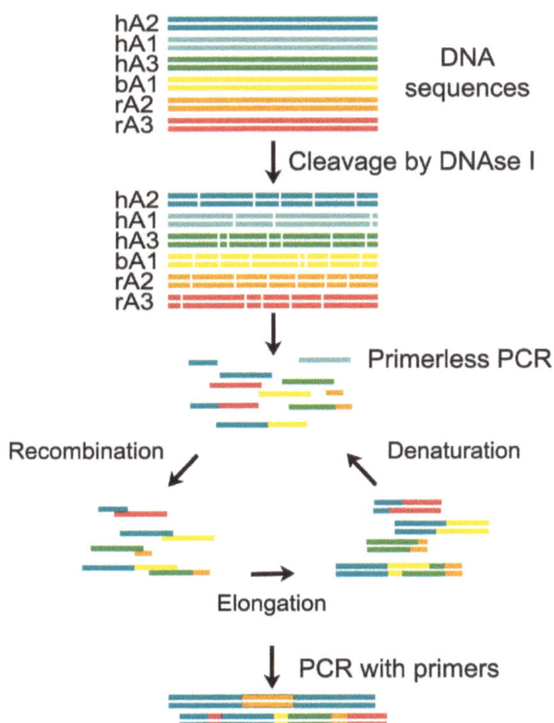

Figure 6.7, Olof Modén and Bengt Mannervik (See Page 224 of this volume.)

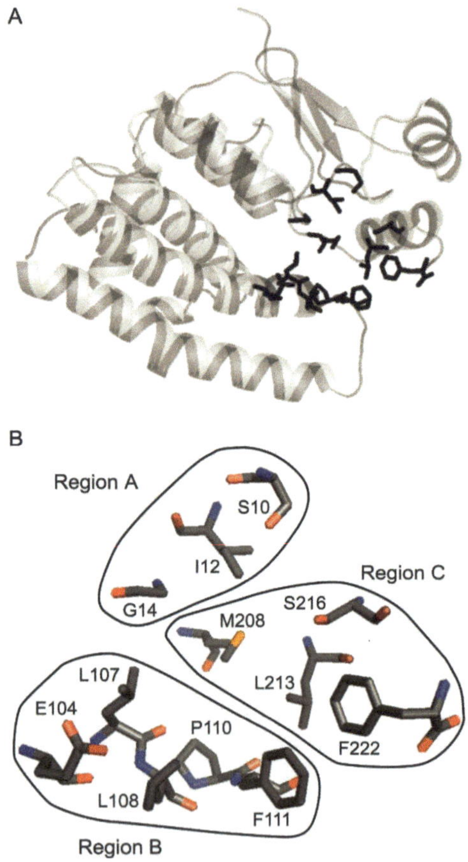

Figure 6.9, Olof Modén and Bengt Mannervik (See Page 227 of this volume.)

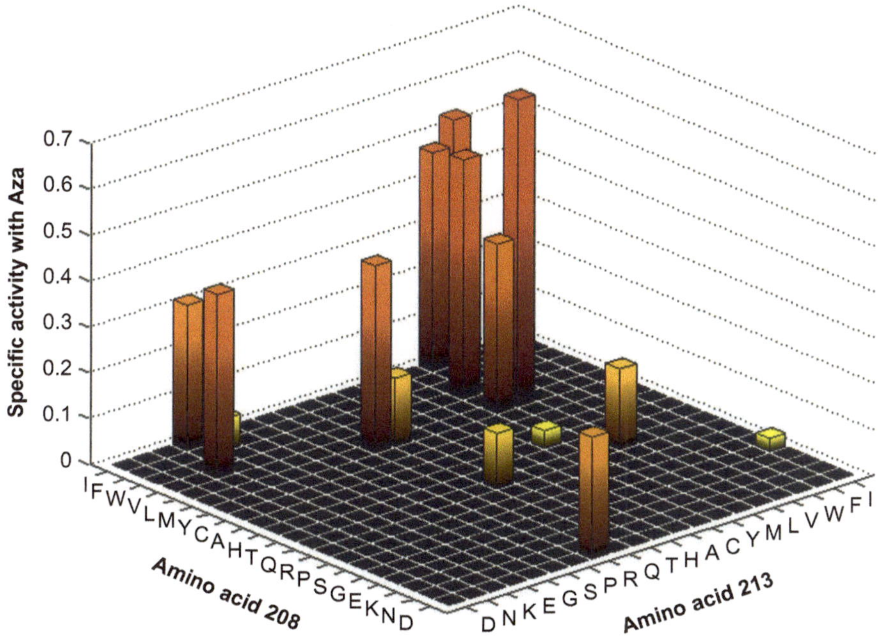

Figure 6.11, Olof Modén and Bengt Mannervik (See Page 231 of this volume.)

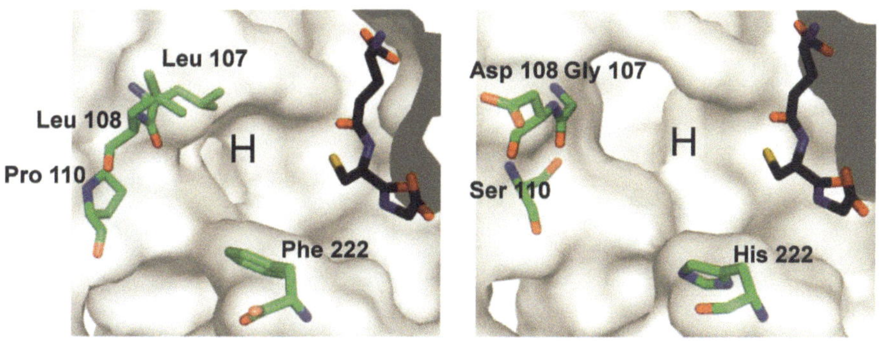

Figure 6.12, Olof Modén and Bengt Mannervik (See Page 235 of this volume.)

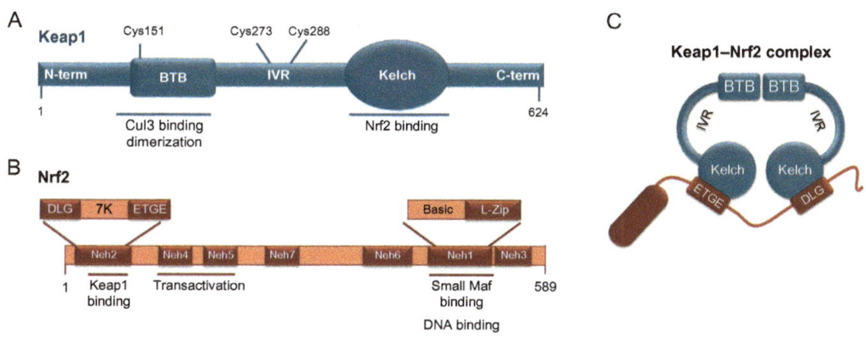

Figure 8.1, Hanna M. Leinonen *et al.* (See Page 283 of this volume.)

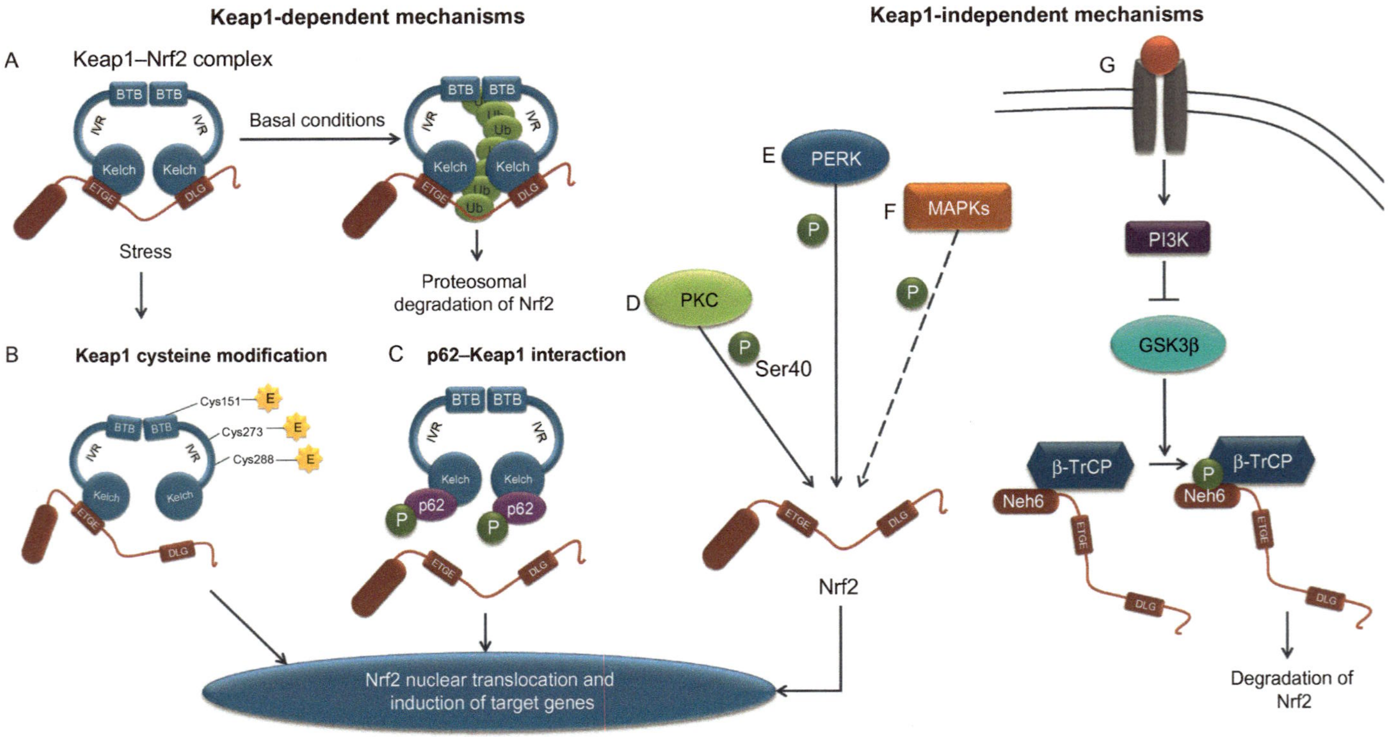

Figure 8.2, Hanna M. Leinonen *et al.* (See Page 285 of this volume.)

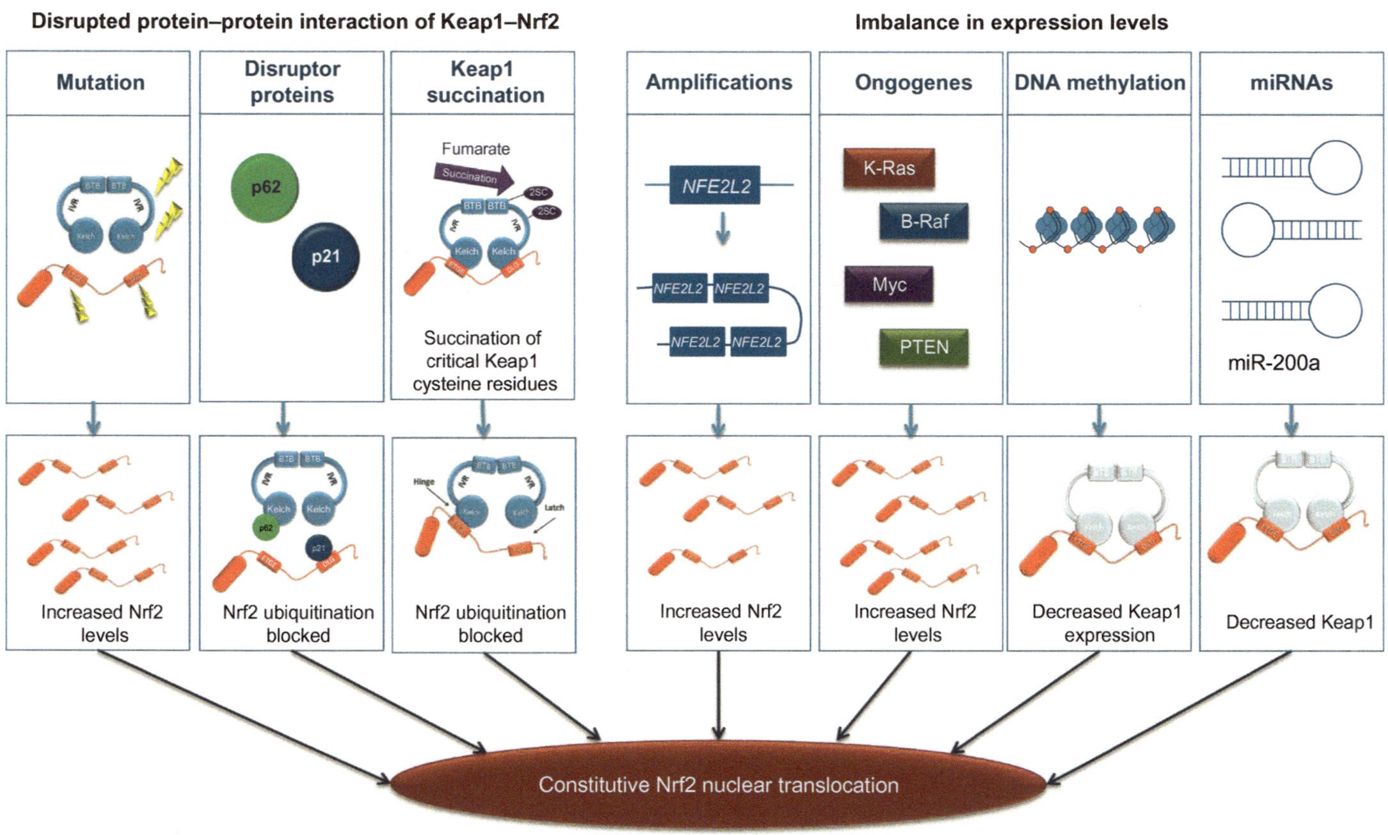

Figure 8.3, Hanna M. Leinonen *et al.* (See Page 291 of this volume.)

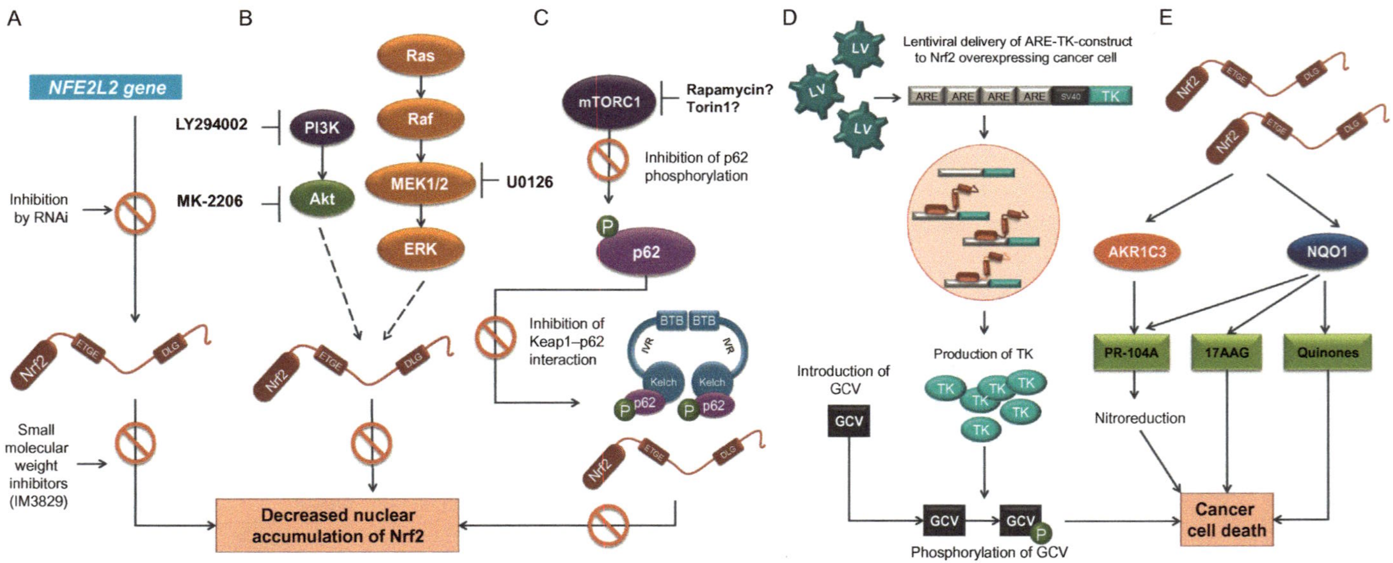

Figure 8.5, Hanna M. Leinonen *et al.* (See Page 303 of this volume.)

FSC
www.fsc.org
MIX
Papier | Fördert
gute Waldnutzung
FSC® C083411

Kolibri 360 GmbH
Ferdinand-Jühlke-Straße 7
99095 Erfurt, Deutschland
produktsicherheit@kolibri360.de